Meyer—*Introduction to Mathematical Fluid Dynamics*
Montgomery and Zippin—*Topological Transformation Groups*
Morse—*Variational Analysis: Critical Extremals and Sturmian Extensions*
Nagata—*Local Rings*
Nayfeh—*Perturbation Methods*
Niven—*Diophantine Approximations*
Plemelj—*Problems in the Sense of Riemann and Klein*
Ribenbolm—*Algebraic Numbers*
Ribenbolm—*Rings and Modules*
Richtmyer and Morton—*Difference Methods for Initial-Value Problems,* 2nd Edition
Rivlin—*The Chebyshev Polynomials*
Rudin—*Fourier Analysis on Groups*
Samelson—*An Introduction to Linear Algebra*
Sansone—*Orthogonal Functions*
Siegel—*Topics in Complex Function Theory*
Volume 1—Elliptic Functions and Uniformization Theory
Volume 2—Automorphic Functions and Abelian Integrals
Volume 3—Abelian Functions and Modular Functions of Several Variables
Stoker—*Differential Geometry*
Stoker—*Nonlinear Vibrations in Mechanical and Electrical Systems*
Stoker—*Water Waves*
Tricomi—*Integral Equations*
Wasow—*Asymptotic Expansions for Ordinary Differential Equations*
Whitham—*Linear and Nonlinear Waves*
Yosida—*Lectures on Differential and Integral Equations*
Zemanian—*Generalized Integral Transformations*

AN INTRODUCTION TO LINEAR ALGEBRA

AN INTRODUCTION TO LINEAR ALGEBRA

HANS SAMELSON

Stanford University
Stanford, California

A WILEY-INTERSCIENCE PUBLICATION

JOHN WILEY & SONS, New York · London · Sydney · Toronto

Library of Congress Cataloging in Publication Data:

Samelson, Hans, 1916–
An introduction to linear algebra.

(Pure and applied mathematics: A Wiley-Interscience series of texts, monographs, and tracts)
"A Wiley-Interscience publication."

QA184.S25 512′.5 74-17001
ISBN 0-471-75170-7

Printed in the United States of America

10 9 8 7 6 5 4 3 2 1

TO NANCY

PREFACE

This book developed from a set of notes for an undergraduate course in linear algebra that I have given several times at Stanford. The audience for the course consisted of mathematicians, physicists, statisticians, operations research majors, engineers, economists, and others, all of whom needed linear algebra, with some sophistication, for their further study and work. The book is addressed to that sort of audience. It is neither an elementary introduction to matrix operations nor an advanced mathematical text on vector spaces and modules. It operates on a middle level, developing the geometric–conceptual background of linear equations (and quadratic forms), emphasizing implicitly all the time that one understands the theory only if one can compute with it. However, it could be used for a more elementary course, since the concrete case of column vectors, row vectors, and matrices is developed for each essential topic.

The "prerequisites" for the book are few: In theory acquaintance with the real and complex numbers and some algebra (up to roots of polynomials and statement of the fundamental theorem of algebra) is all that is needed. In practice most of the students in the course had had a year of calculus, with a fair dose of plane and space analytic geometry; this book is meant for readers or users of about that level of mathematical experience and maturity.

The subject matter of linear algebra is pretty well circumscribed these days. I have not tried to be encyclopedic; this is a textbook, not a handbook. Many individual topics could be developed much further; I have only brought them as far as is necessary for a general understanding. What I have included is best seen from the table of contents. The one big topic omitted (mainly because it did not fit into the two quarters time span of the course) is multilinear algebra, tensors, and exterior algebra.

I have, of course, tried to arrange all the material in the order that seems most natural to me. To some extent the order is a matter of taste; in fact, differences in taste and order constitute one of the main reasons for writing one more book on linear algebra.

For a shorter and/or more elementary course one might leave out the following: Chapter 1, Section 3; Chapter 3, Section 3; Chapter 5, Sections 1 and 3; Chapter 7, Section 4; Chapter 8, Sections 4–11; Chapter 10, all except Section 1; Chapter 11, Sections 5, 7, and 8; Chapter 12, Sections 8–11. On the other hand, the material on the Gram-Schmidt process (Chapter 12, Sections 4, 5, and 6) could be moved to Chapter 4, Section 6, which contains a short discussion of the usual inner product on $\mathbb{R}^n$.

There are both numerical practice problems and more theoretical ones. As in any field of endeavor, to learn the subject one must work (many) problems (preferably without looking up the answer in advance). The numerical problems have, with very few exceptions, small integers as coefficients and solutions. Granting that the problems that one meets in nature are usually not of this type, I still thought it wise not to overload the course with computational matters, important as they are.

The numbering system is as follows: Each chapter contains a number of sections. Definitions, theorems, and so on, are numbered through each section, in the style m.n Item for item n of Section m. They are referred to as Item m.n or Item n of Section m in the chapter in which they appear; in other chapters their chapter number is given. Some formulae are designated by a letter symbol such as (P); they are referred to by that letter, with an indication of section and chapter as needed.

I am indebted to the students in my courses, who were willing to live with a set of notes that came out usually, but not always, in time for class; and I thank Charlotte Austin for a marvelous job of typing and preparing the notes, in a rather hectic atmosphere, from the scribbled sheets that I kept handing her.

HANS SAMELSON

Stanford, California
May 1974

CONTENTS

AN INTRODUCTION TO LINEAR ALGEBRA

INTRODUCTION

We begin by establishing some notation that will be used in this book. The standard symbols of elementary set theory are used: $a \in S$ means "the object a belongs to the set S"; $A \subset B$ means "the set A is a subset of the set B"; $A \cup B$ and $A \cap B$ denote union and intersection. $\{x: P(x)\}$ is the set of those x's that have the property P. The symbols R and C mean the systems of real numbers and complex numbers ($z = a + ib; a, b \in \mathsf{R}; i^2 = -1$) with the usual operations and properties. Finally, the end of a proof, argument, or remark is denoted by a solid square, ■.

A few remarks on origin, scope, and use of linear algebra are in order.

Linear algebra is an extension of analytic geometry of plane and space to "higher dimensions," in particular of that part that has to do with lines and planes as well as with conics (ellipses, etc.) and quadrics (ellipsoids, etc.). It is also an extension to "more variables" of the familiar theory and practice of (systems of) linear equations in two or three variables.

As for its use, there are several, loosely related areas. One is that of (first order) approximation: One attacks a problem by considering it as a slight deviation from a simpler one; this amounts to expanding all functions that enter into the problem by Taylor's theorem and then keeping only the (constant and the) linear terms. In particular, all equations to be solved become linear equations. Thus the concepts of linear algebra can be applied. Second, there are systems that have a natural "linear" behavior: It makes sense to "add" two states of the system; the "superposition principle" holds. The possible forces on a body or the possible electrical fields in a region of space are such systems. Linear algebra is the natural language in such cases. Finally, there is a part of linear algebra in which one studies quadratic expressions (this may sound paradoxical); this finds applications in such areas as maxima and minima of functions (of several variables) and oscillations of systems.

We will review briefly the ideas of analytic geometry and linear equations (this is quite elementary, but in fact most of what appears later is already here in embryonic form). First, we consider the line. One makes it into a number scale by choosing an *origin*, called O, and a point corresponding to 1. This sets up, in the usual way, a one-to-one correspondence between the points on the line and the real numbers.

Next, we consider the plane. One treats it by choosing a point in it, again called the *origin* O, and two lines through O, called the coordinate axes (they are often chosen at right angles to each other, but do not have to be). Each axis is made into a number scale, with O at the origin, by choosing a point "1" on it. Then coordinates of points are introduced: Given a point in the plane, one draws parallels to the axes through the point and finds where they cut the axes, getting a number on each of the axis. These two numbers are the *coordinates* of the point (with respect to the given axes); they are often called x and y coordinates. (See Figure 1.) For specific points the coordinates are numerical values, for example, the point $(2,-3)$. Sometimes one considers points that are not specific; for example, one wants to talk about "*any* point on a certain line." One is then forced to use letters for the coordinates, such as (x,y).

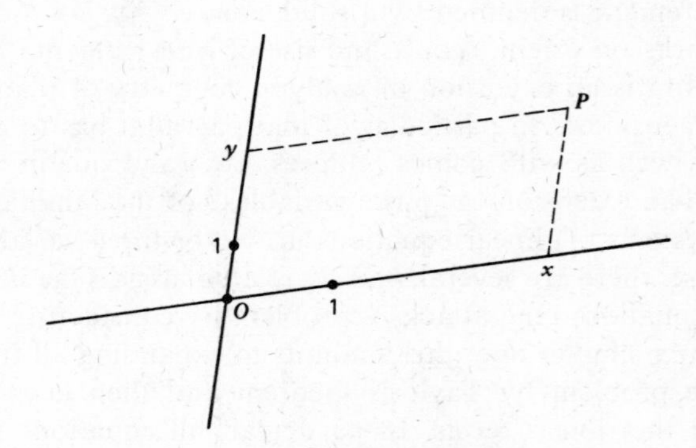

Figure 1.

We take it as known that an equation such as $3x-2y=5$ determines a line in the plane (such equations are called "linear"). (x,y) is the "running" point on the line; numerical values for x and y give a point on the line in question exactly if they satisfy the above equation. Thus $(1,-1)$ is a point on the line since $3\cdot 1-2\cdot -1$ equals 5; but $(1,1)$ is not, since $3\cdot 1-2\cdot 1$ does not equal 5. We also recall that a quadratic equation such as $x^2+2xy+3y^2=2$ determines a conic, which might be an ellipse, a hyperbola, or "degenerate" (two parallel lines); and that there is a way to determine which, from the coefficients a, b, and c of x^2, xy, and y^2 (b^2-4ac does it; negative means ellipse, positive means hyperbola, and zero means degenerate).

Geometrical problems are now translated into algebra; for instance, to find the point of intersection of two lines amounts to solving a *pair* of linear equations for the two unknowns x and y.

Example

$2x+3y=2, x+2y=3$. Recall how one proceeds: One *eliminates* x (or y) by multiplying each equation by a suitable factor and subtracting; the factors are the coefficients of x (or y) in the *other* equation. Thus here we write

$$1\cdot(2x+3y)-2\cdot(x+2y)=1\cdot 2-2\cdot 3.$$

This simplifies to $-y=-4$ or $y=4$.

The main point is that the operation has made x disappear; the factors are carefully chosen to effect precisely that. Of course, once y is known, we can find x from either of the two original equations. ■

We note that two equations might have *no* solution. Take $x+2y=3$ and $x+2y=4$, as an example. No (x,y) can satisfy this, since then 3 would have to equal 4. Geometrically this means, of course, that the two equations represent two *parallel* lines with no point in common.

Finally we consider space. After choosing an origin, one takes three axes (through the origin)—usually, but not necessarily, at right angles—and makes them into number scales. One assigns three coordinates, say x, y, and z, to any point P in space by noting for each axis where it is cut by the plane through P, parallel to the other two axes. An equation like $x+2y-3z=5$ now determines a plane. A quadratic equation like $x^2+4xy+4y^2+2xz+z^2+4yz=16$ determines an ellipsoid, or a hyperboloid of one or two sheets; or it is degenerate—we shall not list the possibilities here.

As for systems of equations, to find a point in common to three given planes amounts to finding a "simultaneous" solution to three linear equations in three unknowns x, y, and z. How does one proceed? By *elimination*; one eliminates x from the first and second equation (by the process described above), as well as from the first and the third (or second and third), obtaining *two* linear equations for y and z. One treats these two as above. One eliminates (say) y, getting *one* equation for z, from which z results immediately. Then one gets y from one of the two equations for y and z, and finally, x from one of the original equations. We note that, while "in general" there is exactly one solution, there might be *no* solution (two of the planes might be parallel to each other, or the three planes might be arranged like the faces of a triangular column), then again there might be many solutions, filling out a whole line (the three planes might go through a common line). We will meet all this again, and make it more systematic, when we talk about linear equation and the "row-echelon form of a matrix."

We will *not* use the notions length, distance, and angle for quite some time. The point is that a lot of geometry does not make use of these concepts. But they are, of course, important. We will come to them later at the proper time, when we introduce the notion of "inner product."

PROBLEMS

1. Solve the three equations

$$2x - y + z = 3,$$

$$x + 2y - z = -1,$$

$$2x - 3y + 2z = -1.$$

2. Construct a "skew" coordinate system in the plane (where the axes are not at right angles and the unit distances on the axes are not equal to each other). Locate the points with coordinates $(1,1)$, $(-1,-4)$, and $(2,4)$. Draw the line with equation $3x - y = 1$. Are the points on the line?

1

VECTOR SPACES

In this chapter we introduce the basic concept of linear algebra, the **vector space**. Vector spaces appear in many different contexts; we describe a number of examples. We also give a preliminary discussion of the natural way for vector spaces to "interact," namely, by **linear transformations**.

1. BASIC DEFINITIONS

We will now consider the plane again. Suppose the origin is chosen. The choice of axes is, of course, quite arbitrary (in analytic geometry this gives rise to the topic "change of coordinates"). What is important is the use of parallel lines as outlined in the introduction; we describe this, a bit differently, through the idea of the *parallelogram*. Given the origin O and two points A and B, we have the two "directed segments" OA and OB, and we can complete them to a parallelogram with a fourth point C such that OA is parallel to BC and OB parallel to AC; the point C and the segment OC are well determined. This construction, which from A and B (with O in the background) produces C, is the basic construction of linear algebra. (See Figure 2.) For instance, the introduction of coordinates can be viewed as saying that, given the axes through O, *any* point P in the plane can be constructed from two points on the axes in this fashion (in fact, in only one way).

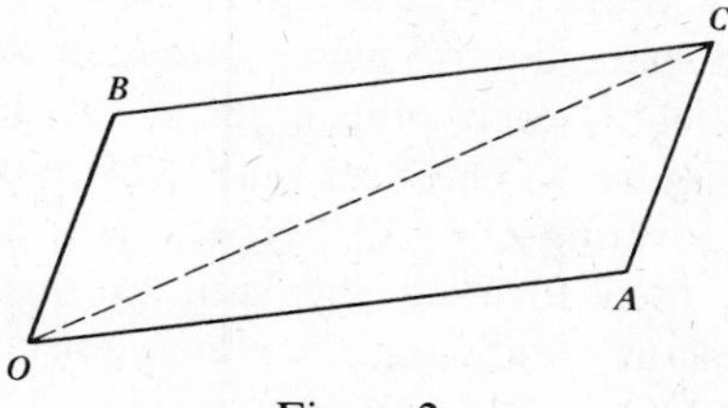

Figure 2.

We visualize a directed segment OA by an arrow, with end at O and tip at A, and call it the *vector* OA. (See Figure 3.) The parallelogram construction, which from OA and OB yields OC, is written as addition: $OA+OB = OC$; "the vector OC is the sum of the vectors OA and OB."

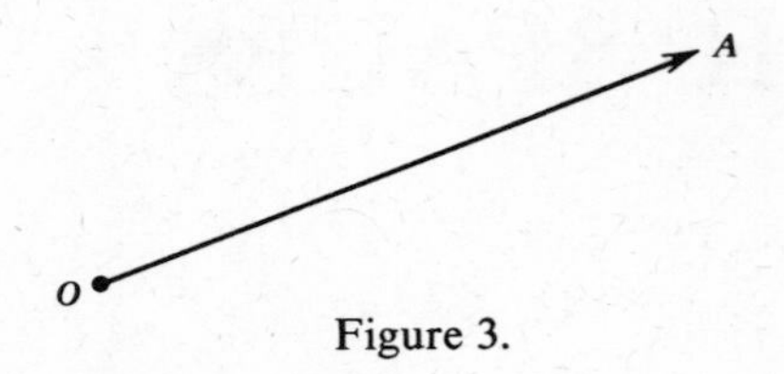

Figure 3.

Actually there is a slight difficulty here. If B happens to lie on the line through O and A, then the parallelogram "collapses." However, intuitively the operation "$+$" still makes sense and yields a C on the same line (see Problem 5). In fact, what we are led to here is the idea of multiplying a vector by a number. Take, for example, B equal to A; then $OA+OB$ would naturally be written as $2\cdot OA$. Similarly, we could construct $3\cdot OA$ or $17\cdot OA$ or $-5\cdot OA$ (going to the other side of O on the line). And this extends clearly to fractional multiples $p/q\cdot OA$, and by approximation to arbitrary real multiples $r\cdot OA$ (if r is irrational, replace it by a section of its decimal expansion). In particular, $0\cdot OA$ is the "degenerate" segment OO. Of course, all this just means that we have made the line through O and A into a number scale, with O and A corresponding to 0 and 1 and with "$+$" of vectors corresponding to "$+$" for numbers.

The basic facts bear recapitulation. Given the plane, with an origin O chosen in it, any two vectors OA and OB can be *added* to give a new vector $OC = OA+OB$ (the parallelogram construction); a vector OA can be *multiplied* by any real number $r(>0, =0,$ or $<0)$ to produce $OD = r\cdot OA$, with D located appropriately on the line through O and A.

This applies equally well to vectors in space (where a vector is again a directed line segment or arrow from the chosen origin to any point A); we can add two vectors, and we can multiply a vector by a number.

Some properties of these operations must be considered. Three vectors OA, OB, and OC in space determine a "parallelepiped" (a skew box with opposite faces parallel to each other); let P be the corner of this box opposite to O. (See Figure 4.) One sees that OP equals $(OA+OB)+OC$; it is obtained by first forming $OA+OB$ (constructing the parallelogram on OA and OB), and then forming the sum of this vector and OC (by constructing the relevant parallelogram, which goes through the interior of the parallelepiped). One could, of course, get OP as $OA+(OB+OC)$

equally well (and there is a third way to get it). In other words, $OA + (OB + OC) = (OA + OB) + OC$ for any three vectors OA, OB, OC in space—addition is "associative." Of course, this law also holds for any three vectors in the plane. (Figure 4 for the parallelepiped is in fact a plane figure!) Another, rather obvious, rule is $OA + OB = OB + OA$—"commutativity." We simply get the same parallelogram and fourth vertex for both sides.

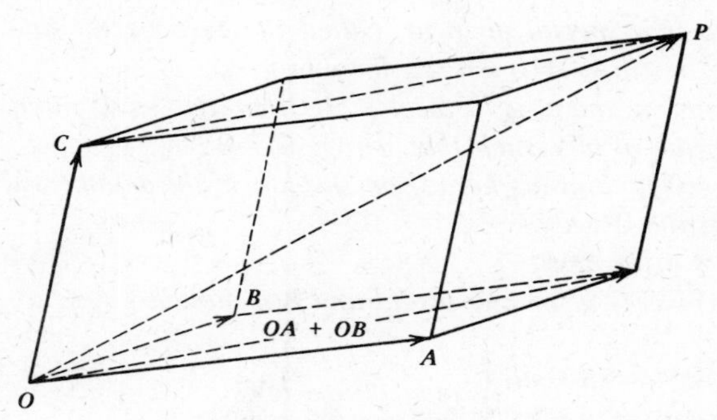

Figure 4.

There are a few more such general rules about addition, for example, $OO + OA = OA$ (the degenerate segment OO acts as "zero" under addition), and also about multiplication by numbers, for example, $r\cdot(OA + OB) = r\cdot OA + r\cdot OB$. Experience has shown that a certain number of these (listed below as axioms VS_1–VS_9) are basic. Indeed, linear algebra is "simply" an elaboration of these rules. The importance of linear algebra comes from the existence "in nature" of many systems that are similar to plane and space in the sense of permitting operations of addition and multiplication-by-numbers that obey just these rules; linear algebra can then be applied to such systems. (We will see examples below.) This sort of structure is called a vector space; vector spaces will be our main object of study.

There is one generalization that is very useful and that we will incorporate into the formal definition of vector spaces: We consider vector spaces where the numbers that can be used to multiply vectors are complex numbers (instead of only real numbers). To handle both cases simultaneously, we shall use F to stand for either R or C. Elements of F are called numbers or *scalars*.

1.1. **MAIN DEFINITION.** *A vector space* (*also linear space*) *over* F *consists of* (a) *a* (*nonempty*) *collection or set of elements, called vectors, and denoted by*

$u, v, w, \ldots$ *(but see Section* 2, *i*), (b) *an operation* $+$, *called addition, that assigns to any two vectors (elements of the given set)* u *and* v *a vector (element of the set)* $u+v$ *as "sum,"* (c) *an operation* $\cdot$, *called (scalar) multiplication, that assigns to any scalar* r *(in* F*) and any vector* u *a vector* $r\cdot u$ *(also written* ru*) as "product," such that the following rules, the vector space axioms, hold*:

VS$_1$: $u+v=v+u$ *for any vectors* u *and* v; *the commutative law.*

VS$_2$: $(u+v)+w=u+(v+w)$ *for any three vectors; the associative law.*

VS$_3$: *there is a special element, called the zero vector and written as* 0, *such that* $u+0=0+u=u$ *for any* u.

VS$_4$: *to any* u *there is a vector, written as* $-u$*("minus* u*" or "the negative of* u*"), such that* $u+(-u)=0$.

VS$_5$: $0\cdot u=0$ *for any* u; *here* 0 *on the left is the number* 0, *and* 0 *on the right the* 0*-vector.*

VS$_6$: $1\cdot u=u$ *for any* u.

VS$_7$: $r(su)=(rs)\cdot u$ *for any* u *and any two numbers* r, s; *an "associative" law.*

VS$_8$: $(r+s)\cdot u=ru+su$

VS$_9$: $r\cdot(u+v)=ru+rv$

(VS$_8$, VS$_9$: *distributive laws.*)

Parts (b) and (c) of the definition are often loosely described by saying that the vector space is *closed* under addition and multiplication-by-scalars. This means that the operations $+$ and $\cdot$ are unambiguously defined and that they yield elements of that same vector space (and not of some other set).

This may look like a long list, but one gets used to it. (Actually, there is some redundancy; Axiom VS$_4$, for instance, can be obtained from the other axioms by taking $-u$ as $(-1)\cdot u$.)

To reiterate, these axioms, or some reformulations of them, are the important properties abstracted from plane and space. We should, of course, check them for plane and space; we forego this, however, and note only that there the zero vector is, of course, the degenerate directed segment OO. (To be quite honest, we should really have given an *axiomatic* description for plane and space; without that, all we would be doing is to verify that the vector space axioms above are compatible with our intuition about plane and space.) We emphasize that all the vectors mentioned in the axioms have to belong to *one* given vector space. Vectors from different vector spaces cannot be combined by $+$; if we want to talk about a vector, we must know to which vector space it belongs.

In generalizing the whole theory, one can replace R or C by any *field*, but we shall make no use of this here. We denote vector spaces by U, V, W, etc.; the operations $+$ and $\cdot$ are usually not mentioned in the notation.

We have to verify a few elementary facts:

a. The 0-vector of any vector space is unique. Suppose there is another vector $0'$ in our space with the property $u+0'=0'+u=u$ for any u. Then we have $0=0+0'$, as well as $0+0'=0'$, and so $0=0'$.

b. Similarly the vector $-u$ in VS_4 is unique. Suppose v is another candidate, that is, a vector satisfying $u+v=0$. Add $-u$ to the equation $u+v=0$, on both sides. This results in the relation $-u+(u+v)=-u+0$. The left side can be written as $(-u+u)+v$ by axiom VS_2; the right side is $-u$ by VS_3. Since $-u+u=0$ by VS_1 and VS_4, the left side reduces to $0+v$, which is v by VS_3. Altogether, $v=-u$. ■

By the way, in Axiom VS_4 there is no intention of considering u as positive or $-u$ as negative. The minus sign only says that the two vectors are "opposite to each other"; that is, their sum is 0. The idea of a positive vector exists only in very special vector spaces.

c. Axioms VS_1 and VS_2 imply that the sum of any k vectors $u_1,\ldots,u_k$ is well defined, regardless of the order in which the terms are arranged or of any bracketing. We write $\Sigma_1^k u_i$ or $u_1+\cdots+u_k$. For example, $(u_1+u_2)+(u_3+u_4)=u_3+(u_1+(u_4+u_2))=u_1+u_2+u_3+u_4$. (On the left we first add u_1 and u_2 to get a vector v_1, say, and also u_3 and u_4 to get a v_2, and then form v_1+v_2.)

By similar arguments one shows (see Problem 3):

i. $-u=(-1)\cdot u$. (By uniqueness of $-u$ this amounts to showing that $u+(-1)\cdot u$ is 0, for any u.)

ii. For given u, w the equation $u+v=w$ has a solution; it is unique and is given by $v=-u+w(=w+(-u))$. Incidentally, for $w+(-u)$ one writes simply $w-u$. (See Figure 5.)

iii. $r\cdot 0=0$, for any scalar r. ■

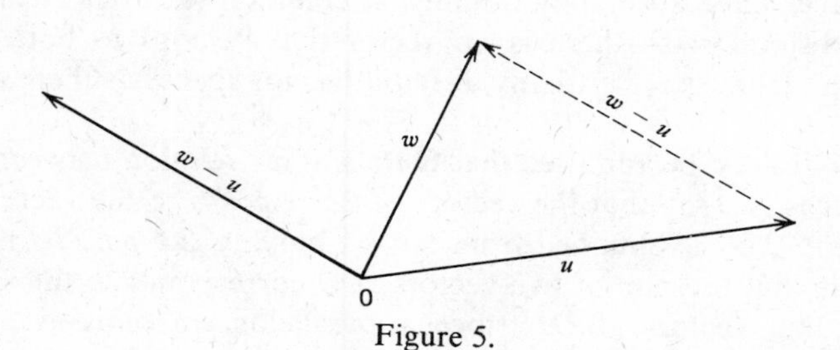

Figure 5.

The above comments and examples show that, once a vector space is given, one can compute with its vectors in a way similar to what one does with numbers. For instance, in an equation one can shift a term from one side to the other (with a minus sign, to be sure); one can divide an equation by a (nonzero) number (this is the same as multiplying by the reciprocal). Be forewarned that it makes no sense to divide by a vector; the one thing that can be done is this: If v is *not* the zero vector, and $r \cdot v = 0$ for some *scalar* r, then one can conclude $r = 0$ (see Problem 2).

PROBLEMS

1. Show that $-ru = r \cdot (-u)$ for any scalar r and vector u.
2. Complete the arguments for (i), (ii), (iii) above.
3. Show: If $r \neq 0$ and $rv = 0$ for some vector v, then $v = 0$.
4. Show: If $v \neq 0$, and $rv = 0$ for some scalar r, then $r = 0$.
5. Invent a geometric construction (using only parallels) for $OA + OB$ in the case where O, A, and B are collinear. (*Hint*. Use some auxiliary vector not on the line.)

2. EXAMPLES

As mentioned before, the need for the study of vector spaces comes from the fact that they occur at many places "in nature"; we consider some examples.

a. We studied already (and thereby got the motivation for the general concept) the plane or space of ordinary geometry; once an origin has been chosen, these become vector spaces with $+$ and $\cdot$ as described.

Some comments on the role of the origin in space follow. As noted, ordinary 3-space becomes a vector space by the choice of an origin, say O; the vectors are then the arrows OA. If we choose another origin O', we get a *different* vector space, whose vectors are the arrows $O'B$. These two vector spaces have absolutely nothing in common (at least at first sight); they are disjoint sets—there is no vector that belongs to both of them. Thus we get from space as many different vector spaces as there are points in space.

But now it must be admitted that there is some relation between all these vector spaces; we say that the vector OA *corresponds* to the vector $O'A'$, if the segment $O'A'$ is obtained from OA by moving OA *parallel* to itself. It is then true that the sum of two vectors at O corresponds to the sum of the corresponding vectors at O' (since a parallelogram remains one under parallel displacement); similarly, if $O'A'$ corresponds to OA, then any multiple $rO'A'$ corresponds to rOA (with the same r). In other words, the vector spaces attached to the different origins in space are faithful copies of each other, under the operation "parallel displacement." Occasionally one "identifies" all these vector spaces; one considers corresponding vectors OA, $O'A'$ as equal, but we shall not do so in this book. All these comments apply, of course, also to the plane (and the line!).

b. In physics, the forces that can act on a given mass point form a vector space; the sum of two forces is the (well-determined) single force with the same effect on the mass point. Multiplying by a (real) number r is equivalent to changing the amount of force without changing the direction (negative r means reversal of direction). That the vector space axioms hold is a matter of experience and is usually tacitly taken for granted. The vector 0, for example, is "the" zero force (which has no direction!). The forces here could be mechanical, electrical, and so on. The possible electrical field forces at a given point in space form a vector space; as in example (a), at each point of space we have here a vector space (of the forces operating there). The different vector spaces have nothing to do with each other (at least at first sight); in an electrical field, for instance, it makes no sense to form the sum of two field vectors that sit at different points of space.

c. Whenever physicists talk about the *superposition principle*, they mean that a vector space is present. Superposition means that for the objects under consideration the operation of sum is defined. Multiplication by reals or complexes is also present, but usually not mentioned, and the axioms are tacitly assumed to hold. The "vectors" here are often states of a system; for example, the stresses at a point in an elastic body, the waves on a string, the electrical fields in a portion of space, or the wave functions describing a particle in quantum mechanics.

d. Consider a (linear) differential equation $y'' + p \cdot y' + q \cdot y = 0$. Here p and q are functions of x, on some interval $[a,b]$. The solutions of this equation form a vector space: The sum of two solutions is again a solution (because of $(y_1 + y_2)' = y_1' + y_2'$, etc.), and so is any constant multiple of a solution. The vector space axioms are easily verified. (The vector 0 is the function 0, that is, the function that is constant equal to 0.) We emphasize the fact that a vector in this space is a function $y(x)$ on $[a,b]$ that satisfies the equation. For example, for $y'' + y = 0$ some such vectors are the functions $\sin x$, $\cos x$, or, more generally, $\sin(x + \alpha)$ for any fixed α.

This example generalizes to linear differential equations of *any* order, not just two. Note a special, but important case: For any $n = 0, 1, 2, \ldots$, the differential equation $y^{(n+1)} = 0$ has for solutions exactly all *polynomials* of degree $\leqslant n$. Thus the collection of these polynomials is a vector space P^n, with the usual $+$ and $\cdot$. Note once more that an individual vector in this space is a polynomial (of degree $\leqslant n$), that is, an expression of the form

$a_0x^n + a_1x^{n-1} + \cdots + a_n$, where the a_i stand for numerical values. For example, $2x^2 + 3x - 1$ is an element or vector in P^2. (It is also one in P^{17}, namely, $0\cdot x^{17} + 0\cdot x^{16} + \cdots + 0\cdot x^3 + 2x^2 + 3x - 1$.) The 0-vector is the polynomial $0 = 0\cdot x^n + 0\cdot x^{n-1} + \cdots + 0\cdot x + 0$.

e. Consider an electrical network consisting of members that carry resistances or inductances or batteries, and are joined at certain nodes. A state of this system is given by arbitrarily assigning a current through each member (whether physically possible or not); two states are added by adding the currents in each member; similarly, we multiply by a real number. The states form a vector space. (That the axioms hold can be checked.)

f. A state of an economic system is characterized by a certain number of numerical parameter (values of items such as different resources, services, productivity, and tax rates). Two states are "added" by adding the values of each parameter; the procedure for multiplication by a number is similar. Thus the states of the system form a vector space and can be treated by linear algebra; this should, of course, be interpreted as a first approximation to what really goes on.

g. The real numbers $\mathbf{R}$ form a vector space over $\mathbf{R}$. We add "vectors" (real numbers) in the obvious way (as numbers), and we multiply a vector by a scalar also in the obvious way (just multiplying the two numbers). The axioms hold. We have to get used to the fact that in this case the vector space and the field of scalars are the same set. Similarly, the complex numbers $\mathbf{C}$ form a vector space over $\mathbf{C}$. The fact that the complex numbers also form a vector space over $\mathbf{R}$ is a little more tricky: $+$ is defined as before (addition in $\mathbf{C}$), and $\cdot$ is also defined as before, except that in forming $r\cdot u$ (with u in $\mathbf{C}$) we only use *real* factors r and simply do not admit any truly complex $r = a + ib$ with $b \neq 0$. This generalizes to the fact that any vector space over $\mathbf{C}$ is also one over $\mathbf{R}$, with the same vectors and the same $+$; we just refuse to multiply any vector by any truly complex number.

h. The **0-vector space** is a vector space that contains exactly *one* vector (which is necessarily the 0-vector 0). Addition is completely described by $0 + 0 = 0$, and multiplication by $r\cdot 0 = 0$ for all r. The axioms hold! There is not very much to such a space, and any two such spaces are pretty much alike ("isomorphic" in later terminology). Even so, the zero space does come up and has to be considered. We write 0 for it. (We now have three meanings for 0: the number, the vector, and the vector space; there will be more.)

i. For the so-called **standard** space (also Euclidean or arithmetic space), real number space $\mathbf{R}^n$ or complex number space $\mathbf{C}^n$ ($\mathbf{F}^n$ is the common symbol), let n be a natural number (1 or 2 or ...). A vector in $\mathbf{R}^n$ (or $\mathbf{C}^n$) is an (ordered) n-tuple of real (or complex) numbers, arranged (perversely at this point, but to fit in with later developments) as a **column**, enclosed by parentheses. For instance, $\begin{pmatrix} 2 \\ -1 \\ \pi \end{pmatrix}$ is a vector of $\mathbf{R}^3$. The "general" vector is

$$\begin{pmatrix} x_1 \\ x_2 \\ \vdots \\ x_n \end{pmatrix}.$$

The number n is called the *dimension* of the vector space $\mathbf{F}^n$; we will see later (Chapter 3, Section 1) what this means. The vectors of $\mathbf{F}^n$ are often called column vectors; in this book we denote column vectors by letters such as $X, Y, \ldots$ instead of $u, v, \ldots$. Also, to save space, we shall print and write them not as columns but as rows, horizontally, within *square brackets*; thus the vector above will be printed as $[2, -1, \pi]$, and the "general" vector as $X=[x_1, x_2, \ldots, x_n]$; we hope this will be useful and not too confusing. The values $x_1, x_2, \ldots$ of $X=[x_1, x_2, \ldots, x_n]$ are called the first, second, etc., components or entries of X. Thus the third component of the vector above is π.

We add two column vectors "componentwise"; the first entry of the sum is the sum of the first entries of the two vectors, and so on. Thus, for $X=[x_1, \ldots, x_n]$ and $Y=[y_1, \ldots, y_n]$ we have $X+Y=[x_1+y_1, x_2+y_2, \ldots, x_n+y_n]$. Similarly, we multiply by a real (or complex) number componentwise: $rX=[rx_1, rx_2, \ldots, rx_n]$. For example, $[2, -1, \pi]+[3, 1, 2]=[5, 0, \pi+2]$, and $-2\cdot[2, -1, \pi]=[-4, 2, -2\pi]$. Check the vector space axioms; it is all very trivial and goes straight back to standard properties of numbers. In particular, the vector 0 is $[0, \ldots, 0]$, that is, has all entries 0. The symbols $\mathbf{R}^0$ and $\mathbf{C}^0$ mean 0-spaces by definition. $\mathbf{R}^1$ is essentially the same as the real number field $\mathbf{R}$: A vector $[x_1]$ here has *one* entry x_1, addition of vectors is just addition of these numbers, and to multiply the vector $[x_1]$ by r is the same as multiplying the number x_1 by r. Similarly, $\mathbf{C}^1$ is essentially $\mathbf{C}$.

We shall also have occasion to consider **row vectors**. A row vector of "length" n, is an ordered sequence of n numbers $a_1, \ldots, a_n$, written *horizontally*, as a row, and enclosed by parentheses $(a_1, \ldots, a_n)$. An example of length 4 is $(1, 0, \pi, -i)$. We use letters like $A, B, \ldots$ for these. Two row vectors of the same length are added by adding corresponding entries or components, just as for column vectors. Similarly, for multiplication by a scalar, we multiply all entries by the scalar. The collection of *all* row vectors of a given length n, with this $+$ and $\cdot$, is, of course, a vector space. We denote it by $(\mathbf{R}^n)'$ or $(\mathbf{C}^n)'$ (note the "prime"), depending on whether we use $\mathbf{R}$ or $\mathbf{C}$ as numbers; the common symbol is $(\mathbf{F}^n)'$. The distinction between $\mathbf{F}^n$ and $(\mathbf{F}^n)'$ is very slight; it is just the rather formal matter of writing the entries of a vector vertically or horizontally. Nevertheless, we are going to keep the two spaces separate; we will see later why (Chapter 4, Section 1). In particular $\mathbf{F}^n$ and $(\mathbf{F}^n)'$ have no vector in common (except for $n=1$!); the two vectors $[2,-1,3]$ and $(2,-1,3)$ are distinct and belong to different vector spaces. $(\mathbf{F}^0)'$ is again the 0-space by definition.

We add a comment about comparison between $\mathbf{R}^2$ and the plane (or $\mathbf{R}^3$ and space).

Consider $\mathbf{R}^2$: A vector there has two components, x_1 and x_2 (one often writes x, y instead); the vector is $[x,y]$, for instance, $[2,3]$ or $[\sqrt{2}\,,1/e]$. This looks very much like "a point in the plane," namely the point with coordinates x, y or 2, 3 or whatever. True, one *can* visualize $\mathbf{R}^2$ in this way as "being the plane"—once coordinates have been introduced in the (geometric) plane. One should understand though that the two are by no means identical: A vector in the plane (with a chosen origin O) is simply given by a point in the plane (or, better, by the arrow from O to the point); what the coordinates of the point are depends on the choice of axes. In $\mathbf{R}^2$, on the other hand, a vector like $[2,3]$ is just that: it has 2 for first and 3 for second component, and that is all there is to it; no changes are possible. Still, we can, and even should, visualize $\mathbf{R}^2$ as the plane *once origin and axes are chosen in the plane*, with the vector $[x,y]$ *represented* by the arrow from O to the point with coordinates x and y. The beauty of this is that $+$ in $\mathbf{R}^2$ now means the same as $+$ in the plane, and the same is true for $\cdot$. For $+$ this can be read off from Figure 6. The same consideration holds for $\mathbf{R}^3$ and (geometrical) space: After choice of origin and axes we can visualize $\mathbf{R}^3$ as space by representing the vector $[x,y,z]$ of $\mathbf{R}^3$ by an arrow from O to the point with coordinates x,y,z and (again, $+$ and $\cdot$ carry over from $\mathbf{R}^3$ to space).

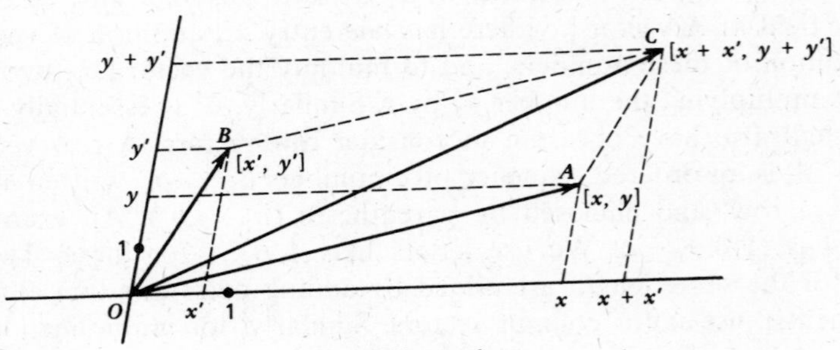

Figure 6.

As a general moral, when we talk in the future about an *arbitrary* vector space, we should be aware of the fact that there are many possibilities. But we should also think about the special cases R^2 and R^3 and their geometric counterparts, ordinary plane and space, and—as a generic example—about F^n. The reason for the former is that they are the ones that one can visualize. The latter, F^n, is in some ways more concrete than all other examples; furthermore, in a certain sense, which we discuss later (Chapter 3, Section 4), all other cases can be reduced to it.

PROBLEMS

1. Are the following sets of polynomials vector spaces (with the usual definition of $+$ and $\cdot$)?
a. All polynomials of degree 10.
b. All polynomials of degree $\leqslant 17$ that vanish at $x=0$.
c. All polynomials of degree $\leqslant 17$ that vanish at $x=1$.
d. All polynomials of degree $\leqslant 17$ that take value 1 at $x=0$.(The question is: Do the vector space properties (a), (b), (c) and axioms VS_1–VS_9 hold?)

2. Let P^n be the set of polynomials of degree $\leqslant n$ (for some arbitrary n). From the usual $+$ and $\cdot$ we construct an artificial "$+$" and "$\cdot$" as follows: $p(x)$"$+$"$q(x)$ $=p(x)+q(x)+1$, r"$\cdot$"$p(x)=2r \cdot p(x)$. Does this yield a vector space? Which of (a), (b), (c), and VS_1–VS_9 hold?

3. For the vectors $X=[1,2,-3,-1]$ and $Y=[2,-1,2,1]$ in R^4 compute $X+Y$, $X-Y$, $2X$, and $3X-2Y$.

4. Forces $u_1,\ldots,u_r$, acting on a mass point P, are *in equilibrium* if their sum $u_1+\cdots+u_r$ is 0. Construct three forces in equilibrium. (Forces are represented by arrows in plane or space, with P as origin, and added and multiplied by scalars as such.)

3. LINEAR MAPS

The concept of vector spaces that we have discussed is only one-half of linear algebra. The other half is that of *linear map* (between vector spaces). We will take up the detailed discussion of this later, but will give the basic ideas here. There are two things that get "generalized": (1) the familiar concept of linear functions, expressions of the form $2x+3y-5z$, with x, y, and z coordinates in space (note that there is no *constant* term in these); (2) the geometric idea of (parallel) projection. Take two planes in space (both through O) and project one onto the other along some chosen direction. Through each point in the first plane draw the line with the given direction and find where that line meets the second plane. (The chosen direction should not be parallel to the second plane.) Think of the projecting lines as light rays from the sun; each figure in the first plane has a shadow in the second plane. What is important to us is that straight lines in the first plane go to straight lines in the second one. In fact, one can see that more is true. The origin of the first plane goes to that of the second (the two origins are actually the same point); any vector (directed segment OA) in the first plane goes to one (OA') in the second; and now, the crucial fact, the sum of two vectors goes to the sum (a parallelogram projects to a parallelogram), and a multiple of a vector goes to the corresponding multiple (clear for double, triple, and true in general). If the projecting direction is, by accident, parallel to the first plane, the situation is "degenerate"; the plane projects to a *line* in the second plane. This is all generalized in the following definition.

3.1. DEFINITION. *Let U and V be two vector spaces ($U=V$ permitted) over* $\mathbf{R}$ *or* $\mathbf{C}$. *A linear map or linear transformation, say T, from U to V (written as T: $U \to V$) is a rule that assigns to each vector u of U a vector Tu (or $T(u)$) of V (called the T-image of u), with the property that sums and scalar multiples are preserved, that is, that the following laws hold*:

L_1: $T(u_1+u_2)=Tu_1+Tu_2$ *for any* u_1, u_2 *in* U; "*additivity*".
L_2: $T(ru)=r\cdot Tu$ *for* u *in* U, r *a scalar*; "*homogeneity*".

It follows, of course, that $T(u_1+u_2+u_3)=Tu_1+Tu_2+Tu_3$, and similar laws. In particular, $T(0)=0$ (take $r=0$ in L_2 with any u). We denote linear maps by $T, S, R, \ldots$. Linear transformations from a vector space to itself ($V=U$) are called *operators*. Occasionally we write $u \overset{T}{\mapsto} v$ or just $u \mapsto v$ to indicate that the T-image of u is v.

Examples

1. $U=P^{17}$ is the vector space of polynomials of degree $\leqslant 17$, $V=P^{16}$; a linear map D is given by *differentiation*: If $p(x)$ is a polynomial (of degree $\leqslant 17$), then $Dp(x)$ is the derivative $p'(x)$ (a polynomial of degree $\leqslant 16$); this is a linear transformation by standard properties of the derivative.

The same rule ($p(x) \mapsto p'(x)$) can also be considered as a linear transformation from P^{17} to P^{17}, P^{18}, and so on. Note that D: $P^{17} \to P^{17}$ and D: $P^{16} \to P^{17}$ are not the same linear transformations since they go to different spaces (even though the D-image of any $p(x)$ is the same in the two cases). This may sound like casuistry. But it is necessary to understand that part of the definition of a linear transformation is the space from which and the space to which the map goes.

Other examples can be constructed from D. For instance (with D^2 meaning second derivative), $T=x^2D^2-xD-3$, which means, in more detail, that the T-image $Tp(x)$ of $p(x)$ is $x^2\cdot p''(x)-x\cdot p'(x)-3p(x)$.

2. A map S from $\mathbf{R}^3$ to $\mathbf{R}^2$ is given by the rule: $S[x_1, x_2, x_3] = [2x_1 - x_2, x_1 + x_2 - x_3]$; this means that the two entries y_1 and y_2 of the S-image SX of any $X = [x_1, x_2, x_3]$ are *computed* from x_1, x_2, and x_3 by the formulae $y_1 = 2x_1 - x_2$ and $y_2 = x_1 + x_2 - x_3$. That this is a linear map depends on the fact that these two formulae are *linear* functions as described above (no constant term and no other functions such as sin or log). L_2 should be clear; multiplying all the x's by r clearly has the effect of multiplying the y's also by r. For L_1, if $X = [x_1, x_2, x_3]$ and $X' = [x_1', x_2', x_3']$, then $X + X' = [x_1 + x_1', x_2 + x_2', x_3 + x_3']$. Then $S(X + X')$, by definition, has first entry $2(x_1 + x_1') - (x_2 + x_2')$, which equals $(2x_1 - x_2) + (2x_1' - x_2')$; that is, the sum of the first entries of SX and SX'. This works similarly for the second component. Thus, $S(X + X') = SX + SX'$.

PROBLEMS

1. Apply the operator T of Example 1 to the vectors x, $5x^4 - x^2 - 2$, and $x^{17} - x^{15}$.

2. Apply the operator S of Example 2 to the vectors $[1, -1, 0] = X$ and $[2, 4, -3] = Y$. Form $X + Y$ and $S(X + Y)$, and verify $SX + SY = S(X + Y)$.

3. For the T of Example 1 try to find a "vector" $p(x)$ with $Tp(x) = 0$. (*Hint*. Try a power of x.)

2

LINEAR COMBINATIONS

We introduce the standard working notions of vector space theory: **linear combinations, subspace, span, linear dependence,** and **independence.** For the purpose of handling these matters computationally, we introduce **matrices** and develop the reduction of a matrix to **column-echelon form** (or **row-echelon form**).

1. LINEAR COMBINATIONS, SUBSPACE, SPAN

Let U be a vector space (over F). The only thing that one can do at first in a vector space is to use the operations $+$ and $\cdot$. Let $u_1,\ldots,u_k$ be vectors in U (a "finite sequence or family of vectors"); let $a_1,\ldots,a_k$ be scalars (elements of F). (Some of the u_i or the a_i might be equal to each other.) We multiply each u_i by the corresponding a_i and add. The resulting vector $a_1u_1+a_2u_2+\cdots+a_ku_k$ or $\Sigma_i a_iu_i$ (well defined by Chapter 1, Section 1.c) is called the **linear combination** of the u_i with coefficients a_i.

Example

In R^3, with $u_1=[1,-1,0]$, $u_2=[2,1,-1]$, $a_1=2$, and $a_2=-1$, we have $2u_1-u_2=[0,-3,1]$.

We also say that a vector u is linear combination of the (family of) vectors $u_1,\ldots,u_k$, if it *can be* written as $\Sigma_i a_iu_i$ for *some* suitable choice of the a_i; in other words, if there *exist* scalars $a_1,\ldots,a_k$ such that $u=\Sigma_i a_iu_i$.

Example

Let U be the vector space of solutions of $y''+y=0$, let u_1 be $\sin x$, and let u_2 be $\cos x$. Then the "vector" $u=\sin(x+\pi/4)$ is linear combination of u_1 and u_2; namely, $u=1/\sqrt{2}\cdot\sin x+1/\sqrt{2}\cdot\cos x$ (using a little trigonometry).

Now a problem arises: Given the family $u_1,\ldots,u_k$, and a vector u, how does one decide whether u is linear combination of the u_i, and if it is, how does one find coefficients a_i that will do the trick? We will eventually answer this question (Chapter 4, Section 5.1), when we will find that the problem amounts to solving a system of linear equations. But before that we have to discuss a number of concepts and facts.

We introduce the notion of subspace of a vector space; this is the analog of a line or plane (through O) in space.

1.1. DEFINITION. *Let U be a vector space (over F). A subset, say V, is called a subspace of U if* (a) *it is nonempty,* (b) *for any two elements u and v of V their sum $u+v$ (an element of U) is also in V, and* (c) *for any u in V and any scalar r the vector $r\cdot u$ (an element of U) is also in V.*

Conditions (b) and (c) are often described by saying that V is *closed* under the operations $+$ and $\cdot$ of U. We usually write $V\subset U$ to indicate a subspace.

Our first proposition is important, although its proof is trivial.

1.2. PROPOSITION. *Let V be a subspace of U, then, with $+$ and $\cdot$ inherited from U, V is a vector space.*

In other words, we define the sum of two vectors in V as their sum in U; by (b) this is a vector in V. In the same vein, we define the product of a scalar and a vector in V as their product in U; by (c) this is a vector in V. It is obvious that the axioms VS_1–VS_9 of Chapter 1, Section 1, hold in V, since they do hold in U. Convince yourself that the 0-vector of U is necessarily in V and is the 0-vector of V. ■

Examples

First, we will consider two trivial examples.

1. The 0-vector of U by itself, or rather the subset of U containing only 0, is a subspace; (a) holds since 0 is in it, (b) holds by $0+0=0$, and (c), by $r\cdot 0=0$. This is a 0-space, called the 0-subspace 0 of U. There is not much to it, but one has to consider it.

2. Take V equal to U; the subset is *all* of U. Clearly, this satisfies (a), (b), and (c).

These two extreme cases are always present and are called the trivial subspaces. We will now consider some nontrivial ones. In the (geometric) plane *any line* through O is a subspace; check (a), (b), and (c). In space *any line* or *any plane* through O is a subspace. (For instance, if two points A and B lie in a given plane, then the whole parallelogram $OABC$ lies in the plane; and thus $OC=OA+OB$ lies in the plane—and so (b) holds.) In talking about subspaces, one should have these simple cases in mind.

In $\mathbf{R}^3$ the subset consisting of all vectors with third component 0 is a subspace; (a) is clear, and (b) and (c) hold by $0+0=r\cdot 0=0$. (Of course, this subspace (all $[x,y,0]$) looks very much like $\mathbf{R}^2$ (all $[x,y]$); $+$ and $\cdot$ have the "same" effect in these two spaces; we will come back to this later on.) In $\mathbf{R}^5$ the subset $\{X\colon x_2=x_3\}$ of all vectors whose second and third entry are equal is a subspace: for (a) take, for instance, $[0,0,0,0,0]$. For (b) note that $x_2=x_3$ and $y_2=y_3$ imply $x_2+y_2=x_3+y_3$; it is similar for (c). The vector $[1,2,3,4,5]$, for instance, is not in this subspace.

Generalizing, any linear equation, such as $2x_1-x_2+3x_3-5x_4=0$ for X in $\mathbf{R}^4$, defines a subspace, as do several equations taken simultaneously. Note that it is important that we have 0 on the right.

In Chapter 1, Section 2.d (the differential equation), the subset of all those solutions that at a given x_0 have $y(x_0)=2y'(x_0)$ forms a subspace. Check (a), (b), and (c).

If $m\leqslant n$, then the space P^m (polynomials of degree $\leqslant m$) is a subspace of P^n. We have $P^m\subset P^n$ (any polynomial of degree at most m is one of degree at most n), and $+$ and $\cdot$ mean the same in both spaces.

Some examples of things that are *not* subspaces.

1. A single vector not equal to 0. Conditions (b) and (c) fail.

2. The set of vectors $[x,y]$ with $x^2+y^2=1$ in $\mathbf{R}^2$ (the "unit circle"). Again, (b) and (c) fail.

3. The set of all vectors on the two axes in the plane. Here (a) and (c) hold, but (b) fails (if u and v are not on the same axis).

4. In $\mathbf{R}^4$, the set of all X with $x_4=1$. Here (b) and (c) fail. (The equation $x_4=0$, instead of 1, does give a subspace.)

5. Any line in the plane that does *not* go through 0. (b) and (c) fail. Similarly, a line or a plane in space not through 0 is not a subspace. These objects are intuitively very much like subspaces; should we just forget them? The answer is, of course, no; they are important, and we will get to them eventually ("linear varieties," Chapter 4, Section 4). ■

We now describe a way to make subspaces. Let M be any nonempty subset (preferably not a subspace) of U. The **span** of M, written in this book as $((M))$, is the set of all possible linear combinations $\Sigma a_i u_i$ with the u_i taken from M (any number!) and arbitrary scalars a_i. This is a subspace: Conditions (a) and (c) should be clear; as for (b), if $u=\Sigma a_i u_i$ and $v=\Sigma b_j v_j$, with the u_i and v_j in M, then $u+v$ is of the same type (the set of u's might overlap that of the v's, but that does not matter).

If $M=\{u_1,\dots,u_k\}$ is a finite family, then we write $((u_1,\dots,u_k))$ for the span, which here is simply the set of all possible $\Sigma_1^k a_i u_i$.

Example

Let M, in $\mathbf{R}^3$, consist just of the two vectors $E^1=[1,0,0]$ and $E^2=[0,1,0]$. Then $((E^1,E^2))$ is the set (subspace) of all $X=[x,y,z]$ with $z=0$. This is true, since E^1 and E^2 have third entry 0, as does any $a_1E^1+a_2E^2$; on the other hand, any $[x,y,0]$ can be written as xE^1+yE^2.

In practice subspaces are very often described as span of some vectors; to "find" a subspace means to find a (finite) set of vectors spanning it. A second, equally important, way of describing subspaces is *by equations*; we shall take this up later in Chapter 4, Section 7. The above example shows both aspects in a very simple situation.

PROBLEMS

1. For the vectors $[2,3,1,-2]=X_1$, $[1,-1,0,1]=X_2$, and $[1,9,2,-7]=X_3$ form the linear combination

a. with coefficients $a_1=1$, $a_2=-1$, $a_3=2$;
b. with coefficients $b_1=2$, $b_2=-3$, $b_3=-1$.

2. Show: If a vector u is linear combination of $v_1,\ldots,v_s$, and each v_i is linear combination of $w_1,\ldots,w_t$, then u is linear combination of the w's.

3. Show: If the vectors $u_1,\ldots,u_r$ are linear combinations of the vectors $v_1,\ldots,v_s$, then the span $((u_1,\ldots,u_r))$ is a subspace of the span $((v_1,\ldots,v_s))$.

4. In $\mathbf{R}^3$ the span of the two vectors $E^1=[1,0,0]$ and $E^2=[0,1,0]$ is the same as that of the two vectors $X_1=[1,1,0]$ and $X_2=[1,-1,0]$.

5. In P^{17}, take $p_1(x)=x^5+x^4-2x^2+1$, $p_2(x)=x^3-3x^2+x-2$, $p_3(x)=x^2-x+2$.

a. Is $x^6-2x^5+3x^4-x^2+2x-1$ linear combination of the p_i?
b. Same for $x^5+2x^4-3x^3+2x^2+x-1$.
c. Same for $x^5+x^4+2x^3-7x^2+x-1$.

6. Show: If S is a subspace of U, then the span $((S))$ equals S itself.

7. For the subspace $\{X\colon x_2=x_3\}$ of $\mathbf{R}^5$ construct a (finite) number of vectors that span it. (Later we will get a procedure for such problems; here we just have to make guesses and pick what seem like promising candidates.)

8. In P^6 let V be the subspace of all *even* polynomials (i.e., satisfying $p(x)=p(-x)$). Find a finite number of polynomials that span V.

2. OPERATIONS ON SUBSPACES

First, we have a piece of mumbo-jumbo.

2.1. PROPOSITION. *A subspace of a subspace of a vector space U is a subspace of U.*

This is trivial, once one sees what it is about. A subspace of U is a vector space in its own right and thus can have subspaces. But any such is automatically also a subspace of U. Complete the details. As an example, consider a line (through 0) in a plane in space.

The next fact is more important.

2.2. PROPOSITION. *Let V and W be two subspaces of the vector space U. The intersection $V \cap W$* (i.e., *the set of vectors in U that are in V and in W simultaneously*) *is a subspace.*

PROOF. For (a) of Definition 1.1, the zero of U is in V and in W, as noted earlier, and thus also in $V \cap W$. For (b), say u and v are in $V \cap W$, then, in particular, u and v are in V and so is then $u+v$ (since V is a subspace). By the same token, $u+v$ is also in W, and so $u+v$ is in $V \cap W$. The same applies for (c). ■

We may take intersection $V_1 \cap V_2 \cap \cdots \cap V_k$ of more than two subspaces; this is associative (meaning $(V_1 \cap V_2) \cap V_3 = V_1 \cap (V_2 \cap V_3)$).

Examples.

1. In the plane the intersection of two lines through 0 is either 0 or a whole line (namely, if the two lines are identical); in either case, it is a subspace.
2. In space the intersection of two planes (through 0) is usually a line, but sometimes it is a plane (namely, if the two planes are identical); in any case it is a subspace.
3. In $\mathbf{R}^5$ let V consist of all vectors with $x_5=0$, and W of all those with $x_4=0$; then $V \cap W$ consists of all X with $x_4=x_5=0$. ■

Now comes an unpleasant surprise. If V and W are subspaces, then $V \cup W$ (the set of all vectors in V or W or both) can fail to be one. For example, let V and W be the axes in the plane; we saw above that the union is not a subspace. To make up for this, we *define* the *sum* $V+W$ for the subspaces V and W of U as the set of all u that *can* be obtained as $v+w$ with v in V and w in W. In brief, $V+W=\{v+w\colon v \in V, w \in W\}$. Check that $V+W$ could also be defined as the *span* $((V \cup W))$ of the union.

2.3. PROPOSITION. *$V+W$ is a subspace; it contains V and W.*

PROOF. For $+$, suppose u_1 can be written as v_1+w_1 (with v_1 in V and w_1 in W), and u_2 can be written as v_2+w_2. Then $u_1+u_2=(v_1+w_1)+(v_2+w_2)=(v_1+v_2)+(w_1+w_2)$; now v_1+v_2 is a vector v_3 in V, and $w_1+w_2=w_3$ is in W. The equation $u_1+u_2=v_3+w_3$ shows that u_1+u_2 is in $V+W$

whenever u_1 and u_2 are. We find the same to be true for $\cdot$. Furthermore, $v = v + 0$, $w = 0 + w$ shows $V \subset V + W$, $W \subset V + W$. ∎

Example

If V and W are two lines (through 0) in space, then $V + W$ is the *plane* spanned by the two lines; that is, *any* point in the plane, and no other, can be obtained as sum of a vector in V and one in W. (Of course, if, by accident, the two lines are identical, then $V + W$ just equals that line.)

The notion "sum of two subspaces of a vector space" extends, of course, to more than two subspaces: If $V_1, \ldots, V_k$ are subspaces of U, we naturally define $V_1 + \cdots + V_k$ as the set of all those vectors u that can be written as $v_1 + v_2 + \cdots + v_k$ with v_i in V_i. One proves that this is a subspace of U just as in the case of two terms. And clearly, the associate and commutative laws hold: $(V_1 + V_2) + V_3 = V_1 + (V_2 + V_3)$ and $V + W = W + V$ for any subspaces of U. ∎

Let once more V and W be two subspaces of U. Is the representation $u = v + w$ for u in $V + W$ unique, or could there be several ways of "splitting" u (several choices of v and w)? If V and W are coordinate axes in the plane, then, as we saw earlier, the splitting is unique. On the other hand, if $V = W = U$ (and $U \neq 0$), the splitting is quite nonunique. In fact, we can take for v any vector of U whatsoever, take w as $u - v$, and we will have $u = v + w$ with v in V and w in W. In general the nonuniqueness depends on the intersection $V \cap W$. We describe the "best" case.

2.4. PROPOSITION. *The representation $u = v + w$ (for any u in $V + W$, with v in V and w in W) is unique exactly if $V \cap W$ is* 0.

PROOF. (a) Suppose $V \cap W = 0$. Let two splittings of a vector u be given; that is, $u = v + w = v' + w'$, with v and v' in V and w and w' in W. Then $v - v' = w' - w$. The left side is in V, the right in W; thus, both are in $V \cap W$. But this is 0. Thus, $v - v' = 0$, that is, $v = v'$, and $w' - w = 0$, that is, $w = w'$.

(b) Conversely if $V \cap W \neq 0$, let z be a nonzero vector in $V \cap W$. Then writing $0 = z + (-z)$ gives a splitting of 0 (since z is in V, and $-z$ is in W) *different* from the splitting $0 = 0 + 0$. Adding $0 = z + (-z)$ to any equation $u = v + w$, we get nonuniqueness for any u in $V + W$; namely, $u = (v + z) + (w - z)$. ∎

Notation. If $V \cap W = 0$, we say that the sum $V + W$ is **direct**, and write it as $V \oplus W$. More generally, the sum of subspaces $V_1, \ldots, V_k$ of U is called **direct** and written $V_1 \oplus \cdots \oplus V_k$, if every vector v in the sum appears in (one and) *only one* way as $v_1 + \cdots + v_k$ with v_i in V_i. (For an equivalent notion see Chapter 3, Section 3, Problem 7.)

Examples

1. Two (different) lines through 0 in the plane, or a plane and a line not in the plane—both through 0 in space—gives examples of unique splitting. The span is the whole plane, in the first case, and all of space, in the second case.
2. Two (different) planes through 0 in space give an example of nonuniqueness. The span is all of space, clearly (argument?), the intersection $V \cap W$ is a line.

3. More abstractly, in $\mathbf{R}^4$, let $V=\{X:\ x_1=x_2=0\}$ and $W=\{X:\ x_3=x_4=0\}$. The intersection is, of course, 0. (Any point in it has $x_1=x_2=x_3=x_4=0$.) The span is the whole space $\mathbf{R}^4$. Any $X=[x_1,x_2,x_3,x_4]$ can be written as $[0,0,x_3,x_4]+[x_1,x_2,0,0]$; the splitting is unique.
4. Again in $\mathbf{R}^4$, let $V'=\{X:\ x_2=0\}$ and $W'=\{X:\ x_3=0\}$. The intersection is $\{X:\ x_2=x_3=0\}$; this is not at all 0. The span $V'+W'$ is again all of $\mathbf{R}^4$: any vector $[x_1,x_2,x_3,x_4]$ can be written as $[0,0,x_3,x_4]+[x_1,x_2,0,0]$, but, for instance, also as $[x_1,0,x_3,0]+[0,x_2,0,x_4]$, exhibiting the nonuniqueness.

The second case of the first example and the second example should be kept in mind, as standard cases of uniqueness and the opposite.

PROBLEMS

1. Show the following:
a. If $W \subset V$, then $V+W=V$.
b. Conversely, if $V+W=V$, then $W \subset V$.

2. Show that $W_1 \cup W_2$ is not a subspace and not equal to W_1+W_2, unless one of the two spaces is contained in the other.

3. Let V, W be subspaces of U. Show that the representation $u=v+w$ for vectors of $V+W$ is nonunique exactly up to $V \cap W$; that is, if also $u=v'+w'$, then $v'=v+z$ and $w'=w-z$ with a certain z in $V \cap W$.

4. In $\mathbf{R}^4$, let V be the subspace formed by those vectors X whose first and second coordinates are related by $x_2=2x_1$; similarly, define W by $x_2=3x_1$. Show that $V \cap W$ is the subspace consisting of the X with $x_1=x_2=0$, and $V+W$ is all of $\mathbf{R}^4$. Choose some X in $\mathbf{R}^4$, and split X along V and W in three different ways.

5. Let r and n be integers with $0<r<n$. In $\mathbf{F}^n$ let V be the set of all those vectors whose last $n-r$ coordinates are 0; similarly, $V'=\{X:\ x_1=x_2=\cdots=x_r=0\}$. Show that V and V' are subspaces and that $\mathbf{F}^n$ is *direct* sum of them.

6. Let W_1 and W_2 be subspaces of U, and let V be a subspace of W_1+W_2.
a. Show with an example, using U as ordinary 3-space, that the relation $V=V \cap W_1+V \cap W_2$ does not (always) hold.
b. Show that the relation does hold (and reads $V=W_1+V \cap W_2$) in case W_1 is contained in V; illustrate with an example in space and describe all spaces clearly.

3. LINEAR DEPENDENCE AND INDEPENDENCE

Again, let U be a vector space over F. Suppose a vector u is known to be linear combination of the vectors $u_1,\ldots,u_k$. It can happen that there are several sets of "suitable" coefficients a_i, that the a_i are *not unique*; for example, for $k=2$, if by chance $u_1=u_2$, then the two formally different linear combinations u_1+u_2 and $2\cdot u_1+0\cdot u_2$ are, in fact, the same vector. With more vectors this can be subtler. Suppose, for instance, that u_5 itself is a linear combination of $u_1,\ldots,u_4$, say $u_5=u_1-u_2+2u_3+\sqrt{2}\,u_4$. Then any v that is linear combination of $u_1,\ldots,u_5$ is also linear combination of $u_1,\ldots,u_4$, by the device of substituting for u_5 its expression in terms of $u_1,\ldots,u_4$ and rearranging the terms. The new formula for u can still be thought of as linear combination of all five u_i, but u_5 having coefficient 0. Thus, we clearly have nonuniqueness. These examples lead to the following definition.

3.1. DEFINITION. *The finite sequence* $\{u_1,\ldots,u_k\}$ *of vectors in* U *is linearly dependent (also simply dependent), if one of the* u*'s is a linear combination of the others.*

A more symmetrical form of this definition, which is very often used, is the following:

The family $u_1,\ldots,u_k$ *is (linearly) independent, if the only way to get the vector* 0 *as linear combination of the* u_i *is to take all coefficients equal to* 0, *or, equivalently, if from an equation* $\Sigma_i a_i u_i=0$ *one can conclude that all* a_i *must be* 0. *And the family is dependent, if it is possible to get the vector* 0 *as* $\Sigma_i a_i u_i$ *with not all* a_i *equal to* 0 *(note that this does not say "all* $a_i\neq 0$," *only that at least one* $a_i\neq 0$*); such an expression for* 0 *is called a nontrivial dependence relation between the* u_i.

The two versions of the definition of dependence and independence amount to the same: If one has a nontrivial dependence relation between the vectors, one can "solve" for any one u_i whose coefficient a_i is not 0, and thus express that u_i as linear combination of the others. Conversely, if one u_i equals a linear combination of the others, then by transferring it to the other side of the equation, we get 0 expressed as linear combination of the vectors with not all coefficients 0, since the coefficient of that u_i is -1.

In particular, if one of the vectors, say u_i, is 0, then the family is dependent: We can take $a_i=1$ and the other $a_j=0$. (Or we can get u_i as linear combination of the other u_j with all coefficients 0.)

The point of the definition is this: If $u_1,\ldots,u_k$ are independent and a vector v is linear combination of them, that is, v *can* be written as $\Sigma a_i u_i$, then the coefficients a_i are *unique*. To show that the definition accomplishes this, we proceed as follows: Assume that the u_i are independent. Suppose $\Sigma a_i u_i=\Sigma b_i u_i$, with two sets of coefficients a_i and b_i which in advance are not known to be equal. (Think of two people, each claiming to construct the same vector v from the u_i as linear combination.) We can rewrite $\Sigma a_i u_i = \Sigma b_i u_i$ as $(\Sigma a_i u_i) - \Sigma(b_i u_i) = 0$ or (by distributivity, etc.) $\Sigma(a_i - b_i)u_i = 0$. The u_i being independent, the coefficients $a_i - b_i$ in this relation must be 0, that is, $a_i = b_i$, for all i. ■

Examples

1. $k=1$, that is, a single vector u. This is independent if $r \cdot u = 0$ can hold only for $r=0$. Clearly, that is equivalent to $u \neq 0$; that is, u is not the 0-vector.

2. $k=2$: Two vectors u_1 and u_2 are dependent if one is a multiple of the other. We cannot say which is the "one" and which is the "other"; for example, if $u_1 = 0$ and $u_2 \neq 0$, we have $u_1 = 0 \cdot u_2$, but $u_2 \neq r \cdot u_1$ for any r; if both are nonzero, then each one is a multiple of the other–in a dependence relation $a_1 u_1 + a_2 u_2 = 0$ neither a_i can now be 0 (reason?), and we can solve for either vector. Geometrically, this means that they lie on the same line (through 0). ■

To prove independence of a family $\{u_1, \ldots, u_k\}$, one always has to go through the following procedure: *Assume* $\Sigma a_i u_i = 0$, with (unspecified) scalars a_i; *prove* that the a_i must all be 0.

Example

The vectors $X_1 = [2,1,0]$, $X_2 = [0,2,-1]$, and $X_3 = [0,0,4]$ of $\mathbf{R}^3$ are independent. Say $a_1 X_1 + a_2 X_2 + a_3 X_3 = 0$. On the left side, the first entry is $a_1 \cdot 2 + a_2 \cdot 0 + a_3 \cdot 0, = 2a_1$. On the right side, the first entry is 0, of course. Thus, a_1 must be 0. The second entry on the left is $a_1 \cdot 1 + a_2 \cdot 2 + a_3 \cdot 0$. This reduces to $2a_2$, since a_1 is 0, as we just saw. On the right side we have 0 again, and so a_2 is 0. The third entry on the left is $a_1 \cdot 0 + a_2 \cdot -1 + a_3 \cdot 4$, which reduces to $4a_3$, since $a_1 = a_2 = 0$. Again we have 0 on the right, and so a_3 is 0. Thus, the only way to get $a_1 X_1 + a_2 X_2 + a_3 X_3$ to be 0 is to take *all three* a_i equal to 0. ■

We consider an example for dependence: The three "vectors" (in P^n) $p_1(x) = 1 + 3x + 4x^2 - x^3$, $p_2(x) = 1 - 3x + 2x^2 + x^3$, $p_3(x) = 1 - 12x - x^2 + 4x^3$ are dependent, because of the nontrivial relation $3p_1(x) - 5p_2(x) + 2p_3(x) = 0$. Check that the relation holds. (We will see later how one goes about *finding* such relations.)

We consider next, the standard example of independent vectors in $\mathbf{R}^n$ or $\mathbf{C}^n$: As *standing notation* we denote by E^i the vector with 1 as ith entry, and 0 for all other entries; we call E^i the ith **standard** vector. Thus we have $E^1 = [1,0,\ldots,0]$, $E^2 = [0,1,0,\ldots,0], \ldots, E^n = [0,\ldots,0,1]$. Strictly speaking, the E^i should carry an index n to tell us the *number* of components; usually the n will be clear. If needed, we write $E^i_{(n)}$; for example, $E^2_{(3)}=[0,1,0]$.

The E^i have the following important property: Each vector $X=[x_1,\dots,x_n]$ is equal to the linear combination $\Sigma x_i E^i$ (using the *components* x_i of X as *coefficients*). For example, x_1E^1 is $x_1\cdot[1,0,\dots,0]=[x_1,0,\dots,0]$; similarly, x_2E^2 is $[0,x_2,0,\dots,0]$, etc. Now we add all this up; for example, in R^3, $[2,-1,4]=2E^1-E^2+4E^3$.

Furthermore, $E^1,\dots,E^n$ are independent in F^n.

PROOF. We just saw that the linear combination $a_1E^1+\cdots+a_nE^n$ is precisely the vector $[a_1,\dots,a_n]$. Therefore, if Σa_iE^i is 0, so is $[a_1,\dots,a_n]$. But that *means* $a_1=a_2=\cdots=a_n=0$. ■

The subspace ("line") spanned by E^i is called the ith coordinate axis of F^n.

PROBLEMS

1. In P^n let $p_1,\dots,p_k$ be polynomials, with no two of the same degree. Show that they are independent.

2. Show that a family $u_1,\dots,u_r$ of nonzero vectors in U is independent exactly if some u_j is linear combination of the "preceding" u_i, that is, of those with $i<j$. Why is the hypothesis "nonzero" needed?

3. Show: If a subfamily $u_{i_1},\dots,u_{i_r}$ of a family $u_1,\dots,u_k$ of vectors is dependent, so is the whole family. (Here $i_1,\dots,i_r$ are some of the indices $1,\dots,k$ in increasing order, with $r\geqslant 1$, of course.)

4. Show in detail: If $u_1,\dots,u_r$ are independent and v is a vector such that $u_1,\dots,u_r,v$ are dependent, then v is a linear combination of the u_i (we used this principle implicitly in the text already.)

5. Show that the vectors $[1,1,0,0]$, $[1,-1,0,1]$, and $[0,0,1,1]$ of R^4 are independent. (Start with "Assume $aX+bY+cZ=0$.")

4. MATRICES, ELEMENTARY OPERATIONS, COLUMN-ECHELON FORM

Naturally, the question arises: How does one recognize dependence or independence? We describe a procedure for this in the most important case F^n (R^n or C^n); we will see later that all other cases reduce to this one. The procedure goes under the name of "reduction of a matrix to column-echelon form."

First, we introduce some general notions. There are three operations, the **elementary operations**, on a family $\{u_1,\dots,u_k\}$ of vectors (in a vector space U) that preserve independence or dependence of the family: (1) interchanging two of the vectors, (2) multiplying a vector by a nonzero scalar, and (3) replacing any one u_i by $u_i'=u_i+r\cdot u_j$, with any scalar r and any index j ($\neq i$). The operations are reversible; that should be clear for (1) and (2). For (3) one should replace u_i' by $u_i' - r \cdot u_j$ to get back to u_i. That the operations preserve independence or dependence should again be clear for (1) and (2). As for (3), suppose there is a (nontrivial) dependence relation for the old family. Substitute $u_i' - ru_j$ for u_i (the two are equal) and collect terms. The result is a nontrivial dependence relation of the new family. If the factor of u_i was not 0, so is now that of u_i' (it is the same factor!); if it was 0, then the substituting does not change *any* of the coefficients, and so there is still at least one nonzero one. The procedure is similar for the other direction; as noted above, there is an operation of type (3) that gets us back from the new family to the old family. ■

All this can be applied, in particular, to families of column vectors. To handle this efficiently, we introduce the concept "matrix": Let $X_1, \ldots, X_k$ be k vectors in F^n. We place the columns X_i next to each other, in order (from left to right), without any brackets around the vectors, and enclose the resulting array of numbers in brackets. The result is called a **matrix**, more specifically an $n \times k$ matrix, or a matrix of n rows and k columns.For example, the vectors [2,5], [1,0], and $[\pi, -7]$ in R^2 give rise to the 2×3 matrix

$$\begin{pmatrix} 2 & 1 & \pi \\ 5 & 0 & -7 \end{pmatrix}.$$

If we have an $n \times k$ matrix, we can also think of it as made up of n row vectors $A_1, \ldots, A_n$, all belonging to F^k, written below each other, in order (from top to bottom). In the example above the matrix is made up of the two row vectors $(2, 1, \pi)$ and $(5, 0, -7)$. An $n \times 1$ matrix is the same as a column vector (of length n); a $1 \times k$ matrix is the same as a row vector (of length k).

In any matrix the ith component of the jth column vector (which is the same as the jth component of the ith row vector) is called the (i,j)-entry or -component. In the example the (2,3)-entry is -7. We denote matrices by capital letters like $A, B, \ldots, L, M, \ldots$. For a matrix A, one writes the (i,j)-entry as a_{ij} (usually with the corresponding lower case letter); the two indices denote row and column. If we call the example above M, then we have, for example, $m_{23} = -7$. The "general" $p \times q$ matrix A will often be written as

$$A = \begin{pmatrix} a_{11} & a_{12} & \cdots & a_{1q} \\ a_{21} & & \cdots & \vdots \\ \vdots & & & \\ a_{p1} & a_{p2} & \cdots & a_{pq} \end{pmatrix},$$

or also $(a_{ij})_{1,1}^{p,q}$ or just (a_{ij}).

The entries of our matrices are numbers or indeterminates (symbols or variables, like a_{ij}). Later on we will meet more general entries, for example, polynomials in some variable. We also allow ourselves the following handy piece of notation: If M is an $n \times k$ matrix with column vectors $X_1,\ldots,X_k$ and row vectors $A_1,\ldots,A_n$, then we write $M=(X_1,\ldots,X_k)$, that is, a $1\times k$ matrix or row vector, whose entries are column vectors, "a row of columns," and, similarly, $M=[A_1,\ldots,A_n]$, that is, an $n\times 1$ matrix or column vector, whose entries are row vectors, "a column of rows."

We now come back to our main topic: How to recognize dependence of a family of vectors.

Let then $X_1,\ldots,X_k$ be vectors in F^n; we form the $n\times k$ matrix $(X_1,\ldots,X_k)$ that has the X_i as columns. The elementary operations ((1), (2), and (3) above) on vectors, now appear as operations on the matrix: (1) interchanges two columns, (2) multiplies a column by r ($\neq 0$), and (3) replaces the ith column X_i by $X_i + rX_j$ (addition as for vectors in F^n!). We now use these operations to *change* the given matrix into a new one with a specific form, the **column-echelon form.**

A matrix in column-echelon form looks as follows: The first nonzero entry (if any), the "leading term," of any column is 1; and for each column this 1 occurs at least one step lower than the leading term in the preceding column. The "general" shape is then this:

$$\begin{pmatrix}
0 & 0 & 0 & \cdots \\
\vdots & \vdots & \vdots & \\
0 & & & \\
1 & \vdots & \vdots & \\
* & & & \\
 & 0 & & \\
\vdots & 1 & \vdots & \\
 & * & & \\
 & & 0 & \\
\vdots & \vdots & 1 & \\
 & & * & \\
 & * & & \\
 & \vdots & & \cdots \\
* & * & * &
\end{pmatrix}$$

Here the $*$'s mean arbitrary scalars. There may be one or several 0-columns at the right end. (For a $p \times q$ matrix with $q > p$ in this form the columns number $p+1, p+2, \ldots$ are necessarily 0.)

We describe now the procedure, using the column operations, for changing a given matrix, say A, to column-echelon form. Consider the first row that has a nonzero entry (usually the top row). By (1) bring the column of some such nonzero element all the way to the left; by (2) make that element equal to 1; by (3) make the other elements in that row equal to 0 (by adding or subtracting multiples of the first column). The first column is left alone from now on. Use the same process on the remaining columns to generate a suitable second column, and so on. Clearly this will end with the matrix in column-echelon form.

Example

$$M = \begin{pmatrix} 0 & 2 & 3 \\ 2 & -1 & 1 \end{pmatrix} \overset{(1)}{\rightarrow} \begin{pmatrix} 2 & 0 & 3 \\ -1 & 2 & 1 \end{pmatrix} \overset{(2)}{\rightarrow} \begin{pmatrix} 1 & 0 & 3 \\ -1/2 & 2 & 1 \end{pmatrix}$$

$$\overset{(3)}{\rightarrow} \begin{pmatrix} 1 & 0 & 0 \\ -1/2 & 2 & 5/2 \end{pmatrix} \overset{(2)}{\rightarrow} \begin{pmatrix} 1 & 0 & 0 \\ -1/2 & 1 & 5/2 \end{pmatrix} \overset{(3)}{\rightarrow} \begin{pmatrix} 1 & 0 & 0 \\ -1/2 & 1 & 0 \end{pmatrix}.$$

The type of operation used is indicated for each step. In the third step we subtracted $3 \cdot$ first column from the third one; in the fourth we multiplied the second column by $1/2$. The reader is advised to work out a couple of examples, and to try some large matrix.

Note that the shape of the matrix does not change in this process. The 0-columns (or rows) *cannot* be dropped (although in practice, as a short cut, we will sometimes do just that). By the way, in practice one often does not reduce the leading terms to 1.

What we can do with columns, we can also do with rows. There is the analogous concept of a matrix in row-echelon form: The leading (first nonzero) term of any row is 1, and the leading term of any occurs later than (= to the right of) the leading term of the preceding row. One can bring any matrix to row-echelon form by the three elementary row operations: interchanging of two rows, multiplying a row by a nonzero scalar, and adding a multiple of a row to another row.

Remark. The column(or row)-echelon form is not quite unique. Starting from a given matrix one can end up with slightly different matrices. However, the location of the leading 1's is always the same (Section 5, Problem 8). One gets uniqueness if one uses the *reduced* column(or row)-echelon form, in which in the row (or column) of any leading 1 *all* other entries have been reduced to zero.

PROBLEMS

1. Bring the following matrices into column-echelon form:

i.
$$A=(0\ 0\ 2\ 4\ -1\ 3).$$

ii.
$$B=\begin{pmatrix} 0 & 2 & 1 \\ 1 & 1 & 1 \\ 2 & 3 & 1 \end{pmatrix}.$$

iii.
$$C=\begin{pmatrix} 1 & 3 & 2 \\ 2 & -2 & 1 \\ -3 & 5 & 0 \\ 1 & -1 & 1 \end{pmatrix}.$$

iv.
$$M=\begin{pmatrix} 1 & 1 & 2 & 1 \\ 2 & 2 & 4 & 2 \\ 1 & -1 & 1 & 3 \\ -4 & 2 & -5 & -10 \end{pmatrix}.$$

2. Bring the following matrices into row-echelon form:

i.
$$N=\begin{pmatrix} 4 & 3 & 2 & -1 \\ 3 & 2 & 2 & 1 \\ -2 & -2 & -1 & 2 \\ 9 & 4 & 3 & 3 \end{pmatrix}.$$

ii.
$$D=\begin{pmatrix} 0 & 3 & 2 & 3 \\ 2 & 2 & 1 & 5 \\ 1 & 3 & 2 & 5 \end{pmatrix}.$$

5. USE OF THE COLUMN-ECHELON FORM

We now come to the promised method of testing a family $\{X_1,\ldots,X_k\}$ of vectors in F^n for independence. We form the matrix $M=(X_1,\ldots,X_k)$ and bring it into column-echelon form, say M', with columns $Y_1,\ldots,Y_k$. By the discussion in Section 4, the X's are dependent precisely if the Y's are so. And for the Y's this is very easy to decide: For a matrix in column-echelon form, the column vectors are dependent exactly if one (or several) is 0. Of course, if one is 0, so are all the ones to the right of it.

PROOF. If one vector is 0, then the family is dependent (see Chapter 2, Section 3; coefficient 1 for the 0-vector, 0 for the others). On the other hand, suppose none of the vectors is 0. Given a relation $\Sigma a_i Y_i = 0$, we have to show that *all* a_i must be 0. Consider the leading 1 of Y_1. The corresponding entries of the other Y_i are 0, and so that entry of $\Sigma a_i Y_i$ is $a_1 (= a_1 \cdot 1 + a_2 \cdot 0 + \cdots + a_k \cdot 0)$. Since $\Sigma a_i Y_i$ is supposed to be 0, we must have $a_1 = 0$. Now look at the leading term of Y_2. In $\Sigma a_i Y_i$ we get a_2 $(a_2 \cdot 1 + a_3 \cdot 0 + \ldots)$ as the corresponding entry; thus a_2 must be 0, etc. This goes all the way to a_k since column Y_k has a leading 1. ■

Example

$$X_1 = [1\ 1\ -1\ 2], \qquad X_2 = [2\ 0\ 3\ 2], \qquad X_3 = [3\ 1\ 2\ 4].$$

$$\begin{pmatrix} 1 & 2 & 3 \\ 1 & 0 & 1 \\ -1 & 3 & 2 \\ 2 & 2 & 4 \end{pmatrix} \xrightarrow{(3)} \begin{pmatrix} 1 & 0 & 0 \\ 1 & -2 & -2 \\ -1 & 5 & 5 \\ 2 & -2 & -2 \end{pmatrix} \xrightarrow{(2)} \begin{pmatrix} 1 & 0 & 0 \\ 1 & 1 & -2 \\ -1 & -\frac{5}{2} & 5 \\ 2 & 1 & -2 \end{pmatrix}$$

$$\xrightarrow{(3)} \begin{pmatrix} 1 & 0 & 0 \\ 1 & 1 & 0 \\ -1 & -\frac{5}{2} & 0 \\ 2 & 1 & 0 \end{pmatrix} = M'.$$

In the first step we subtracted $2X_1$ from X_2, and $3X_1$ from X_3.

Answer: X_1, X_2, and X_3 are dependent.

Remarks: After the first step two equal columns appear; this already shows dependence (clear?). Also, in retrospect, we might have seen at once that $X_3 = X_1 + X_2$, giving dependence. The moral that we can derive is that there are often some shortcuts to the general procedure.

One general result follows from this.

5.1. **PROPOSITION.** *A family of more than n vectors in F^n is necessarily dependent.*

PROOF. In column-echelon form there must be 0-columns, since the leading 1's move down at least one step from any column to the next. ■

Question: What does the column-echelon form for n independent vectors in F^n look like?

Here is an important operation that one can perform with the column-echelon form: "Reduce" a family $\{X_1,\ldots,X_k\}$ of vectors to an independent one. This means that one wants to find a subfamily (by omitting, or dropping, some of the vectors) that is (a) independent and (b) such that any omitted vector is linear combination of the ones one keeps (so that one "does not lose anything"—the span does not change). To accomplish this, one goes through the vectors in succession and removes ("crosses out") any X_i that is 0 or a linear combination of the preceding ones (see the last remark of this section). Another method is to bring the matrix $(X_1,\ldots,X_k)$ into column-echelon form, and to drop from the list any X_i whose column has been reduced to 0 (keeping track, of course, of any interchanges).

Example

Reduce the family $[1,0,1,1]$, $[-1,1,0,2]$, $[0,1,1,3]$, $[2,-1,2,1]$, and $[2,-2,1,-2]$ (of vectors in R^4) to an independent one.

$$\begin{pmatrix} 1 & -1 & 0 & 2 & 2 \\ 0 & 1 & 1 & -1 & -2 \\ 1 & 0 & 1 & 2 & 1 \\ 1 & 2 & 3 & 1 & -2 \end{pmatrix} \to \begin{pmatrix} 1 & 0 & 0 & 0 & 0 \\ 0 & 1 & 1 & -1 & -2 \\ 1 & 1 & 1 & 0 & -1 \\ 1 & 3 & 3 & -1 & -4 \end{pmatrix}$$

$$\to \begin{pmatrix} 1 & 0 & 0 & 0 & 0 \\ 0 & 1 & 0 & 0 & 0 \\ 1 & 1 & 0 & 1 & 1 \\ 1 & 3 & 0 & 2 & 2 \end{pmatrix} \to \begin{pmatrix} 1 & 0 & 0 & 0 & 0 \\ 0 & 1 & 0 & 0 & 0 \\ 1 & 1 & 1 & 0 & 0 \\ 1 & 3 & 2 & 0 & 0 \end{pmatrix}.$$

Answer. X_3 and X_5 should be dropped. In the last step we interchanged the third and fourth columns, and subtracted the (new) third column from the fifth. (The first step consisted in subtracting multiples of the first column from the other columns to produce the zeros on top.) Minor variations are possible: After the first step we see that X_3 should be dropped because the second column is the same as the third column; therefore we could have *omitted* the third column from the rest of the computation (although removing a column is not one of the elementary operations).

Implicit in what we said is the following important fact: The span of a finite family $X_1,\ldots,X_k$ in F^n or, more generally, the span of a family $u_1,\ldots,u_k$ in any vector space U does not change if one applies the elementary operations to the family (once, or repeatedly). For (1) and (2) this is obvious; for (3) we see that because of $u_i = u_i' - ru_j$ any vector in the span of either family is also in that of the other one. ■

The column-echelon form also makes it possible to decide whether a given vector Z is linear combination of given vectors $X_1, \ldots, X_k$. Bring $(X_1, \ldots, X_k)$ into column-echelon form $(Y_1, \ldots, Y_k)$; the span of the X_i equals that of the Y_i, as just noted. Now decide whether Z can be reduced to 0 by successively subtracting multiples of $Y_1, \ldots, Y_k$. By keeping track of the moves one can even find the a_i in $Z = \Sigma a_i X_i$ (if such a relation exists, i.e., if Z is in the span of the X_i). We shall replace this whole procedure with a better one later (Chapter 4, Section 5.1).

Example

Express $Z = [2, -5, 1]$ as linear combination of $X_1 = [2, 1, -1]$ and $X_2 = [1, 2, -1]$ (if possible). We start with the column-echelon process:

$$\begin{pmatrix} 2 & 1 \\ 1 & 2 \\ -1 & -1 \end{pmatrix} \to \begin{pmatrix} 1 & 2 \\ 2 & 1 \\ -1 & -1 \end{pmatrix} \to \begin{pmatrix} 1 & 0 \\ 2 & -3 \\ -1 & 1 \end{pmatrix}.$$

The new column vectors are $Y_1 = X_2$ and $Y_2 = X_1 - 2X_2$. Now we try to reduce Z to 0. $Z - 2Y_1 = [0, -9, 3]$ (the first entry has become 0), and then $Z - 2Y_1 - 3Y_2 = [0, 0, 0] = 0$ (we made the second entry 0, and the third happened to turn into 0 also). Thus $Z = 2Y_1 + 3Y_2 = 2X_2 + 3(X_1 - 2X_2) = 3X_1 - 4X_2$.

PROBLEMS

1. Are the following vectors of F^4 independent:

$$[1,0,1,1], \quad [1,1,-1,1], \quad [2,1,2,2], \quad \text{and} \quad [0,0,1,0]?$$

2. Is $[2, 1, -1, -1]$ a linear combination of

$$[2,-1,-2,-8], \quad [-1,0,2,3], \quad \text{and} \quad [1,-2,1,-3]?$$

3. Reduce to an independent family:

$$[2,0,1,0,1], \quad [0,0,1,1,-1], \quad [0,1,1,0,2], \quad \text{and} \quad [2,1,1,-1,4].$$

4. Reduce to an independent family:

$$[1,2,-1,1,0,2], \quad [2,4,-1,4,1,3], \quad [1,2,-2,-1,-1,3], \quad \text{and} \quad [-1,-2,2,3,0,0].$$

5. Reduce to an independent family (in C^4):

$$[i,1+i,2,-1], \quad [1-i,1,i,2], \quad [-i,i,3i,2-i], \quad \text{and} \quad [2+i,2+2i,1+i,1+2i].$$

6. Are the following vectors of P^3 independent:

$$4x^3+3x^2+2x-1, \quad 5x^3+4x^2+3x-1, \quad -2x^3-2x^2-x+2,$$

$$\text{and} \quad 11x^3+6x^2+4x+1?$$

(*Hint*. Each polynomial can be replaced by a column vector.)

7. Use the method sketched in the text to show that $Y=[4,2,-1]$ is linear combination of $X_1=[2,0,1]$ and $X_2=[1,-1,2]$ and to find the coefficients.

8. Prove the statement in the Remark at the end of Section 4. (*Hint*. If M' and M'' are two reductions of M, then every column of M' is a linear combination of columns of M'', and the same holds with M' and M'' interchanged.)

3

DIMENSIONS; BASIS

The intuitive idea of **dimension** of a vector space is made precise through the notion of **basis** and the main theorem about bases. We collect a number of facts pertaining to these concepts, among them the **modular law**, which has to do with the dimensions of the sum and the intersection of two subspaces. Finally, we describe the main device for computing in vector spaces—**representing** a vector by its **components** relative to a basis.

1. THE NOTIONS BASIS AND DIMENSION

One "knows" that a line has one dimension, the plane has two, space has three. If one tries to explain, one usually says: one can go in one (or two or three) "different" directions, but not more. We make this more precise in vector language: That we go in a direction means that we look at a vector; that we can go in so many "different" directions, means that we are given so many *independent* vectors; that there are no more directions to go, means that any other vector is dependent on the given ones (or, equivalently, that the given ones span the space). This suggests the following definition.

1.1. DEFINITION. *A family* $\{u_1,\ldots,u_k\}$ *of vectors, in a vector space* U, *is a basis for* U, *if* (*a*) *it is independent and* (*b*) *it spans* U.

Another way to express this is to say that $u_1,\ldots,u_k$ are a basis of U if every vector u of U *can* be written as linear combination of the u_i, with *unique* coefficients. The "can be written..." corresponds to (b), and the "unique..." to (a) (see Chapter 2, Section 1). Bases will sometimes be denoted briefly by letters like α, β,... (rather than by writing out the vectors in the basis).

Examples

1. The vectors $Z_1=[1,1]$ and $Z_2=[1,-1]$ form a basis for $\mathbb{R}^2$. They are independent (neither is a multiple of the other), and they span $\mathbb{R}^2$, since we can manufacture $E^1=1/2Z_1+1/2Z_2$ and $E^2=1/2Z_1-1/2Z_2$, and, thus, any $X=x_1E^1+x_2E^2$ from Z_1 and Z_2.

2. P^n = vector space of polynomials of degree $\leqslant n$. *One* basis is given by the very special polynomials $1,x,x^2,x^3,\ldots,x^n$; this just states the obvious fact that any polynomial can be written as $\Sigma_0^n a_i x^i$ with unique a_i. *Another* basis is $\{1,x-1,(x-1)^2,(x-1)^3,\ldots,(x-1)^n\}$. In fact, any polynomial can be written as $b_0+b_1(x-1)+b_2(x-1)^2+\cdots+b_n(x-1)^n$ with unique b's (Taylor expansion at $x_0=1$). (Using $x-a$, with any a instead of 0 or 1, we get many other bases; this does *not* exhaust the possible bases of P^n.) One should realize that for a given polynomial the a's in $\Sigma a_i x^i$ and the b's in $\Sigma b_i(x-1)^i$ are not the same numbers but that they determine each other.

3. We note an important example. The family $\{E^1,\ldots,E^n\}$ of standard vectors forms a basis σ, called the **standard basis**, for $\mathbb{F}^n$; the vectors span $\mathbb{F}^n$ (any $X=[x_1,\ldots,x_n]$ can be written as $\Sigma x_i E^i$), and they are independent (see Chapter 2, Section 3; this is quite simple). Similarly, the vectors $E_1=(1,0,\ldots,0)$, $E_2=(0,1,0,\ldots,0),\ldots,E_n=(0,\ldots,0,1)$ form the **standard basis** for the space $(F^n)'$ of row vectors. (Note the position of the indices on the E_i.)

4. Let $X_1,\ldots,X_t$ be vectors in $\mathbb{F}^n$; bring the matrix $(X_1,\ldots,X_t)$ to column-echelon form, say $(Y_1,\ldots,Y_t)$. The *non-zero ones* of the Y's are a basis for the span $((X_1,\ldots,X_t))$; (a) and (b) hold by Chapter 2, Section 5. ■

As suggested by the introductory remarks, we should say that U has dimension k, if it has a basis of k vectors. But for this to make sense, we have to make sure that U has at least one basis, *and* that any two bases have the same number of elements. We shall do this now. Since we want bases to be finite sets, we need one more definition.

1.2. DEFINITION. *A vector space U is finitely generated if it is the span of some finite subset. Note that then there might be (and will be) many different finite subsets that span U.*

Examples

Most of the examples in Chapter 1 are finitely generated.

1. The plane is spanned by two vectors; space, by three.

2. For P^n (polynomials in x of degree less than or equal to n), convince yourself that a y (finite) set of polynomials of degree less than or equal to n, such that each degree from 0 to n occurs at least once, spans P^n.

such that each degree from 0 to n occurs at least once, spans P^n.

We should give at least one example of a *not* finitely generated vector space. Let P^∞ be the vector space consisting of *all* polynomials in a variable x, of any degree whatever. No finite number of polynomials can generate this, clearly, because the degree of the polynomials generated could not go beyond the maximal degree of the given ones.

Now the main theorem that justifies the idea of dimension.

1.3. THEOREM (AND DEFINITION). *Let U be a finitely generated vector space different from* 0. *Then U has* (*many*) *bases, and all bases have the same number of elements. This common number is called the dimension of U, denoted by* $\dim U$.

The 0-space is a bit exceptional since one cannot find any independent family of vectors in it (0 is the *only* vector in 0), and so there cannot be any basis. This is annoying since in many arguments one has to say "unless $U=0$" and then make special provisions for 0; that case is usually completely trivial, and we will often skip this detail. At any rate, the dimension of the space 0 is 0 *by definition*.

From now on, any vector space is automatically assumed finitely generated.

For Theorem 1.3 we first show *existence* of bases. This rests on a simple idea that we met already in Chapter 2, Section 1 ("reduction to independence").

1.4 PROPOSITION. *Let $\{u_1,\ldots,u_m\}$ be a finite family in U (with not all u_i equal to* 0), *then there is an independent subfamily with the same span.*

PROOF. We go through the family from u_1 to u_m and drop any u_i that is 0 or a linear combination of the earlier vectors. What remains, clearly satisfies the proposition. The condition "some u_i not 0" insures that something is left over. (To be a bit more fancy, let us consider all possible independent subfamilies (the number of such is, of course, finite and positive). Of these, take one, say $\{u_{i_1},\ldots,u_{i_k}\}$, with the *largest* number of elements. Any other u_i is a linear combination of the u_{i_r}; otherwise we would get a larger independent subfamily. And so this subfamily has the same span as the original one.) ■

1.5 PROPOSITION. *If U is finitely generated, it has at least one basis.*

PROOF. Apply Proposition 1.4 to any finite family that spans U. ■

Next, to show the uniqueness of the number of elements in a basis, we first prove a slightly weaker statement.

1.6 PROPOSITION. *Let U be spanned by the finite family $\{u_1,\ldots,u_m\}$ (not* *assumed independent). Let $v_1,\ldots,v_m$ be m independent vectors in U (same m). Then the v_i also span U.*

1.7. COROLLARY. *If U is spanned by m vectors, then any $m+1$ or more vectors are dependent; equivalently, if $v_1,\ldots,v_k$ are independent, then the inequality $k \leqslant m$ holds.*

PROOF OF COROLLARY 1.7. Let $v_1,\ldots,v_{m+1}$ be given. Then $v_1,\ldots,v_m$ are either dependent or independent. In the latter case, $v_1,\ldots,v_m$ span U by Proposition 1.6, but then v_{m+1} is linear combination of them. In either case, $v_1,\ldots,v_{m+1}$ are dependent. ■

PROOF OF PROPOSITION 1.6. The proof is very general and, therefore, somewhat abstract. The idea is sometimes called the "(Steinitz) exchange principle." It consists in exchanging one u_i at a time for a v_i, without changing the span.

We start with v_1. We have $v_1 = a_1u_1 + \cdots + a_mu_m$ with suitable a_i, since the u_i span U. Not all the a_i are 0, since v_1 is not 0 by independence of the v_i. For simplicity of writing, we shall assume that a_1 is not 0 (otherwise we renumber the u_i; this trick will be used repeatedly). Then we can solve for u_1 in terms of $v_1, u_2,\ldots,u_m$; we get an expression $u_1 = r_1v_1 + r_2u_2 + \cdots + r_mu_m$. It follows that the span of $v_1, u_2,\ldots,u_m$ is the same as that of $u_1,\ldots,u_m$ (since from each set we can get the vectors in the other set as linear combinations); therefore, $((v_1, u_2,\ldots,u_m)) = U$.

We turn to v_2. The last equation shows that it can be written as linear combination of $v_1, u_2,\ldots,u_m$; we have $v_2 = b_1v_1 + b_2u_2 + \cdots + b_mu_m$. The coefficients $b_2,\ldots,b_m$ cannot all be 0; otherwise, v_2 would be a multiple of v_1, contradicting independence. We may assume $b_2 \neq 0$. We can then solve for u_2 and get $u_2 = s_1v_11 + s_2v_2 + s_3u_3 + \cdots + s_mu_m$. It follows that the spans of $v_1, u_2,\ldots,u_m$ and of $v_1, v_2, u_3,\ldots,u_m$ are equal, and so $((v_1, v_2, u_3,\ldots,u_m)) = U$.

We turn to v_3. We write it as $c_1v_1 + c_2v_2 + c_3u_3 + \cdots + c_mu_m$. The coefficients $c_3,\ldots,c_m$ cannot all be 0, and we can "exchange" v_3 for one of the u_i, say u_3, obtaining $((v_1, v_2, v_3, u_4,\ldots,u_m)) = U$. We continue, until at the end we find $((v_1, v_2,\ldots,v_m)) = U$. Try to see clearly how independence of the v_i is used in this argument. ■

We come to the main consequence of this, which will finish the proof of Theorem 1.3, namely that any two bases of U have the same *number* of elements. In Corollary 1.7, each of the two bases can take the role of the first (spanning) or second (independent) family (since a basis spans *U and* is independent). Thus each basis has *at most* as many elements as the other, and the two numbers must be equal.

The notion of dimension of a (finitely generated) vector space is now established.

Main Example

$\mathbf{R}^n$ and $\mathbf{C}^n$ have dimension n. (If this were not so, our definition would be defective.)

PROOF. We saw in Section 1, Example 3, that $\mathbf{F}^n$ has a basis of precisely n elements, namely the standard basis.

Second Example

P^n, the vector space of polynomials of degree $\leqslant n$, has dimension $n+1$, since, as we saw in Section 1, Example 2, it has bases consisting of $n+1$ vectors each (of course, one basis is enough to establish the dimension).

More Examples

The vector space of solutions of $y''+y=0$. One knows that any solution can be written as $a\cdot \sin x + b\cdot \cos x$, and that $\sin x$ and $\cos x$ are independent (neither is a multiple of the other). Thus, $\sin x$ and $\cos x$ form a basis; and the dimension is 2.

The vector space $\mathbf{C}$ over $\mathbf{R}$ (see Chapter 1, Section 2.g). Here 1 and i are a basis, since any complex number z can be written uniquely in the form $a+b\cdot i$, or better $a\cdot 1+b\cdot i$, with a and b in $\mathbf{R}$. Thus the dimension is 2. However, the dimension of $\mathbf{C}$ *over* $\mathbf{C}$ is 1 since 1 by itself is a basis. (Any z can be written uniquely as $c\cdot 1$, with c in $\mathbf{C}$, namely with $c=z$!)

As "generic" example, suppose a family $\{u_1,\ldots,u_k\}$ of vectors can be reduced to the independent subfamily $\{u_{i_1},\ldots,u_{i_r}\}$. Then the dimension of the span $((u_1,\ldots,u_k))$ is r, since the subfamily is clearly a basis for it.

PROBLEMS

1. Find the dimensions of the spaces spanned by the vectors given in Problems 3–5 of Chapter 2, Section 5.

2. Show that the vectors $[0,1,1]$, $[1,0,1]$, and $[1,1,0]$ form a basis for $\mathbf{R}^3$. (*Suggestion*. Show that they are independent and that one can generate the standard vectors E^1, E^2, and E^3 from them.)

3. Find a basis for $\mathbf{F}^4$ containing the vector $[1,1,1,1]$.

4. Show that $X_1=[1,2]$ and $X_2=[1,3]$ form a basis for $\mathbf{R}^2$ by showing that any $X=[x,y]$ can be written, and uniquely so, as aX_1+bX_2. (Show that one can solve for a and b uniquely in terms of x and y.)

5. Express $3+2x+x^2$ as $\Sigma b_i(x-1)^i$.

6. The vector space $\mathbf{C}^n$ (over $\mathbf{C}$) is also a vector space over $\mathbf{R}$ (by "forgetting to use i," see Chapter 2, Section 2.g). Show that the vectors $E^1, iE^1, E^2, iE^2, \ldots, E^n, iE^n$ form a basis over $\mathbf{R}$. What is the dimension of $\mathbf{C}^n$ over $\mathbf{R}$?

2. FACTS ABOUT DIMENSION AND BASIS

1. In a vector space U of dimension n, fewer than n vectors cannot span the space, and more than n vectors are automatically dependent.

PROOF. Suppose $u_1,\dots,u_p$ with $p<n$, span U; by reducing to an independent set we get a basis for U with *at most p* elements, which is impossible. The second part is simply Corollary 1.7. If $\dim U = n$, then U is spanned by n elements (take any basis), and then any independent family has at most n elements. (A special case of this occurred already in the discussion of the column-echelon form of matrices (Chapter 2, Section 1).

We rephrase this as follows. The *dimension* of a (finitely generated) vector space U has three equivalent meanings: (a) number of vectors in any basis; (b) maximum number of vectors in an independent set; and (c) minimum number of vectors in a spanning set. ■

2. Let $u_1,\dots,u_n$ be n vectors in a vector space U of dimension n(same n!), then if they span U, they are independent, and conversely. In either case they form a basis. Thus, if one has the right number of vectors, one needs only one of the two properties that define a basis.

PROOF. Suppose they span U. If they were dependent, we could reduce to fewer than n vectors that still span U and get a basis of fewer than n elements, contradicting the main result about dimension.

Suppose they are independent. Let v be any vector in U. Then the family $u_1,\dots,u_n$, v is dependent by our Fact 1. But then v is linear combination of the u_i (in a nontrivial relation between $v,u_1,\dots,u_n$ the factor of v must be nonzero, since otherwise the u_i would be dependent; and so we can solve for v, see Problem 2 of Chapter 2, Section 3). Thus the u_i span U and so form a basis. ■

3. Any independent family $u_1,\dots,u_r$ can be **extended** to a basis of U; that is, there exist vectors $u_{r+1},\dots,u_n$ such that the whole set $u_1,\dots,u_r,\dots,u_n$ is a basis for U. (In particular, if $r=n$, then the given set is already a basis.)

PROOF. Suppose $u_1,\dots,u_r$ are not yet a basis for U; since by assumption they are independent, this means that they do not span U. Thus there are vectors in U that are *not* linear combinations of $u_1,\dots,u_r$. Take any such vector and call it u_{r+1}. The argument used in (2) shows that then the set $u_1,\dots,u_{r+1}$ is independent. (Otherwise, we could get u_{r+1} as linear combinations of $u_1,\dots,u_r$.) If this set still does not span U, we can find a vector u_{r+2} that is *not* linear combination of $u_1,\dots,u_{r+1}$; the argument of Fact 2 just used shows that the $r+2$ vectors $u_1,\dots,u_{r+2}$ are independent. We continue this way. When we get to n vectors, they form a basis for U by our Fact 2. ■

4. If V is a subspace of U, the V is also finitely generated and the inequality $\dim V \leqslant \dim U$ holds; if "=" holds, then in fact V equals U.

PROOF. For vectors in V independence in V means the same as independence in U, clearly. Therefore, according to our Fact 1, there can be *at most n* vectors in an independent set in V. Now let $v_1,\ldots,v_r$ be a set with as many independent vectors in V as possible (we just saw $r \leqslant n$). The argument used in our Fact 3 shows that the v_i must span V, and so they form a basis for V. Therefore, $\dim V = r$, and we know already $r \leqslant n$. If $r = n$, then by our Fact 2 the v_i are a basis for U, and it follows that $V = U$. ■

5. Let V be a subspace of U. A **complement** to V is any other subspace W, such that $V \cap W = 0$ and $V + W = U$, that is, such that the sum is direct and equal to U. (This is *not* the set complement.)

Examples

1. In space, a plane and any line not in the plane are complementary subspaces.
2. In R^5, $((E^1, E^2))$ and $((E^3, E^4, E^5))$ are complementary subspaces.

Now, any subspace V of U always has a complement (in fact, infinitely many possibilities except in the cases $V = 0$ or $V = U$).

PROOF. Let $u_1,\ldots,u_k$ be a basis for V. Extend it to a basis $u_1,\ldots,u_n$ of U, by (2). Then $((u_{k+1},\ldots,u_n))$ is clearly a complement to V. The different possibilities come from different ways of extending the basis. ■

If $\{v_1,\ldots,v_r\}$ is any family of vectors, independent or not, one calls **rank** of the family the dimension of the subspace $((v_1,\ldots,v_r))$ spanned by the vectors. In F^n we know how to find the rank of a family $\{X_1,\ldots,X_r\}$ by Chapter 2, Section 5. We bring the matrix $A = (X_1,\ldots,X_r)$ into column-echelon form; the rank is the number of *nonzero* columns appearing there.

PROBLEMS

1. Let $u_1,\ldots,u_r$ be independent. Let $\{v_1,\ldots,v_m\}$ be any family spanning U (independent or not), form the family $\{u_1,\ldots,u_r,v_1,\ldots,v_m\}$ consisting of the u's *and* v's, and reduce it to an independent set by the usual procedure. Prove that the set so obtained is an extension of $\{u_1,\ldots,u_r\}$ to a basis. (In practice, for F^n, we take the standard basis for the v's.)

2. Extend the independent set $\{[1,0,1], [1,1,1]\}$ to a basis of R^3.

3. Extend $\{[2,1,-1,0], [2,2,0,1]\}$ to a basis of F^4.

4. Do $[1,0,1,0]$, $[0,1,1,1]$, $[1,0,2,1]$, and $[1,1,0,-1]$ form a basis for F^4? (Using our Fact 2, are they independent?)

5. $X_1=[1,i]$ and $X_2=[i,-1]$. Is this a basis for $\mathbf{C}^2$?

6. Do $p_1=x^2+x+1$, $p_2=x^2+2x+1$, and $p_3=x^2+2x+2$ form a basis of P^2?

7. Let $\dim U=n$.

a. Show that for any k with $0\leqslant k\leqslant n$ there is at least one subspace of U whose dimension is k.

b. Show that for any k with $0<k<n$ there are infinitely many subspaces of U with dimension equal to k.

8. Find the ranks of the families of vectors in Problems 3–5 of Chapter 2, Section 5.

9. In $\mathbf{F}^5$ find a complement to the subspace $V=\langle\langle[1,1,1,1,1],[2,1,1,2,1]\rangle\rangle$. In fact, find two different complements.

3. THE "MODULAR LAW"

3.1. THEOREM. *Let V and W be subspaces of U; they determine the subspaces $V\cap W$ and $V+W$. Then*

$$\dim V+\dim W=\dim(V\cap W)+\dim(V+W).$$

Before proving it, we note an important special case.

3.2. THEOREM. *If $V\cap W=0$* (i.e., *if the sum is direct*), *then* $\dim(V+W)=\dim V+\dim W$.

Examples

1. Look at two different planes (through 0) in space. The intersection is a line (of dim = 1), the span is the whole space (proof?) (of dim = 3); the equation reads $2+2=1+3$.

2. Examine a plane and a line in space, where the line is *not* in the plane. Then $V\cap W=0$ and $V+W$ is space; $1+2=0+3$.

3. As in Example 2, look at a line and a plane where the line is in the plane. Now $V\cap W$ is the line, and $V+W$ is the plane; $1+2=1+2$.

Before we give the proof we note a general moral: Since dimension is defined in terms of bases, questions about dimension usually have to be treated by considering bases.

We do Theorem 3.2 separately, even though it is a special case of Theorem 3.1; it has a slightly simpler proof. Let $v_1,\ldots,v_p$ be a basis for V; let $w_1,\ldots,w_q$ be one for W.

Claim. Because of $V\cap W=0$, the set $\{v_1,\ldots,v_p,w_1,\ldots,w_q\}$ is a basis for $V+W$ (this will prove the theorem, since these are $p+q$ vectors).

Reason. (a) The set spans $V+W$ since the v_j's give any vector v in V, and the w_j's any vector w in W as linear combination, and so together they give any $v+w$.

(b) They are independent: Can we have a relation $a_1v_1+\cdots+a_pv_p+b_1w_1+\cdots+b_qw_q=0$ with not all a's and b's equal to 0? Transpose the w's to get $a_1v_1+\cdots+a_pv_p=-b_1w_1-\cdots-b_qw_q$. The left-hand side is in V, the right-hand side in W, so the common value is a vector in $V\cap W$. But $V\cap W$ is 0! We conclude that $a_1v_1+\cdots+a_pv_p$ and $b_1w_1+\cdots+b_qw_q$ are both 0! (One should understand clearly how the *one* original equation has led to *two* separate equations.) And now since the v's are independent, all the factors a_i must be 0; and since the w's are independent, all the b_j must be 0. ∎

Now, we will look at the modular law: Taking bases for V and W is not quite enough now, since together they will not be a basis for $V+W$ (this would contradict the result, e.g., Example 2). Instead, we start from a basis $\{u_1,\ldots,u_r\}$ of $V\cap W$. By Fact 2 of Section 2 we can extend it to a basis of V; write $v_1,\ldots,v_s$ for the additional vectors (so that $r+s=\dim V$). Similarly, we can extend it to a basis of W, obtaining additional vectors $w_1,\ldots,w_t$, with $r+t=\dim W$. We claim that the set $\{u_1,\ldots,u_r,v_1,\ldots,v_s,w_1,\ldots,w_t\}$ is a *basis* for $V+W$. If this is so, we are done, since dim $V\cap W=r$, dim $V+W=r+s+t$, and $r+r+s+t=r+s+r+t=\dim V+\dim W$. Now the u's and v's are a basis for and so span V; the u's and w's span W. It is clear then that the u's, v's and$w$'s together span $V+W$. Independence is a little tricky. Suppose we have $a_1u_1+\cdots+a_ru_r+b_1v_1+\cdots+b_sv_s+c_1w_1+\cdots+c_tw_t=0$. We have to show that all a's, b's, and c's vanish. We rewrite the equation as $a_1u_1+\cdots+a_ru_r+b_1v_1+\cdots+b_sv_s=-c_1w_1-\cdots-c_tw_t$. The left side is in V, the right side in W; thus both are in $V\cap W$. But if the left side is in $V\cap W$, then all the b's must be 0 (since a vector in $V\cap W$ *can* be written in the form $d_1u_1+\cdots+d_ru_r+0\cdot v_1+\cdots+0\cdot v_s$, and the coefficients, with respect to the basis $u_1,\ldots,u_r,v_1,\ldots,v_s$ of V, are uniquely determined). The equation reduces now to $a_1u_1+\cdots+a_ru_r=-c_1w_1-\cdots-c_tw_t$. Since the set $\{u_1,\ldots,u_r,w_1,\ldots,w_t\}$ is independent, all the a's and c's must vanish (we transpose the right-hand side of the equation to the left to see this), and we are done.

PROBLEMS

1. In R^6 describe two subspaces V and W of dimensions two and three that satisfy $V\cap W=0$, and verify the claim of Theorem 2 by constructing suitable bases.

2. In F^{10} describe

a. two subspaces V and W with $\dim V=3$, $\dim W=5$, and $V\cap W=0$,

b. two subspaces V' and W' with $\dim V'=5$, $\dim W'=7$, and $\dim V'\cap W'=3$,
c. two subspaces V'' and W'' with $\dim V''=5$, $\dim W''=7$, and $\dim V''\cap W''=1$.

3. In $\mathbf{R}^4$ let $V=(([1,-1,1,1],[1,-2,3,4]))$ and $W=(([1,2,3,6],[1,-1,0,0]))$. Find $\dim V\cap W$. (*Hint*. Find the dimensions of V, W, and $V+W$ first.)

4. Show by examples that there *cannot be* a formula expressing $\dim(V_1+V_2+V_3)$ in terms of the dimensions of $V_1, V_2, V_3, V_2+V_3, V_1+V_3, V_1+V_2, V_2\cap V_3, V_1\cap V_3, V_1\cap V_2$, and $V_1\cap V_2\cap V_3$.

5. Show that $\dim(V_1+V_2+V_3)=\dim V_1+\dim V_2+\dim V_3-\dim V_1\cap V_2-\dim V_3\cap(V_1+V_2)$. (This is the correct form of the modulator law for three subspaces; note the "mixed" term $V_3\cap(V_1+V_2)$.)

6. Extend Problem 5 to k summands.

7. Subspaces $V_1, V_2,\ldots,V_k$ of U are called independent if a relation $v_1+v_2+\cdots+v_k=0$, with v_i taken from V_i for $1\leqslant i\leqslant k$, can hold only if all the v_i are 0 separately. Show that subspaces $V_1,\ldots,V_k$ are independent precisely if the equation $\dim(V_1+\cdots+V_k)=\dim V_1+\dim V_2+\cdots+\dim V_k$ holds. Give an example of three independent subspaces in $\mathbf{R}^{10}$; also give an example of three subspaces such that any two are independent, but all three are not independent.

4. COMPONENTS RELATIVE TO A BASIS

We introduce a very important notion, that of components or coordinates of a vector relative to (or with respect to) a basis. Let U be a vector space, and let $\beta=\{u_1,\ldots,u_n\}$ be a basis. Then any vector u in U can be written uniquely as linear combination $\Sigma_1^n x_i u_i$; the x_i are certain scalars, called the **components** or **coordinates** of u relative to β, and "uniquely" means that for a given u there is only one choice of the x_i. We write the x_i as a column vector $X=[x_1,\ldots,x_n]$, and say that u corresponds to X, via β, or that u is **represented**, relative to β, by X; we write $u\underset{\beta}{\leftrightarrow}X$. Conversely, to any X (in $\mathbf{F}^n$)$=[x_1,\ldots,x_n]$ corresponds a vector u, namely simply the vector $x_1u_1+\cdots+x_nu_n$. The vector 0 in U corresponds to the column 0; this expresses the obvious fact $0=0\cdot u_1+\cdots+0\cdot u_n$.

Example

1. In P^2 the vectors ($=$polynomials) p_1,p_2,p_3 of Section 2, Problem 6, form a basis. Consider now, for instance, the polynomial x^2-2. What is its component vector with respect to the above basis? After some experimentation one finds the equation $x^2-2=2\cdot p_1+2\cdot p_2-3\cdot p_3$ (check!). Therefore, $x^2-2\underset{\beta}{\leftrightarrow}[2,2,-3]$.

2. This example is very special. $U=\mathbf{F}^n$, with basis σ, the standard basis $\{E^1,\ldots,E^n\}$. Any $X=[x_1,\ldots,x_n]$ can be written (as we know) as Σx_iE^i.

This means that the components of X with respect to σ are precisely the entries x_i of X. In other words, the component vector of X with respect to σ is X itself; $X \underset{\sigma}{\leftrightarrow} X$.

We repeat the main point. Although vectors in U are not column vectors (unless U happens to be F^n), but are whatever they are (polynomials, line segments OA, forces, etc.), once we choose a basis in U, we can *represent* them by column vectors. For plane or space this is the familiar process of "introducing coordinates," via choice of axes, and so on. It is *not* true that space (3-space) is R^3; a line segment OA is not a triple $[x,y,z]$ of numbers, but, after choice of a basis, we can *assign* to each vector OA a triple $[x,y,z]$ and conversely, so that the two vector spaces do "look alike."

In fact, in the general situation $u \underset{\beta}{\leftrightarrow} X$, we have the additional pleasant fact that $+$ and $\cdot$ carry over. That is, if $u \underset{\beta}{\leftrightarrow} X$, and $v \underset{\beta}{\leftrightarrow} Y$ for a second vector v, then $u+v$ is represented by the column vector $X+Y$. We can write this in symbols as $u+v \underset{\beta}{\leftrightarrow} X+Y$. For $\cdot$ this reads $r \cdot u \underset{\beta}{\leftrightarrow} r \cdot X$.

PROOF. From $u = \Sigma_1^n x_i u_i$ and $v = \Sigma_1^n y_i u_i$ we get $u+v = \Sigma_1^n (x_i + y_i) u_i$; the work is similar for $r \cdot u = r \cdot (\Sigma x_i u_i) = \Sigma r \cdot x_i u_i$. Thus, once a basis is chosen, U behaves as if it were F^n. ■

The point of all this is that in practice *all numerical* computations in a vector space are always done with the components of the vectors with respect to some suitable basis.

The only flaw in all this is that so far we have no good way to *find* the coordinates of a vector u with respect to a basis $\{u_i\}$, even in the simple case $U = \mathsf{F}^n$, where u and the u_i are column vectors (except when the u_i are the E^i). We saw a suggestion how to go about this in Chapter 2, Section 5. The systematic approach is by way of linear equations; this will be developed in the next chapter (Chapter 4, Section 5.1). The main idea is described in Problem 1.

PROBLEMS

1. Prove that $X_1 = [1,2]$ and $X_2 = [1,3]$ form a basis for R^2. Find the components of $Y_1 = [3,-1]$, $Y_2 = [-1,1]$, and $Y_3 = [2,0]$ with respect to X_1 and X_2. Check that the component vector for Y_3 is the sum of those for Y_1 and Y_2. (Note that $Y_3 = Y_1 + Y_2$.) (*Hint.* $Y_1 = aX_1 + bX_2$ gives two equations for a and b from the two components of the vectors; solve for a and b.)

2. Find the components of the vector $p(x) = 2x^2 - x - 1$ with respect to the basis of P^2, formed by the vectors p_1, p_2, p_3 of Section 2, Problem 6. (Approach this problem by expressing the polynomials x^2, x, and 1 in terms of p_1, p_2, and p_3 and substituting into $p(x)$.)

4

LINEAR FUNCTIONALS AND LINEAR EQUATIONS

In this chapter we consider the theory and practice (but only "by hand" practice, not computer practice) of systems of linear equations, homogeneous (right sides equal to 0) and nonhomogeneous. Linear equations and their solutions are studied from a somewhat more general point of view, introducing the concepts **linear functional**, **kernel (annihilator)**, **linear variety (flat)**, and the **duality** between equations and solutions. The **rank** appears, a number that describes an important aspect of a matrix.

The idea of **dot product** or **inner product** in Euclidean space is presented in a brief discussion; this topic will be taken up in more detail in Chapter 12.

1. LINEAR FUNCTIONALS

As noted in the introduction, part of the business of linear algebra is solving linear equations. Let us consider such an equation, say $2x+3y-5z=0$. What sort of thing is the expression $2x+3y-5z$ that is being "put equal to 0" here? First of all, it is a *function* on R^3, in the sense that it takes a *value* at each (numerical) $[x,y,z]$; for example, for $[1,-2,-1]$ the value is $2\cdot 1+3\cdot -2-5\cdot -1,=1$. For most $[x,y,z]$ this value is not 0; our job in solving the equation is to find those $[x,y,z]$ for which the value is 0. Of course, the function at hand is a rather special (and simple) function; there are no squares, no roots, no log, no products like xy, just each variable multiplied by a scalar, and these terms added together. The general form on F^n is $a_1x_1+a_2x_2+\cdots+a_nx_n$, with given scalars $a_1,\ldots,a_n$. This is what one calls a "linear" function. The name has been chosen because of the relation of these functions, in the case of two variables, to lines in the plane (known from analytic geometry). The description of these functions just given is, of course, rather naive and simple minded. What one needs, is a description in terms of characteristic properties that distinguishes this class of functions from others. The following *definition* accomplishes this (for any vector space U instead of $\mathbf{R}^3$).

Let U be a vector space over $\mathbf{F}$. A **linear functional** (also called linear function or linear form) on U is a function, say φ, from U to $\mathbf{F}$ (we use function in the usual sense of a rule that assigns to each u in U unambiguously a *value*, denoted by $\varphi(u)$, in $\mathbf{F}$) with the *linearity* properties

L_1: $\varphi(u_1+u_2)=\varphi(u_1)+\varphi(u_2)$

for any u_1, u_2 in U ("sum goes to sum"; additivity),

L_2: $\varphi(ru)=r\cdot\varphi(u)$

for u in U and r in $\mathbf{F}$ ("multiple goes to multiple"; homogeneity).

Note that $\varphi(0)=0$ necessarily (take $r=0$ in L_2, with any u). Simple formal arguments show that L_1 and L_2 generalize to the following: If $u=\sum a_i v_i$ is a linear combination of vectors v_i with coefficients a_i, then $\varphi(u)=\sum a_i\cdot\varphi(u_i)$. One says that φ "preserves linear combinations."

We note explicitly that to say that two linear functionals φ and ψ are equal means exactly the same as "$\varphi(u)=\psi(u)$ for every u in U."

A trivial, but not unimportant, example that comes with every U is the linear functional that is constant equal to 0; its value at *whatever* u is 0. L_1 and L_2 are trivially satisfied. This linear functional is denoted by 0 (so that one has $0(u)=0$ for every u); this is a fourth use of the symbol 0.

Examples

1. A very simple example on $\mathbf{R}^1$, $\theta(x)=5x$ (i.e., the value assigned to any x by the rule θ is $5x$).
2. On $\mathbf{R}^2$, $\psi(x,y)=2x+3y$.
3. On $\mathbf{R}^3$, $\varphi(x,y,z)=2x+3y-5z$.
4. On $\mathbf{R}^n$, any $\varphi(X)=a_1x_1+a_2x_2+\cdots+a_nx_n$ with given scalars a_i ("linear expression in the x_i"). (Here $X=[x_1,\ldots,x_n]$.)

We could have used $\mathbf{C}$ instead of $\mathbf{R}$ in these examples.

For something a bit different, and more abstract, we consider these examples.

5. On the vector space P^n of polynomials the linear functional η given by $\eta(p(x))=\int_0^1 p(x)\,dx$. For instance, $\eta(x^2-1)=(1/3)-1=-2/3$.

6. On the vector space of solutions $y(x)$ of the differential equation $y''+f\cdot y'+g\cdot y=0$, the function ω given by $\omega(y(x))=y'(1)$, the value of the derivative at $x=1$.

7. On the vector space of forces on a mass point, the work done against the force by moving the point to a second location (this one is a bit tricky; one has to imagine each force acting not only at the original position, but at each point of the path taken).

We consider L_1 for the second example; with $X_1=[x_1,y_1]$ and $X_2=[x_2,y_2]$, we have $X_1+X_2=[x_1+x_2,y_1+y_2]$. Thus we have to verify $\psi(x_1+x_2,y_1+y_2)=\psi(x_1,y_1)+\psi(x_2,y_2)$ for additivity; that is $2(x_1+x_2)+3(y_1+y_2)=(2x_1+3y_1)+(2x_2+3y_2)$. But this is obvious from standard properties of numbers.

Comment: One reason why linear functions, especially as in Example 4, are important is this: In calculus of several variables one learns that a function $f(x_1,\ldots,x_n)$ is *approximately* equal to $f(0)+a_ix_1+\cdots+a_nx_n$, with $a_i=\partial f(0)/\partial x_i$ [Mean Value Theorem]. Thus, apart from the not very important constant $f(0)$ (the 0th order approximation), the first-order approximation to f is of the form of Example 4; "f is approximately a linear functional." (Of course, in "real life" this holds only for small values of the x_i.)

The following description of linear functionals with the help of a basis is important.

1.1. PROPOSITION. ("*Construction Principle*"). *Let* $\{u_1,\ldots,u_n\}=\beta$ *be any basis of* U. (a) "*Uniqueness.*" *A linear functional* φ *is completely determined by its values* $\varphi(u_1),\ldots,\varphi(u_n)$ *on these basis vectors.* (b) "*Existence.*" *Conversely, given any scalars* $a_1,\ldots,a_n$, *there is a (unique) linear functional, say* ψ, *with* $\psi(u_1)=a_1,\ldots,\psi(u_n)=a_n$.

The first part says that a linear functional is determined, once we know its values at a few (n) suitable points; for instance, a linear functional θ on the line is determined by its value $\theta(x)$ at any one point $x(\neq 0)$. The second part describes the totality of such functions; it says that one can prescribe the values of a linear functional on a basis arbitrarily.

PROOF. (a) Any u in U can be written as Σx_iu_i, with unique x_i. But then, by linearity, we have $\varphi(u)=\varphi(\Sigma x_iu_i)=\Sigma x_i\varphi(u_i)$; this shows that $\varphi(u)$ is determined for any u, once the values $\varphi(u_1),\ldots,\varphi(u_n)$ are known. To put it differently, if two linear functionals φ and ψ agree at $u_1,\ldots,u_n$ (i.e., $\varphi(u_1)=\psi(u_1)$, $\varphi(u_2)=\psi(u_2),\ldots,\varphi(u_n)=\psi(u_n)$), then they agree at every u (i.e., $\varphi(u)=\psi(u)$ for all u), and so $\varphi=\psi$.

For (b), we *define* $\psi(u)=\Sigma x_ia_i$ (since we want $\psi(u_i)=a_i$, this is suggested by the relation $\varphi(u)=\Sigma x_i\varphi(u_i)$ for linear functionals). Note that u determines the x_i and that the a_i are given. Thus to each u we get a number $\psi(u)$; we have a *function* ψ. Now, we show that ψ is linear. For L_2, if we multiply u by r, we have to multiply the components x_i by r, and then the sum $\Sigma x_i a_i$ also multiplies by r. For L_1, say $v=\Sigma y_i u_i$ is another vector, then, of course, $u+v=\Sigma(x_i+y_i)u_i$. By our definition we have $\psi(v)=\Sigma y_i a_i$ and $\psi(u+v)=\Sigma(x_i+y_i)a_i$. The distributive law $(x_i+y_i)a_i=x_ia_i+y_ia_i$, for *numbers!*, shows then $\psi(u)+\psi(v)=\psi(u+v)$. Finally, $\psi(u_j)=a_j$, since the expansion of u_j is $0\cdot u_1+0\cdot u_2+\cdots+1\cdot u_j+0\cdot u_{j+1}+\cdots+0\cdot u_n$; the jth component x_j is 1, the other x_i are 0, and so $\Sigma x_i a_i$ reduces to a_j. ■

Example

$Z_1=[1,1]$ and $Z_2=[1,-1]$ form a basis for R^2 (see Section 1.1). The conditions $\varphi(Z_1)=-1$ and $\varphi(Z_2)=2$ determine, by Proposition 1.1, a well-defined linear functional φ. What is $\varphi([1,3])$? We find, one way or another, that $[1,3]=2Z_1-Z_2$. The recipe of Proposition 1.1 says that then $\varphi([1,3])=2\cdot-1-1\cdot 2=-4$.

Proposition 1.1 shows that once a basis has been chosen in U (so that any vector u has components $x_1,\ldots,x_n$, with respect to this basis), then a linear functional φ is simply of the form $a_1x_1+\cdots+a_nx_n$, with certain constants $a_1,\ldots,a_n$. It "looks like" the example $2x+3y-5z$ above, except that there are more variables. Each a_i is, of course, just $\varphi(u_i)$. In particular, in F^n, using the standard basis $\{E^i\}$, if φ is a linear functional on F^n, then there are definite numbers $a_1,\ldots,a_n$ $(=\varphi(E^1),\ldots,\varphi(E^n))$ such that for any $X=[x_1,\ldots,x_n]$ the value $\varphi(X)$ is $a_1x_1+\cdots+a_nx_n$ or $\Sigma_1^n a_ix_i$. This is true because the x_i are precisely the components of X relative to the E^i, since $X=\Sigma x_iE^i$. We see that our linear functionals on F^n are just the usual "linear expressions" described above.

Example

The linear functional on R^5 whose values on $E^1,\ldots,E^5$ are $1,2,-1,0,3$ is $x_1+2x_2-x_3+3x_5$. (Note the $0\cdot x_4$.)

We come to an important idea; we can describe a linear functional $a_1x_1+\cdots+a_nx_n$ on F^n simply by listing the coefficients $(a_1,\ldots,a_n)$. What we have then is $1\times n$ matrix, in other words, a *row vector* (see Chapter 1, Section 2.i); we shall use letters like A and B for these. Thus we may and shall think of a linear functional on F^n simply as *being* a row vector $A=(a_1,\ldots,a_n)$; the *value* of A on a vector $X=[x_1,\ldots,x_n]$ is $a_1x_1+\cdots+$

$a_n x_n$; we write this also as $A \cdot X$ or plain AX. For instance, the linear functional $(2,0,-1)$ on $\mathbf{R}^3$ (i.e., the function $2x_1 - x_3$) takes at the vector $[1,-1,3]$ the value $2\cdot 1 + 0\cdot -1 + -1\cdot 3, \; = -1$; we can write this as $(2,0,-1)\cdot[1,-1,3] = -1$. ■

We digress for a moment: In analytic geometry or calculus one learns to represent functions by their graphs. The function $5x$ is represented by a line in the (x,y)-plane $y = 5x$, and the functional $2x + 3y$ is represented by a plane in (x,y,z)-space $(z = 2x + 3y)$. The general case is not easy to visualize; U itself is abstract, and we need *one more* dimension or axis to represent the values of the function. A somewhat symbolic picture is shown in Figure 7. The horizontal "axis" has to be understood as U, having any dimension. One can visualize the cases $\dim U = 1$ or 2, as described above.

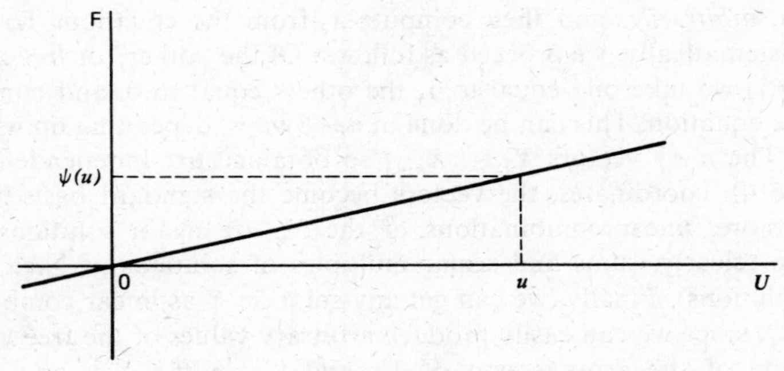

Figure 7.

PROBLEMS

1. Complete the proofs of linearity for Examples 1–4.

2. Let A be the row vector $(2,0,-1,3,1)$ in $(\mathbf{R}^5)'$. Compute the value AX for the following X: $[0,1,2,-4,-3]$, $[3,-1,2,2,0]$, and $[2,0,-1,3,1]$.

3. Find two different row vectors A and B in $(\mathbf{R}^4)'$ that satisfy $AX = BX = 1$ with $X = [2,-1,0,3]$.

4. Suppose u is a nonzero vector in the vector space U. Show: There exists a linear functional φ on U that satisfies $\varphi(u) = 1$. Is φ unique, or can there be several such φ? (*Hint*. Extend u to a basis of U.)

5. Suppose $u_1, \ldots, u_k$ are independent in U and $b_1, \ldots, b_k$ are any scalars. Show that there exists a linear functional ψ on U that satisfies $\psi(u_i) = b_i$, for $i = 1, \ldots, k$. (*Hint*. Use the hint of Problem 4.)

2. LINEAR EQUATIONS; KERNEL, NULL SPACE

We now come to our main topic—solving linear equations. In general, one has several equations (a *system* of equations); we begin here with the case of *one* equation. Abstractly, we are given a linear functional φ on a vector space U, and we are asked to find or describe all vectors u with $\varphi(u)=0$. In concrete cases, like $2x+3y$ in R^2 or $2x+3y-5z$ in R^3, we know, of course, that the answer is a line or a plane, respectively.

To study the question, we begin with the concrete case of one equation for F^n. Thus, we are given a linear functional A, with $A\cdot X=\Sigma a_i x_i$, and we are to find all vectors X that satisfy the equation $AX=0$, that is, $\Sigma a_i x_i=0$. To do this, we start by picking an index, say i, such that a_i is not 0 (we exclude the trivial case $A=0$, where all the a_i are 0 and all X are solutions); usually one takes the first such term, with the smallest index i, called the *leading* term of A. To find solutions X, one can now presecibe all other x_j, for $j\neq i$, *arbitrarily*, and then compute x_i from the equation. To do this more systematically, we proceed as follows: Of the "other" or *free* x's (with index $\neq i$) we take one equal to 1, the others equal to 0, and compute x_i from the equation. This can be done in $n-1$ ways, depending on where the 1 goes. The $n-1$ vectors $X_1,\ldots,X_{n-1}$ so obtained are independent (if we omit the ith coordinates, the vectors become the standard basis in F^{n-1}). Furthermore, linear combinations of the X_k are again solutions of the equation (clearly, sums and scalar multiples of solutions of $\Sigma a_i x_i=0$ are again solutions). Finally, we can get any solution X as linear combinations of the X_k, since we can easily produce arbitrary values of the free variables (by virtue of the arrangement of 1's and 0's in the X_k), and the last variable, x_i, is determined by the equation. We see that the set of solutions of our equation is a *subspace* of F^n (called null space or solution space or "kernel" of A and denoted by $\ker A$), and that its dimension is $n-1$ (since the X_k are a basis). This generalizes what we noted above for R^2 and R^3.

Example

$2x_2-x_3+x_4=0$ in R^5. The second entry is leading; x_1, x_3, x_4, and x_5 are free; we find solution vectors by taking one of them equal to 1, to others equal to 0, and computing x_2 from the equation: For X_1 we take $[1,x_2,0,0,0]$ and determine x_2 from $2x_2-0+0=0$ as 0; thus $X_1=[1,0,0,0,0]$. Then $X_2=[0,x_2,1,0,0]$, with $2x_2-1+0=0$; thus $X_2=[0,1/2,1,0,0]$. Similarly, $X_3=[0,-1/2,0,1,0]$, $X_4=[0,0,0,0,1]$. And X_1,X_2,X_3,X_4 are a basis for $\ker\varphi$. We can write $\ker\varphi=((X_1,X_2,X_3,X_4))$. One also says that $aX_1+bX_2+cX_3+dX_4$ is the *general solution* of our problem, meaning that one can get any particular solution vector by choosing suitable values of the a, b, c, and d. Note that the general solution here works out to $X=[a,1/2(b-c),b,c,d]$, and that a, b, c, and d are just the values of the free variables. This shows one more that the X_i are independent (if $X=0$, then $a=b=c=d=0$) and span the kernel (by taking a, b, c and d equal to the values of the free variables of any given solution vector; x_2 will automatically be correct because the given equation is satisfied).

We turn now to the abstract version. Let φ be a linear function on U.

2.1. DEFINITION AND PROPOSITION. *The set of zeros of* φ, *that is, the set* $\{v\colon \varphi(v)=0\}$, *is a subspace of* U, *called the kernel of* φ *and denoted by* $\ker\varphi$. *In the extreme case* $\varphi=0$ *this is all of* U; *if* $\varphi\neq 0$, *then* $\ker\varphi$ *is of dimension one less that* U *itself* (*of "codimension one"*). (*Other terms for kernel are null space and solution space.*)

PROOF. From $\varphi(u)=0$ and $\varphi(v)=0$ we get $\varphi(u+v)=0$ and $\varphi(rv)=0$, by L_1 and L_2 of Section 1. Thus $\ker\varphi$ is closed under $+$ and $\cdot$. It is also nonempty, since 0 is in it. It follows that $\ker\varphi$ is a subspace.

If φ is 0, then, of course, $\ker\varphi$ is all of U. We take now a nonzero φ. Let $u_1,\ldots,u_k$ be a basis for $\ker\varphi$. Since φ is not 0, there are vectors u with $\varphi(u)\neq 0$, and even with $\varphi(u)=1$. We choose such a vector and call it u_{k+1}. We will prove below that the vectors $u_1,\ldots,u_{k+1}$ form a basis for U; this will establish Proposition 2.1, since we have $\dim\ker\varphi=k$, and $\dim U=k+1$.

We show first that $u_1,\ldots,u_{k+1}$ are independent. Otherwise, u_{k+1} would be a linear combination of $u_1,\ldots,u_k$ (by a familiar argument, using the known independence of $u_1,\ldots,u_k$); but then u_{k+1} would be in $\ker\varphi$, which it is not.

Finally we show that the vectors span U (and then form a basis). Take any v in U, and form $w=v-\varphi(v)u_{k+1}$. (Note that $\varphi(v)$ is a scalar.) Then $\varphi(w)=\varphi(v)-\varphi(\varphi(v)u_{k+1})=\varphi(v)-\varphi(v)\cdot\varphi(u_{k+1})=\varphi(v)-\varphi(v)\cdot 1=0$. Thus w is in $\ker\varphi$; it is therefore a linear combination of $u_1,\ldots,u_k$, and then $v(=w+\varphi(v)u_{k+1})$ is in $((u_1,\ldots,u_{k+1}))$. ∎

Proposition 2.1 agrees, of course, with what we found above for the special case $U=\mathsf{F}^n$ and $\varphi=A$.

We come to the case of *several* equations $A_1\cdot X=0,\ldots,A_k\cdot X=0$. Abstractly, let $\varphi_1,\ldots,\varphi_k$ be k linear functionals on U. We write $\ker(\varphi_1,\ldots,\varphi_k)$ for the **kernel**, i.e., the set of **common zeros** or **solutions** of the φ_i, the set $\{v\colon \varphi_1(v)=\varphi_2(v)=\cdots=0\}$.

2.2 PROPOSITION. $\ker(\varphi_1,\ldots,\varphi_k)$ *is a subspace.* (*We will determine the dimension later.*)

PROOF. Clearly $\ker(\varphi_1,\ldots,\varphi_k)$ is just the intersection of $\ker\varphi_1,\ldots,\ker\varphi_k$, and thus a subspace. ∎

PROBLEMS

1. Find a basis for the solution space V of the equation $x_1 - 2x_2 + x_3 - 3x_4 + 2x_6 = 0$ in F^6.

2. The two equations $x_1 - x_2 + 2x_3 - 3x_4 = 0$ and $2x_1 - x_2 - 3x_3 + x_4 = 0$ determine a subspace W or R^4. Find two *other* equations that also have W as solution space.

3. As the converse of Proposition 2.1, let V be a subspace of U of codimension 1. Show that there is a linear functional φ whose kernel is exactly V. (*Hint*. Take a basis of V, and extend it to one of U; use Proposition 1.1.)

4. The vectors $X_1 = [1,1,0]$ and $X_2 = [0,1,1]$ span a subspace V of F^3 of dimension two. Find a row vector $A = (a,b,c)$ so that $\ker A$ is V. (*Hint*. A must satisfy $AX_1 = 0, AX_2 = 0$.) How much freedom is there in the choice of A?

3. USE OF THE ROW-ECHELON FORM

We now come to the "concrete" version of Proposition 2.2: Given k linear functionals $A_1X, A_2X, \ldots, A_kX$ on F^n, how does one go about *finding* the common zeros? Proposition 2.2 says that the vectors X with $A_1X = 0, \ldots, A_kX = 0$ form a subspace of F^n; how does one construct a basis for this space? (Such a set of equations is called a **homogeneous system** of linear equations, "homogeneous" meaning that the right sides of the equations are all 0.)

The basic idea (generalizing what one does in the case of two or three variables) is to use the operations of addition and multiplication by scalars to simplify the system. For this purpose we make first a rather trivial observation: If AX and BX are two linear functionals on F^n, then $AX + BX$ equals the linear functional $(A+B)X$; written out this says

$$(a_1x_1 + \cdots + a_nx_n) + (b_1x_1 + \cdots + b_nx_n) = (a_1 + b_1)x_1 + \cdots + (a_n + b_n)x_n.$$

Similarly, $r\cdot(AX) = (rA)X$; written out this says $r\cdot(a_1x_1 + \cdots + a_nx_n) = ra_1x_1 + \cdots + ra_nx_n$.

Now we apply this: If for our given homogeneous system of equations we replace, for instance, the second equation $A_2X = 0$ by the sum of the first and the second, that is, by $(A_1 + A_2)X = 0$ (not changing any other equation!), then we get exactly the same solutions as before. For any X with $A_1X = 0$ the new second equation says exactly the same as the old one; and the other equations have not changed anyway. Similarly, if we replace an equation $A_iX = 0$ by $rA_iX = 0$ with a *nonzero* factor r, we clearly get exactly the same solutions.

Instead of the linear functionals $A_1X, \ldots, A_kX$ let us consider the corresponding row vectors $A_1, \ldots, A_k$ (elements of the space $(F^n)'$ of row vectors, see Chapter 1, Section 2). We recognize that the two operations on the equations just described amount to two of the elementary operations (Chapter 2, Section 4) on the vectors A_i. The remaining operation, interchange of two vectors or, correspondingly, interchange of two equations, again does not change the solutions of the system, obviously.

We now develop a procedure for finding the solutions of our system, specifically for finding a basis of the solution space, with the help of these three elementary operations. To have a handy notation, we write

$$A_1=(a_{11},a_{12},\ldots,a_{1n}),\ A_2=(a_{21},a_{22},\ldots,a_2n),\ldots,A_k=(a_{k1},a_{k2},\ldots,a_{kn}).$$

(The terms carry two indices; the first one tells which row vector we are looking at and the second is the number of the component of the vector.) The equations, written out, are then

$$\begin{aligned} a_{11}x_1+a_{12}x_2+\cdots+a_{1n}x_n&=0\\ a_{21}x_1+a_{22}x_2+\cdots+a_{2n}x_n&=0\\ &\vdots\\ a_{k1}x_1+a_{k2}x_2+\cdots+a_{kn}x_n&=0. \end{aligned}$$

We read off a matrix M with the A_i as rows

$$M=\begin{pmatrix} a_{11}\cdots a_{1n}\\ \vdots\\ a_{k1}\cdots a_{kn}\end{pmatrix}.$$

(We write, symbolically, $M=[A_1,\ldots,A_k]$, i.e., the A_i are arranged in a column.) Conversely, any $k\times n$ matrix gives rise to k linear equations for $x_1,\ldots,x_n$, with the rows of the matrix as row vectors for the equations.

The operations (1), (2), and (3) on linear functionals or row vectors mentioned above appear now as operations on the rows of M. With their help we change M (without changing the solutions of the corresponding equations) to a more convenient form, namely into **row-echelon** form (Chapter 2, Section 4), getting, say, N, of type

$$\begin{pmatrix} 0 & \cdots & 0\,1* & & & & \cdots & * \\ 0 & & \cdots & 0\,1* & & & \cdots & * \\ 0 & & & \cdots & & 0\,1* & \cdots & * \\ & & & \vdots & & & & \end{pmatrix}.$$

Each row of N has a **leading** 1 (unless the whole row is 0). The leading 1 moves to the right as we go from a row to the one below it. The equation corresponding to any row has a *leading* x_i, followed by terms with larger indices. (If the row is 0, the equation is $0\cdot x_1+0\cdot x_2+\cdots+0\cdot x_n=0$ and can be discarded.) To solve such a set of equations is easy: The *nonleading* or *free* x_j can be assigned arbitrarily; the leading ones are then determined by the equations, working from the bottom up (the value of the leading x_i in the last equation, once found, is used in the next-to-last equation, etc.) To do this systematically, and, in fact, to get a basis for the space of solutions ($\ker(A_1,\ldots,A_k)$), we proceed similarly as in the case of one equation: We take one of the free x_j equal to 1, the others equal 0, and determine the leading x_i, thereby getting a solution vector. Denoting by r the number of nonzero rows in N (equal the number of leading x_i), there are $n-r$ ways of doing this (namely of deciding which free x_j we take equal to 1). In this manner we get a basis for the solution space: The vectors obtained are independent because of the arrangement of the 1's and 0's for the free variables, and they span the solution space, since by forming linear combinations we can get any values for the free x_j, and the leading ones are determined by the equations.

Example

"Solve"

$$x_1-x_2-x_3+2x_4+x_5=0.$$

$$2x_1-2x_2-x_3+3x_4+x_5=0.$$

$$x_1-x_2+2x_3-x_4-2x_5=0.$$

$$M=\begin{pmatrix}1 & -1 & -1 & 2 & 1\\ 2 & -2 & -1 & 3 & 1\\ 1 & -1 & 2 & -1 & -2\end{pmatrix}\rightarrow\begin{pmatrix}1 & -1 & -1 & 2 & 1\\ 0 & 0 & 1 & -1 & -1\\ 0 & 0 & 3 & -3 & -3\end{pmatrix}$$

$$\rightarrow\begin{pmatrix}1 & -1 & -1 & 2 & 1\\ 0 & 0 & 1 & -1 & -1\\ 0 & 0 & 0 & 0 & 0\end{pmatrix}.$$

The new equations are

$$x_1-x_2-x_3+2x_4+x_5=0,$$

$$x_3-x_4-x_5=0.$$

The leading variables are x_1 and x_3. We can assign x_2, x_4, and x_5 at will. As a first choice we select $x_2=1$, $x_4=0$, and $x_5=0$; we find $x_3=0$ (from the second equation); and then $x_1=1$ (from the first equation). Thus $X_1=[1,1,0,0,0]$. Our next choice is $x_2=0$, $x_4=1$, and $x_5=0$; this leads to $x_3=1$, $x_1=-1$; $X_2=[-1,0,1,1,0]$. Finally, $x_2=x_4=0$, $x_5=1$; this leads to $X_3=[0,0,1,0,1]$. X_1, X_2, and X_3 are a basis for the solution space $(=\ker(A_1,A_2,A_3)$, where A_1, A_2, and A_3 are the rows of the matrix M). Writing out the 5×3 matrix (X_1,X_2,X_3), we see from the nature of rows number two, four, and five

$$
\begin{array}{ccc}
1 & -1 & 0 \\
\hline
\multicolumn{1}{|c}{1} & 0 & \multicolumn{1}{c|}{0} \\
\hline
0 & 1 & 1 \\
\hline
\multicolumn{1}{|c}{0} & 1 & \multicolumn{1}{c|}{0} \\
\hline
\multicolumn{1}{|c}{0} & 0 & \multicolumn{1}{c|}{1} \\
\hline
\end{array}
$$

that the columns are independent (the second, fourth, and fifth entries of $aX_1+bX_2+cX_3$ are just a, b, and c; thus, if this vector is 0, so are a,b,c); and that by taking linear combinations we can get *any* values as second, fourth, and fifth entry.

We note a consequence of our development, the counterpart to Proposition 2.5.1.

3.1. PROPOSITION. *A homogeneous system with more variables than equations always has nontrivial solutions.*

PROOF. After reduction of the $m\times n$ matrix to row-echelon form there are at least $n-m$ free variables present. (Nontrivial means not the 0-vector.) ■

PROBLEMS

1. Find the solution space in $\mathbf{R}^4$ of the equations

$$x_1+x_2-x_3+x_4=0$$

$$2x_1+3x_2+5x_3-x_4=0$$

$$4x_1+5x_2+3x_3+x_4=0$$

(This means find a basis.)

2. Find the null space in $\mathbb{R}^4$ of the homogeneous system of equations with matrix

$$\begin{pmatrix} 1 & 2 & 5 & 0 \\ 4 & 12 & 24 & 2 \\ 3 & 6 & 15 & -3 \end{pmatrix}.$$

3. Find the space in $\mathbb{R}^5$ consisting of the common zeros of the four linear functions $(1,2,1,-4,1)$, $(1,2,-1,2,1)$, $(2,4,1,-5,0)$, and $(1,2,3,-10,1)$.

4. Solve the homogeneous system of equations with the following matrices:

a.
$$\begin{pmatrix} 3 & 6 & 1 & 1 & 1 \\ 2 & 4 & 0 & 1 & 1 \\ 4 & 5 & 1 & -3 & 0 \end{pmatrix}.$$

b.
$$\begin{pmatrix} 1 & -1 & 1 \\ 2 & 6 & 6 \\ 1 & 7 & 5 \end{pmatrix}.$$

c.
$$\begin{pmatrix} 1 & 2 & 3 & 4 \\ 2 & 3 & 4 & 0 \\ 3 & 4 & 0 & 1 \\ 4 & 0 & 1 & 2 \end{pmatrix}.$$

d.
$$\begin{pmatrix} 2 & -2 & -3 & 6 & -1 \\ 1 & 2 & 0 & -3 & 1 \\ 5 & 3 & -3 & -3 & 1 \end{pmatrix}.$$

e.
$$\begin{pmatrix} 2 & -1 & -3 & 0 \\ 7 & -5 & 0 & -3 \\ 1 & -1 & 2 & -1 \end{pmatrix}.$$

f.
$$\begin{pmatrix} 3 & -1 & -4 & 7 & 0 \\ 1 & 1 & -2 & 1 & 2 \\ 2 & 4 & -5 & 0 & 7 \end{pmatrix}.$$

g.
$$\begin{pmatrix} 8 & 0 & -5 & -3 & 2 & -2 \\ 0 & -6 & 5 & 2 & -2 & 1 \\ 11 & -4 & -2 & -5 & 2 & -2 \\ 3 & -7 & 5 & 0 & -2 & 1 \\ 6 & -1 & -2 & -3 & 1 & -1 \end{pmatrix}.$$

4. NONHOMOGENEOUS SYSTEMS; LINEAR VARIETIES

Naturally, the study of linear equations should be extended to the case of **nonhomogeneous** equations, that is, to the problem of finding and describing all solutions X of a system $A_1X=c_1,\ldots,A_kX=c_k$ (in $\mathbf{R}^n$ or $\mathbf{C}^n$), where $c_1,\ldots,c_k$ are given numbers, possibly (and probably) not equal to 0. The first thing to realize is that now there might not be *any* solutions: The system $x+y=1$, $x+y=2$ obviously has no solution. The second thing to realize is that the solutions now do *not* form a subspace of $\mathbf{F}^n$. For instance, the "line" $2x+3y=5$ in $\mathbf{R}^2$ does not go through the origin, and so it is not a subspace.

In the general case, if we have two solutions X' and X'', and if, say, c_1 is not 0, then from $A_1X'=c_1$ and $A_1X''=c_1$ we get $A_1(X'+X'')=2c_1$, but not $=c_1$; thus $X'+X''$ is not a solution; the set of solutions is not closed under $+$, and similarly not under $\cdot$. However, it is true that for any two solutions X' and X'' the *difference* vector $X'-X''$ is a solution of the *homogeneous* system $A_1\cdot X=\cdots=A_k\cdot X=0$: namely, from $A_1\cdot X'=c_1$ and $A_1\cdot X''=c_1$ we get $A_1\cdot(X'-X'')=0$. (Look at this also with $A_1\cdot X'$ written out as $a_{11}x_1'+a_{12}x_2'+\cdots+a_{1n}x_n'$.) The same simple argument shows: If X' is a solution of the *nonhomogeneous* system, and X'' is a solution of the *homogeneous* system, then $X'+X''$ is again a solution of the *nonhomogeneous* system. We know already that the solutions of the homogeneous system form a subspace of $\mathbf{F}^n$.

This leads us to a new concept.

4.1. DEFINITION. *Let V be a subspace of U, and let u_0 be a vector in U. The translate of V by u_0 is the set $V+u_0 \overset{\text{def}}{=} \{v+u_0\colon v\in V\}$; such a set is also called a linear variety or flat of type V. (We could equally well write u_0+V for $V+u_0$.) We define the dimension of u_0+V to be that of V. (See Figure* 8.)

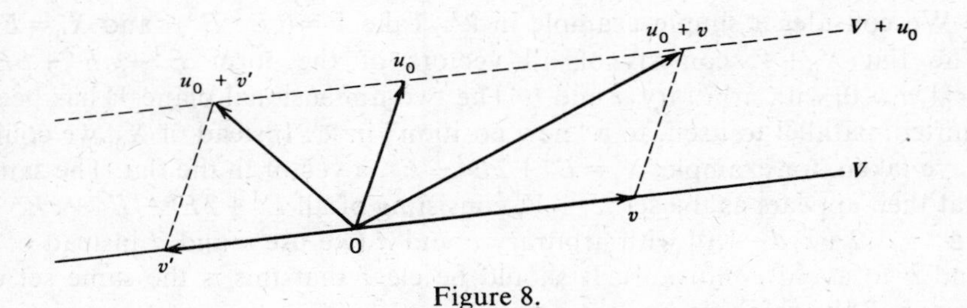

Figure 8.

Thus $V+u_0$ is "V moved over by u_0." Intuitively, what we have in mind are things like lines in the plane, *not* necessarily through 0, or lines or planes in space, *not* necessarily through 0. Note that V itself is one of the linear varieties associated to V (take $u_0=0$). A flat of codimension one (that is of dimension $n-1$, if $\dim U=n$) is called a **hyperplane**. A somewhat more concrete description of a flat is as follows: Let $u_1,\ldots,u_r$ be a basis for V. The flat u_0+V consists of all vectors of the form $u_0+a_1u_1+\cdots+a_ru_r$, with arbitrary scalars $a_1,\ldots,a_r$.

An important fact to note is that two translates of V, by vectors u_0 and u_1, are identical exactly if u_0 and u_1 differ by a vector in V, that is, if $u_0-u_1\in V$.

PROOF. If the two translates are equal, then in particular u_0 (which is in $V+u_0=V+u_1$) must be of the form $v+u_1$ for some v in V. Conversely, if $u_0=v+u_1$ with v in V, then any $w+u_0$ can be written as $(w+v)+u_1$; thus the set $V+u_1$ is contained in $V+u_0$, and similarly with u_0 and u_1 interchanged. (See Figure 9.)

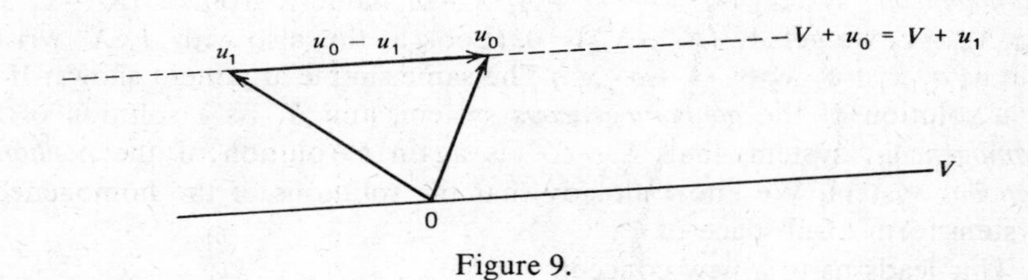

Figure 9.

We can describe part of this as follows: Let $\mathrm{L}=u_0+V$ be a flat of type V, and let u_1 be *any* vector in L; then the flat u_1+V is identical with L. As a consequence, if two linear varieties of type V have a vector in common, then they are in fact identical (if w is a common vector, then both of them equal $w+V$). Thus, the flats of type V fill out U without overlap, like the family of lines in the plane parallel to a given one.

We consider a simple example in $\mathbf{R}^4$: Take $V=((E^2,E^3))$ and $X_0=E^1$. The flat X_0+V consists of all vectors of the form $E^1+aE^2+bE^3=[1,a,b,0]$ with arbitrary a and b. The two-dimensional plane V has been shifted, parallel to itself, to a "new position" in $\mathbf{R}^4$. Instead of X_0, we could have taken, for example, $X_1=E^1+2E^2-E^3$, a vector in the flat. The same flat then appears as the set X_1+V, consisting of all $E^1+2E^2-E^3+cE^2+dE^3=[1,2+c,d-1,0]$ with arbitrary c and d (we use c and d instead of a and b to avoid confusion). It should be clear that this is the same set of vectors ("the same figure") as before.

With the notion of linear variety we can now restate briefly what we worked out before.

4.2. PROPOSITION. *Given an inhomogeneous system* $A_1X = c_1, \ldots, A_kX = c_k$; *the solutions (if there are any at all) form a translate of the space of solutions of the homogeneous system. If there are no solutions, we call the system inconsistent.*

We restate this in more detail: Let X_0 be a solution of the inhomogeneous system (thus assuming it consistent), and let $X_1, \ldots, X_r$ be a basis for the solution space of the homogeneous system (replacing all c_i by 0). The full solution set is then the flat $X_0 + ((X_1, \ldots, X_r))$; one says that X_0 is a "particular" solution, and that $X_0 + \Sigma_1^r a_i X_i$ is the "general" solution.

Of course, what we did here for the concrete F^n is true in any vector space: Given k linear functional $\varphi_1, \ldots, \varphi_k$ and k scalars $c_1, \ldots, c_k$, we look for *all* u in U that satisfy $\varphi_1(u) = c_1$, $\varphi_2(u) = c_2, \varphi_k(u) = c_k$. Either there are none, or they form a linear variety of type $\ker(\varphi_1, \ldots, \varphi_k)$.

The row-echelon form gives us a *procedure* to find the solution (in the F^n case): Let there be given the equations $A_1 \cdot X = c_1, \ldots, A_k \cdot X = c_k$. We form the matrix M with the A_i as rows, *and then* the new matrix M' (the **augmented** matrix) obtained by adding the *column* $C = [c_1, \ldots, c_k]$ as last column; we write M' as (M, C). Now we bring the matrix M' into row echelon form, say N'. Each row of N' gives rise to a linear equation with the last entry of the row as right side, and the first n entries as coefficients of the x_i.

Now the elementary operators (1), (2), and (3) on M' are so designed that the original equations and the new equations have precisely the same solutions (it is for this reason that we included the c_i in the matrix, so that they are transformed along with the left sides; the idea is simply that, for example, $A_1 \cdot X = c_1$ and $A_2 \cdot X = c_2$ imply $(A_1 + A_2) \cdot X = c_1 + c_2$). And, just as for the homogeneous system, once the matrix is in row-echelon form, it is easy to solve the equations. First, if there is a row that has all zeros except for the last one (that one being *not* 0), then there is *no* solution; the original equations are inconsistent. This is so because the equation for this row reads $0 \cdot x_1 + 0 \cdot x_2 + \cdots + 0 \cdot x_n = c \neq 0$, and there are no x's to satisfy this. If there is no such row, then, as in the homogeneous case, we can assign the nonleading or "free" x's at will and compute the leading ones, working upwards through the rows. To get the "flat" description of the solution set (Proposition 4.2), we construct (a) a *basis* $X_1, \ldots, X_r$ for the solutions of the *homogeneous* system (with the right sides equal to 0), as learned earlier, and (b) a "particular" solution X_0 of the nonhomogeneous system (for instance, by putting all free x's equal to 0, and computing the leading ones, from the bottom up). The solution set is then $X_0 + ((X_1, \ldots, X_r))$. ($r$ could be 0, then there would be exactly *one* solution.) The dimension of the linear variety of solutions always equals the number of free variables. In particular, it follows from Proposition 2.1 that the solution set for *one* equation of the form $A \cdot X = c$ (or $\varphi(u) = c$), with $A \neq 0$ (or $\varphi \neq 0$) is a hyperplane.

Examples

1.

$$x_1 + x_2 + 2x_3 = 8.$$

$$2x_2 + x_3 + x_4 = 6.$$

$$x_1 + 2x_2 + x_3 + 2x_4 = 2.$$

$$x_1 + x_2 + x_3 + x_4 = 2.$$

$$\begin{pmatrix} 1 & 1 & 2 & 0 & & 8 \\ & 2 & 1 & 1 & : & 6 \\ 1 & 2 & 1 & 2 & \cdot & 2 \\ 1 & 1 & 1 & 1 & & 2 \end{pmatrix} \rightarrow \begin{pmatrix} 1 & 1 & 2 & 0 & & 8 \\ & 2 & 1 & 1 & : & 6 \\ 0 & 1 & -1 & 2 & : & -6 \\ 0 & 0 & -1 & 1 & & -6 \end{pmatrix}$$

$$\rightarrow \begin{pmatrix} 1 & 1 & 2 & 0 & & 8 \\ 0 & 1 & -1 & 2 & : & -6 \\ 0 & 0 & 3 & -3 & \cdot & 18 \\ 0 & 0 & -1 & 1 & & -6 \end{pmatrix} \rightarrow \begin{pmatrix} 1 & 1 & 2 & 0 & & 8 \\ 0 & 1 & -1 & 2 & : & -6 \\ 0 & 0 & -1 & 1 & \cdot & -6 \\ 0 & 0 & 0 & 0 & & 0 \end{pmatrix}.$$

(The dots only serve to remind us that we are using the augmented matrix; they are not part of the matrix and are not really necessary.)

The new equations are $x_1 + x_2 + 2x_3 = 8$, $x_2 - x_3 + 2x_4 = -6$, and $-x_3 + x_4 = -6$. We leave it to the reader to reconstruct the steps taken. x_4 is free. "The" particular solution comes from $x_4 = 0$: $x_3 = 6$, $x_2 = 0$, and $x_1 = -4$. To find a basis (one vector!) for the solutions of the *homogeneous* system, take $x_4 = 1$, and get $x_3 = 1$, $x_2 = -1$, $x_1 = -1$. Thus $X_0 = [-4, 0, 6, 0]$ and $X_1 = [-1, -1, 1, 1]$. The linear variety of solutions is $X_0 + ((X_1))$ of dimension one; it consists of the vectors $X_0 + aX_1$, any a.

2. This example illustrates a slight variant of the procedure:

$$x_1 + 2x_2 + x_3 - x_4 = 2.$$

$$2x_1 + 4x_2 - x_3 - 5x_4 = 7.$$

$$x_1 + 2x_2 + 2x_3 = 1.$$

$$x_3 + x_4 = -1.$$

$$\begin{pmatrix} 1 & 2 & 1 & -1 & 2 \\ 2 & 4 & -1 & -5 & 7 \\ 1 & 2 & 2 & 0 & 1 \\ 0 & 0 & 1 & 1 & -1 \end{pmatrix} \to \begin{pmatrix} 1 & 2 & 1 & -1 & 2 \\ 0 & 0 & -3 & -3 & 3 \\ 0 & 0 & 1 & 1 & -1 \\ 0 & 0 & 1 & 1 & -1 \end{pmatrix}$$

$$\to \begin{pmatrix} 1 & 2 & 1 & -1 & 2 \\ 0 & 0 & 1 & 1 & -1 \\ 0 & 0 & 0 & 0 & 0 \\ 0 & 0 & 0 & 0 & 0 \end{pmatrix}.$$

The new equations are $x_1+2x_2+x_3-x_4=2$ and $x_3+x_4=-1$. x_2 and x_4 are free. We write a and b for them, for brevity, and solve the equations for x_1 and x_3 *in terms of* a and b beginning with the *last* equation: $x_3=-1-b$, and then (using the value of x_3 just found) $x_1=2-2a-x_3+b$ $=3-2a+2b$. Thus *all* solutions are obtained as

$$X=[3-2a+2b,a,-1-b,b]$$

with arbitrary a and b. We can rewrite this as

$$X=[3,0,-1,0]+a[-2,1,0,0]+b[2,0,-1,1],=X_0+aX_1+bX_2$$

(this defines X_0, X_1, and X_2). The set of all solutions is recognized as the flat $X_0+((X_1,X_2))$. In the earlier procedure we would have gotten X_0, "the" particular solution, by solving the nonhomogeneous equations for x_1 and x_3, after putting the free variables x_2 and x_4 equal to 0; and we would have found X_1 and X_2, basis for the kernel of the homogeneous equations, by solving the homogeneous equations with the choices $[1,0]$, respectively $[0,1]$ for $[x_2,x_4]$.

Remarks

1. One should realize that all the machinery developed here for solving equations is just the "practical" or "concrete" aspect of Propositions 2.2 and 4.2 (and of the abstract version of the latter).

2. We have chosen to solve linear equations by use of the row-echelon form (this is also called Gaussian elimination). There are other procedures to perform the same task. One does not have to stick slavishly to any one method; on the other hand, it is a good idea to get well-acquainted with a particular method. In special cases one often uses special tricks or short cuts.

3. Finding the solution (space or flat) of one or more equations, homogeneous or inhomogeneous, is to be regarded as part of a standing

repertoire, as a task that can be performed whenever it comes up, without hesitation or even much thinking. This is important since almost all problems of linear algebra reduce in the end to solving some linear equations. Needless to say (?), one should acquire facility in this by practicing with a number of examples.

4. The whole process makes sense even if some of the coefficients of the system are not specified numerically but are left as variables (this happens quite often for those on the right-hand side); of course, the answers then involve those variables. Consistency becomes then a *condition* on the variable coefficients (see Problem 6).

PROBLEMS

1. Solve the system

$$\begin{aligned} x_1+2x_2+3x_3+x_4&=1.\\ 2x_1+3x_2-x_3-x_4&=1.\\ 3x_1+7x_2+4x_3-4x_4&=2.\\ x_2+x_3-2x_4&=0. \end{aligned}$$

2. Solve the systems with the following augmented matrices:

a. $$\left(\begin{array}{rrrrr|r} 1 & 3 & -1 & 0 & 2 & 2\\ 2 & 6 & 1 & 6 & 4 & 13\\ -1 & -3 & 0 & -2 & -2 & -5 \end{array}\right).$$

b. $$\left(\begin{array}{rrr|r} 2 & 1 & 5 & 4\\ 3 & -2 & 2 & 2\\ 5 & -8 & -4 & 1 \end{array}\right).$$

c. $$\left(\begin{array}{rrrr|r} 1 & 1 & 1 & 1 & 1\\ -1 & 0 & 2 & 1 & 1\\ 3 & 2 & -1 & 0 & 1\\ 1 & 1 & 2 & 2 & 1 \end{array}\right).$$

d. $$\left(\begin{array}{rrrr|r} 1 & -1 & 1 & 1 & -1\\ 3 & -2 & 1 & 1 & 0\\ -2 & 1 & -1 & 0 & 1 \end{array}\right).$$

3. Given the system

$$x_1 + x_3 = a.$$

$$x_2 + x_3 + x_4 = b.$$

$$x_1 + 2x_3 + x_4 = 0.$$

$$x_1 + x_2 - x_4 = 1.$$

a. Find a (linear) condition on a and b that makes the system consistent;
b. Choose values for a and b that make the system consistent, and solve the system;
c. With a and b left variable (but assumed to satisfy the condition in (a)) solve the system. (See Remark 4 above.)

4. Solve the systems for the following augmented matrices:

a.
$$\begin{pmatrix} 1 & 1 & 1 & \vdots & 3 \\ 1 & -1 & 0 & \vdots & 1 \\ 0 & 2 & 1 & \vdots & 2 \end{pmatrix}.$$

b.
$$\begin{pmatrix} 1 & 2 & 1 & 2 & & 3 \\ 2 & 0 & -1 & 3 & \vdots & 2 \\ -1 & 6 & 5 & 0 & \vdots & 5 \\ 1 & -2 & -2 & 1 & & -2 \end{pmatrix}.$$

c.
$$\begin{pmatrix} 4 & -4 & -1 & 1 & \vdots & -1 \\ 8 & -7 & -2 & 3 & \vdots & 3 \\ 4 & -5 & -1 & 0 & \vdots & -6 \end{pmatrix}.$$

d.
$$\begin{pmatrix} 2 & -1 & 0 & 2 & & 4 \\ 1 & 0 & 1 & 1 & \vdots & 1 \\ 1 & -1 & 1 & -2 & \vdots & 4 \\ 1 & 0 & -1 & 4 & & 1 \end{pmatrix}.$$

5. Treat

$$\begin{pmatrix} 5 & -4 & 1 & 0 & & a+b+1 \\ 4 & -1 & 1 & -1 & \vdots & 2b+3 \\ 3 & 2 & 1 & -2 & \vdots & -a+3b+5 \\ 2 & 5 & 1 & -3 & & -2a+4b+7 \end{pmatrix}.$$

like the matrix in Problem 3.

5. APPLICATIONS

Solving systems of linear equations comes up at several other points.

1. Given vectors $Z_1,\ldots,Z_k$ and a further vector Y in F^n, we saw in Chapter 2 how to decide whether Y is linear combination of the Z's. *But* we did not get a good method for finding the coefficients c_i in $Y=\Sigma c_i Z_i$ (assuming that they exist). We remedy this now: The main point is that the vector relation $\Sigma c_i Z_i = Y$ amounts to n linear equations for the quantities c_i, one equation for each component of the vectors. We follow this in detail: Let M be the matrix $(Z_1,\ldots,Z_k)$ with the Z_i as columns. Denote the rows of M by $A_1,\ldots,A_n$, and put $C=[c_1,\ldots,c_k]$ (a vector in F^k). Then the components of $\Sigma c_i Z_i$ are precisely the "products" $A_1 \cdot C, A_2 \cdot C,\ldots,A_n \cdot C$; for instance, for the first component of $\Sigma c_i Z_i$ we multiply each c_i by the first component of Z_i and add. But that is just $A_1 \cdot C$, since the first components of the Z_i constitute the row vector A_1 (strictly speaking, we used the commutative law for multiplication of numbers; in $A_i \cdot C$ the c's are on the right, and in $\Sigma c_i Z_i$ they are on the left). In other words, the relation $\Sigma c_i Z_i = Y$ is identical with the (nonhomogeneous) system of n equations $A_1 \cdot C = y_1, A_2 \cdot C = y_2,\ldots,A_n \cdot C = y_n$, with the c's as unknowns and the y's as right sides.

We now solve for the c_i as we learned above. The augmented matrix has the Z_i and Y as columns: $M' = (Z_1,\ldots,Z_k,Y)$; we bring it to *row*-echelon form N' and use N' to form the new equations for $c_1,\ldots,c_k$. If the equations are inconsistent, then Y is *not* linear combination of the Z_i; that is, it is not in $((Z_1,\ldots,Z_k))$. If they are consistent, then any solution $(c_1,\ldots,c_k)$ will satisfy $Y=\Sigma c_i Z_i$. (Of course, if the Z_i are independent, there can be at most one solution; the coefficients, if they exist at all, will be unique.)

Note the special case $Y=0$. Here we are trying to find the coefficients c_i for all possible *linear relations* $\Sigma c_i Z_i = 0$ (of course, if the Z_i are independent, the only solution is $c_1 = \cdots = c_k = 0$).

Example

$Z_1=[1,0,1,1]$, $Z_2=[1,2,2,1]$, $Z_3=[2,1,1,1]$, $Z_4=[0,1,2,1]$, and $Y=[8,6,2,2]$. The equations $\Sigma c_i Z_i = Y$ yield the four equations (one for each component of the vectors):

$$c_1 + c_2 + 2c_3 = 8.$$

$$2c_2 + c_3 + c_4 = 6.$$

$$c_1 + 2c_2 + c_3 + 2c_4 = 2.$$

$$c_1 + c_2 + c_3 + c_4 = 2.$$

By accident, these are the same equations as in the first example in Section 4 (except for the names of the unknowns); thus we know they are consistent, and we know the solutions. Let us take *any* solution, for example, $X_0=[-4,0,6,0]$; it tells us that we must have $Y=-4\cdot Z_1+0\cdot Z_2+6\cdot Z_3+0\cdot Z_4$. (Check!)

Example

Let M be an $n\times n$ (square) matrix whose columns Z_i are *independent*. Then the nonhomogeneous system $A_1\cdot X=y_1,\ldots,A_n\cdot X=y_n$, formed with the rows of M, has a solution (is consistent), no matter what the y_i are. This is true because the Z_i are a *basis* for F^n, by Chapter 2, Section 2.2; they span F^n; therefore, x_i with $\Sigma x_i Z_i = Y$ exist.

We interpret this as saying that the x_i are the components of Y relative to the basis $\{Z_1,\ldots,Z_n\}=\beta$ of F^n; X is the column vector *representing* Y with respect to β: $Y\underset{\beta}{\leftrightarrow}X$. We see that, as hinted in Chapter 3, Section 4, the problem of finding components with respect to a basis amounts to solving a nonhomogeneous system of n equations in n unknowns.

2. A problem that we did not finish in Chapter 2 was to find the intersection of two subspaces V and W of U, each given as span of some vectors. Say $V=((v_1,\ldots,v_r))$ and $W=((w_1,\ldots,w_s))$. Vectors in $V\cap W$ are those that can be written as $\Sigma a_i v_i$ and also as $\Sigma b_j w_j$. Thus we get $V\cap W$ by finding all a_i and b_j that satisfy $\Sigma a_i v_i=\Sigma b_j w_j$. Again, let us specialize to F^n, writing X_i for v_i and Y_j for w_j. Each component of the vector relation $\Sigma a_i X_i-\Sigma b_j Y_j=0$ gives us one equation for the unknowns $a_1,\ldots,a_r,b_1,\ldots,b_s$; there are n such equations. Thus we are back to solving a linear system of equations; which we do, of course, by bringing the matrix to row-echelon form. Note that the solution vectors have $r+s$ components, and so lie in F^{r+s}. Let $Z_1,\ldots,Z_t$ be a basis for the space of solutions. For each Z_p we form the linear combination of the X_i with the first r entries of Z_p (the a's) as coefficients (or the linear combinations of the Y_j with the last s entries (the b's) as coefficients—the two are equal by the meaning of our equations). The t vectors so obtained will clearly span $V\cap W$. (But there is no guarantee that they are independent.)

Example

$$U=\mathsf{R}^4.$$

$$V=(([1,2,3,6],[4,-1,3,6],[5,1,6,12]))=((X_1,X_2,X_3)).$$

$$W=(([1,-1,1,1],[2,-1,4,5]))=((Y_1,Y_2)).$$

We note $X_3 = X_1 + X_2$; thus we can *omit* X_3. From $a_1X_1 + a_2X_2 = b_1Y_1 + b_2Y_2$ we get four equations (one for each component):

$$\begin{aligned} a_1 + 4a_2 - b_1 - 2b_2 &= 0.\\ 2a_1 - a_2 + b_1 + b_2 &= 0.\\ 3a_1 + 3a_2 - b_1 - 4b_2 &= 0.\\ 6a_1 + 6a_2 - b_1 - 5b_2 &= 0. \end{aligned} \quad \text{with matrix} \begin{pmatrix} 1 & 4 & -1 & -2\\ 2 & -1 & 1 & 1\\ 3 & 3 & -1 & -4\\ 6 & 6 & -1 & -5 \end{pmatrix}.$$

The row-echelon form is

$$\begin{pmatrix} 1 & 4 & -1 & -2\\ 0 & -9 & 3 & 5\\ 0 & 0 & 1 & 3\\ 0 & 0 & 0 & 0 \end{pmatrix}$$

(we should really divide the second row by 9); the new equations are

$$\begin{aligned} a_1 + 4a_2 - b_1 - 2b_2 &= 0,\\ -9a_2 + 3b_1 + 5b_2 &= 0,\\ b_1 + 3b_2 &= 0. \end{aligned}$$

There is one independent solution $[\frac{7}{9}, \frac{-4}{9}, -3, 1]$; the first two entries are the a_i, the second two the b_j. We know from the general development that $\frac{7}{9}X_1 - \frac{4}{9}X_2$ must equal $-3Y_1 + Y_2$; check that both equal $[-1,2,1,2]$; and so the latter vector, say Z, lies in $V \cap W$. In fact, since there are no other independent solutions of the equations for the a's and b's, $V \cap W$ is $((Z))$ and has dimension one. If in the beginning we had not noticed that X_3 is not needed, we would have had to use three a's: a_1, a_2, and a_3. The matrix of the equations for the a's and b's would be 4×5, and the computation would be a little longer. At the end we must, of course, get the same $V \cap W$.

3. How does one in the concrete case F^n find a linear functional A with given values $b_1, \ldots, b_k$ on independent vectors $X_1, \ldots, X_k$? (This is the concrete version of Problem 5 in Section 1.) This amounts again to a system of linear equations $A \cdot X_1 = b_1, A \cdot X_2 = b_2, \ldots, A \cdot X_k = b_k$, with the entries a_i of A as unknowns. For solving the system, note that the matrix M has the X_i as *rows*; the augmented matrix M' has $b = [b_1, \ldots, b_k]$ as last column.

4. Let $A_1 \cdot X = b_1, \ldots, A_n \cdot X = b_n$ be n equations for n unknowns with *independent* row vectors A_i. Let $M' = (M, B)$ be the augmented matrix. Let us bring M' into row-echelon form, say N'. Then N' has 1's *all along* the "main diagonal" ($1-1-, 2-2-, \ldots,$entries), and 0's below it (for $i > j$), since the A_i are independent (otherwise we would find a 0-row at the bottom). It is easy now to apply further row operations to get 0's also *above* the 1's on the main diagonal; let $b'_1, \ldots, b'_n$ be the entries that appear then as last column. The new *equations*, corresponding to this new augmented matrix, are very simple: the ith one reads $x_i = b'_i$. In other words, there are no free variables, the solution is unique, and the last column is precisely the solution vector. The first n columns of the new matrix are $E^1, \ldots, E^n$; the matrix formed by them (1's on the main diagonal, 0 everywhere else) is called the $n \times n$ **identity matrix** I_n, or I. Thus solving n independent equations in n unknowns amounts to taking the (augmented) matrix (M, B) and changing it by row operations to the form (I, B'); B' is then the solution vector. I_n is also called **unit** matrix.

Example (2×2 for simplicity)

Solve $x_1 + 2x_2 = 1, x_1 + 3x_2 = 2$. Augmented matrix

$$\begin{pmatrix} 1 & 2 & 1 \\ 1 & 3 & 2 \end{pmatrix} \to \begin{pmatrix} 1 & 2 & 1 \\ 0 & 1 & 1 \end{pmatrix} \to \begin{pmatrix} 1 & 0 & -1 \\ 0 & 1 & 1 \end{pmatrix}.$$

(The last step was subtracting 2.2nd row from 1st row.) We read off the solution (last column): $x_1 = -1, x_2 = 1$. (Check!) The final matrix could be written $(I_2, [-1, 1])$.

5. "Lines"; convexity. A one-dimensional flat is often called a line, even if it does not go through 0 (i.e., even if it not a subspace). For given u_0 and u_1, the line $u_0 + ((u_1))$ consists of all $u_0 + tu_1$, with any t. This is called "parametric representation" of the line through u_0, parallel to u_1, with t as "parameter" (running variable). A variant is the "line through two given vectors u and v"; this is the same as the line through u, parallel to $v - u$, consisting of the vectors $u + t(v - u), = tv + (1 - t)u$, with arbitrary scalar t. Note that both u and v belong to this line, for $t = 0$ and $t = 1$. In the case $\mathsf{F} = \mathsf{R}$ the set of vectors $u + t(v - u)$ on this line with $0 \leqslant t \leqslant 1$ is called the **segment** from u to v, often denoted by $[u, v]$. (It is usually understood that $u \neq v$. See Figure 10.)

We describe briefly an important geometric concept based on the notion of segment: Let U be a vector space over R. A subset U is called **convex** if, whenever u and v belong to S, the whole segment $[u, v]$ also belongs to S.

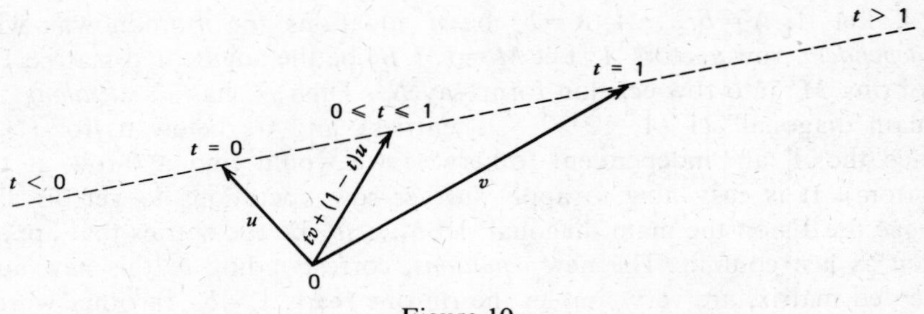

Figure 10.

We will consider some simple examples. In $\mathbf{R}^2$ the set S_1 of all $[x,y]$ with $y \geqslant 0$, or the set S_2 of all $[x,y]$ with $x \geqslant 0$ *and* $y \geqslant 0$, or the set S_3 of all $[x,y]$ with $x^2+y^2 \leqslant 1$ (upper half plane, first quadrant, unit disk). Proof of convexity is simple for the first two (if y_1 and y_2 are $\geqslant 0$, so is $ty_2+(1-t)y_1$ with $0 \leqslant t \leqslant 1$), but not quite so simple for the third case.

We will consider another example of a convex set: Let A be a row vector of n components with $A \neq 0$; let c be any real number. We define a set S in $\mathbf{R}^n$ as $\{X\colon AX \leqslant c\}$, consisting of all vectors X that satisfy the inequality $\Sigma a_i x_i \leqslant c$; it consists of all points "to one side" of the $(n-1)$-dimensional flat with equation $AX=c$. Such a set is called a (closed) **half space**. Our example is furnished by the following statement: Every half space is convex.

PROOF. If $AX_1 \leqslant c$ and $AX_2 \leqslant c$, and $0 \leqslant t \leqslant 1$, then $A\cdot(tX_2+(1-t)X_1) = tAX_2+(1-t)AX_1 \leqslant tc+(1-t)c=c$. (We used that both t and $1-t$ are nonnegative.) (See Figure 11.)

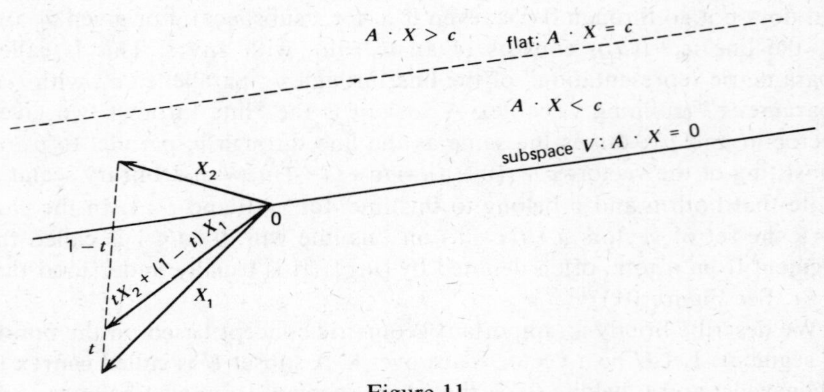

Figure 11.

One also talks about the *open* half space $\{X: AX<c\}$, obtained from the closed one by removing the "boundary" flat.

PROBLEMS

1. Express the vector $Y=[7,3,-1,2]$ as linear combination of $X_1=[1,0,1,1]$, $X_2=[2,-1,2,1]$, and $X_3=[3,1,-1,0]$, if possible.

2. $V_1=(([2,1,0,1],[1,3,-5,3]))$ and $V_2=(([2,1,-1,2],[1,1,0,2],[2,0,-1,1]))$ are two subspaces of $\mathbf{R}^4$. Find (a basis for) the intersection $V_1\cap V_2$.

3. Given $V=(([1,2,1,0],[4,-7,-2,6],[2,-1,0,2]))$ and $Z=[7,4,3,6]$, decide whether Z is in V.

4. $V_1=(([1,2,3,6],[4,-1,3,6],[5,1,6,12]))$ and $V_2=(([1,-1,1,1],[2,-1,4,5]))$ are two subspaces of $\mathbf{F}^4$. Find $V_1\cap V_2$.

5. Find a linear function A on $\mathbf{R}^2$ (i.e., an element of $(\mathbf{R}^5)'$) that takes the values 2, -1, and 1 at the vectors $[1,2,-1,-3,1]$, $[2,3,1,0,2]$, and $[2,2,1,-2,1]$. In fact, find two different such A's.

6. Solve the equations

$$4x_1-6x_2+3x_3=2,$$

$$5x_1-3x_2+2x_3=1,$$

$$x_1-2x_2+x_3=3$$

by reducing the matrix to the identity matrix I_3 (see point 4) above).

7. Find the point where the line through $[-3,-2,3,2]$ and $[-1,0,2,1]$ meets the flat with equation $2x_1+x_2-x_3+x_4=3$.

8. Prove that the unit disk $\{[x,y]: x^2+y^2\leqslant 1\}$ is convex.

9. Let U be a vector space, with two subspaces V and W. Let $L=V+v_0$ and $M=W+w_0$ be linear varieties of types V and W. Show that the intersection $L\cap M$ is nonempty if and only if the vector v_0-w_0 lies in the span of V and W.

10. In $\mathbf{R}^4$ let L be the flat $[2,0,3,2]+(([2,1,0,1],[1,-1,2,1]))$; similarly, $M=[0,6,5,2]+(([1,4,1,2],[2,-1,-3,1]))$. Determine whether L and M have a point in common.

11. With the data of Problem 9 show that the intersection $L\cap M$, if nonempty, is a linear variety of type $V\cap W$. Illustrate with geometric examples in ordinary space.

12. With the data of Problem 10, determine $L\cap M$.

6. DIGRESSION: "DOT PRODUCT"

So far we have always insisted that the coefficients of a linear expression $\Sigma a_i x_i$ form a row vector $A=(a_1,\ldots,a_n)$, and that the linear expression itself be considered as a "product" of the row vector A and a column vector X. However, row and column vectors are not really terribly different from each other, and so we extend our earlier construction: Given two column vectors $X=[x_1,\ldots,x_n]$ and $Y=[y_1,\ldots,y_n]$ in $\mathbf{R}^n$, we form the number $\Sigma x_i y_i$ and regard it as a product of X and Y, called the **dot product** or **standard inner product** and denoted by $X\cdot Y$. For example,

$$[1,2,0]\cdot[-3,2,1]=1\cdot-3+2\cdot2+0\cdot1=1.$$

We note two obvious properties of this construction:

(D_1) $X\cdot Y=Y\cdot X$ for any X and Y; (symmetry).

(D_2) $\left.\begin{array}{l}(X'+X'')\cdot Y=X'\cdot Y+X''\cdot Y\\ (rX)\cdot Y=rX\cdot Y, \text{ for any scalar } r\end{array}\right\}$ linearity in X.

By symmetry (D_2) holds, of course, also on the Y-side; the two (D_2)'s together are called "bilinearity."

The dot product is closely connected with the notions length, distance, and angle. We shall develop this in Chapter 12; here we make only a preliminary contact, in order to introduce some widely used terminology. To begin with perhaps the simplest concept, we say that two vectors X and Y are **orthogonal** (or perpendicular or at right angles) to each other if $X\cdot Y$ is 0. (For the plane or for space, this corresponds to the usual geometric notion of "at right angles," as we shall see in Chapter 12.)

Example

$[1,2,-1,2]$ and $[1,1,5,1]$ are orthogonal to each other (in $\mathbf{R}^4$).

The point of this is the following: If X is a given numerical vector, then the relation $X\cdot Y=0$ is a linear equation for the vector $Y=[y_1,\ldots,y_n]$, and its solution space (of dimension $n-1$, assuming $X\neq0$) can now be interpreted as consisting of all vectors at right angles to X (in space, e.g., all vectors perpendicular to a given vector form a plane). Conversely, a linear equation $\Sigma a_i x_i$ can be interpreted as saying that the vector $X=[x_1,\ldots,x_n]$ is orthogonal to the *column* vector $[a_1,\ldots,a_n]$.

More generally, a homogeneous system of linear equations can be written as $X_1\cdot Y=0,\ldots,X_k\cdot Y=0$, with column vectors $X_1,\ldots,X_k$; the solution space can be thought of as consisting of all vectors Y that are orthogonal simultaneously to $X_1,\ldots,X_k$. In space, for example, the vectors orthogonal to two given (independent) vectors form a line (one-dimensional subspace). Of course, as for actually finding the solutions, nothing has changed!

We discuss briefly two related concepts, distance and angle. For a vector X of $\mathbf{R}^n$ the number $X \cdot X$ ($= \Sigma x_i^2$, or written out, $x_1^2 + x_2^2 + \cdots + x_n^2$, the sum of the squares of the components) is, by definition, the square of the length of X (generalizing the relation $c^2 = a^2 + b^2$ for a right triangle); thus the length itself, denoted by $|X|$ and also called the **norm** of X, is $\sqrt{X \cdot X}$. For example, the norm of $[3, 1, -3, 4, -1]$ is $(9+1+9+16+1)^{1/2}$, which is equal to 6. Extending this a little, the **distance** from (the tip of) X to (the tip of) Y is, again by definition, given by $|Y - X|$ (the length of the "vector from X to Y"). We note that the square of this distance is then $(Y - X) \cdot (Y - X)$, which works out to $Y \cdot Y - 2Y \cdot X + X \cdot X$. Finally, one introduces the **angle**, say ϕ, formed by two vectors X and Y. One defines it from the law of cosines ($c^2 = a^2 + b^2 - 2ab \cdot \cos \phi$). With $a = |X|$, $b = |Y|$, and $c = |Y - X|$, this works out to the prescription:

$$\cos \phi = \frac{X \cdot Y}{|X| \cdot |Y|},$$

with ϕ taken to be the value between 0 and π determined by this formula. (Of course, X and Y should not be 0.) Thus, if $X \cdot Y$ is positive, the angle is acute; if $X \cdot Y$ is negative, the angle is obtuse; and if $X \cdot Y$ is 0, the angle is $\pi/2$, in agreement with the earlier definition of orthogonal.

Remark. Dot product, orthogonality, norm, and distance can also be defined for $\mathbf{C}^n$. But one has to modify the definition slightly, to $X \cdot Y = \Sigma \bar{x}_i y_i$, where the "bar" means "complex conjugate"; otherwise, $|X| = (X \cdot X)^{1/2}$ could be 0 or even complex for nonzero X.

For the moment this is all we shall say about this whole topic. As noted above, we shall take it up in earnest in Chapter 12. (We shall show in Chapter 12 that the right-hand term of the formula for $\cos \phi$ automatically lies between $+1$ and -1, and therefore determines a well-defined angle ϕ.)

PROBLEMS

1. Verify that the two vectors of the example are orthogonal.

2. Find a vector orthogonal to $[1, 2, -3, 2, -1]$.

3. Show that the 0-vector is perpendicular to every vector, in particular to itself.

4. Show that a vector that is perpendicular to itself must necessarily be the 0-vector.

5. Find all the vectors in $\mathbf{R}^4$ orthogonal to $[1, 2, 1, 2]$ and $[1, 1, 2, 2]$.

6. Suppose Y is a solution of the system $2y_1 + y_2 - 3y_3 + y_4 = 0$ and $-y_1 - y_2 + 4y_3 + 2y_4 = 0$. Describe three vectors that are orthogonal to Y.

7. Find the distance from (the tip of) $[4, 1, 0, -2, 3, -5]$ to (the tip of) $[1, 2, 2, -3, 0, -4]$.

8. Find the angle between the two vectors $X=[-18,1,2,20]$ and $Y=[9,7,-4,-4]$ in R^4.

9. Show that two vectors X and Y are orthogonal, if and only if they satisfy the relation $|X+Y|^2=|X|^2+|Y|^2$ (Pythagoras!).

10. Let $Y\neq 0$, and X_0 be given. Show that the vectors X that satisfy the relation $Y\cdot(X-X_0)=0$ form a linear variety through X_0, whose type is the space of all vectors orthogonal to Y.

7. RANK OF MATRIX; ANNIHILATORS

Let M be an $m\times n$ matrix. As noted earlier, we can look at M in two ways: The columns give us n vectors $Z_1,\ldots,Z_n$ in F^m, and the rows give us m vectors $A_1,\ldots,A_m$ in $(\mathsf{F}^n)'$. The span $((Z_1,\ldots,Z_n))$, a subspace of F^m, is called the **column space** of M; the span $((A_1,\ldots,A_m))$, a subspace of $(F^n)'$, is called the **row space** of M. The dimension of the column space is the **column rank** of M (maximum number of independent columns); similarly, we have the **row rank**. We note that the column space does not change under the elementary column operations (Chapter 2, Section 4). Also, for a matrix in column-echelon form the column rank clearly equals the number of nonzero columns (see Chapter 2, Section 5). By obvious symmetry there are the analogous statements about row space, row operations, and row-echelon form. (These remarks give a method to compute column or row rank.)

The most interesting fact about the two ranks is that they coincide.

7.1. THEOREM. *Column rank and row rank of any matrix are equal to each other.*

The common value is called the **rank** of the matrix and denoted by ρ_M or just ρ. We prove Theorem 7.1 together with the next result, which describes the dimension of the solution space of a homogeneous system of equations (compare Section 3). With M as before, let $X=[x_1,\ldots,x_n]$ be a variable vector in F^n.

7.2. THEOREM. *(The Rank-Nullity Law). The dimension of the kernel of the homogeneous system* $A_1\cdot X=\cdots=A_m\cdot X=0$ *(formed with the rows of the matrix* M*) is* $n-\rho_M$.

This dimension is often called the **nullity** (strictly speaking, the row-nullity) of M and denoted by ν_M or just ν; Theorem 7.2 then reads $n=\rho_M+\nu_M$.

PROOF OF THEOREMS 7.1 AND 7.2. Let r and s be the row rank and column rank, respectively, of M. We will prove $r\geqslant s$. By symmetry this also gives $s \geqslant r$, and so $r = s$. The row space of M is spanned by r vectors; we may assume that these are $A_1, \ldots, A_r$. We form the $r \times n$ matrix $M' = [A_1, \ldots, A_r]$. Its column space, a subspace of F^r, is spanned by some t vectors, with $t \leqslant r$ by Proposition 5.1 of Chapter 2. We may assume that these are the first t columns of M'. We will show that then the first t columns of M span the column space of M; from this we find $s \leqslant t$ and therefore also $s \leqslant r$. Let Z be any of the Z_j with $j > t$. The properties of M' imply that there are scalars $c_1, \ldots, c_t$ such that the first r components of $c_1 Z_1 + \cdots + c_t Z_t$ equal those of Z. For any $i > r$ the row A_i is a linear combination of $A_1, \ldots, A_r$. This means that for any Z_j, and therefore also for any vector in the column space of M, the ith component is determined by the first r components. Consequently, $c_1 Z_1 + \cdots + c_t Z_t = Z$. This proves Theorem 7.1. We know that, with M in row-echelon form, the dimension of the solution space of $MX = 0$ is $n - r$, where r is the number of nonzero rows. It is clear that r is the row rank of M and therefore equals ρ_M, and Theorem 7.2 follows. ■

Now, for the dimension of the kernel, we saw already (Section 3) that it is equal to the number of free x's in the row-echelon form, that is, equal to n minus the number of nonzero rows; and we also know that for a matrix in row-echelon form the rank equals the number of nonzero rows. ■

A standard piece of notation involving rank: A square matrix ($n \times n$) is called **singular** if its rank is less than n, **nonsingular** (or, occasionally, **regular**) otherwise.

"Rank" is a notion that is present explicitly or implicitly at many places; the following comment is an example: Consistency of a nonhomogeneous system means that the matrix M of the homogeneous system and the augmented matrix M' have the same rank. Namely, writing the system as $\Sigma c_i Z_i = Y$ (as in 5.1), we see that existence of a solution means that Y is a linear combination of the Z_i. Clearly, this is equivalent to $\dim((Z_1, \ldots, Z_k)) = \dim((Z_1, \ldots, Z_k, Y))$.

We come to a more abstract version of the rank-nullity law. Any fixed solution X of $A_1 \cdot X = \cdots = A_m \cdot X = 0$ clearly also satisfies $A \cdot X = 0$ for any A in $((A_1, \ldots, A_m))$ (we can "add equations" and "multiply equations by scalars"), and conversely, if $A \cdot X = 0$ for all A in $((A_1, \ldots, A_m))$, then in particular $A_1 \cdot X = 0, A_2 \cdot X = 0, \ldots, A_m \cdot X = 0$. This remark leads to a rephrasing of Theorem 7.2.

7.3. DEFINITION AND THEOREM. *Let Q be a subspace of $(\mathsf{F}^n)'$. The set of those X in F^n that satisfy $A \cdot X = 0$ for every A in Q is a subspace of F^n, called the annihilator of Q and denoted by $Q^\perp$ ($\perp$ = perpendicular). The dimensions satisfy the "duality relation"* $\dim Q + \dim Q^\perp = n$.

PROOF. $Q^\perp$ is just the kernel of any system $A_1 \cdot X = \cdots = A_m \cdot X = 0$, where the A_i span Q; that is, annihilator is just another term for kernel. The rank ρ_M of the matrix M with the A_i as rows is $\dim Q$ by definition; Theorem 7.2 says $\dim Q^{\perp} = \nu_M = n - \rho_M = n - \dim Q$. ■

Note the moral: The more (independent) equations, the fewer (independent) solutions.

We can interchange the roles of F^n and $(\mathsf{F}^n)'$ because of the obvious symmetry between the two.

7.3′. THEOREM. *Let $V = ((X_1, \ldots, X_k))$ be a subspace of F^n. The set of all A in $(\mathsf{F}^n)'$ that satisfy $A \cdot X = 0$ for every X in V is a subspace of $(\mathsf{F}^n)'$, called the annihilator of V and written $V^{\perp}$, and* $\dim V + \dim V^{\perp} = n$.

For a fixed A, the requirement "$A \cdot X = 0$ for every X in V" amounts to the same as "$A \cdot X_1 = A \cdot X_2 + \cdots = A \cdot X_k = 0$," analogously to what we saw above for fixed X. We might say that we now prescribe the *solutions* $(X_1, \ldots, X_k)$ and are trying to find the *equations* (the A's in $V^{\perp}$). Note that the relations $A \cdot X_1 = 0, \ldots, A \cdot X_k = 0$ are k linear equations for the components a_i of A. We find the A's as we found the X's earlier: We write out the matrix that goes with the equations $A \cdot X_1 = \cdots = A \cdot X_k = 0$ (with the X_i as *rows* now!), bring it into row-echelon form, and so on.

We can now "iterate": From Q in $(\mathsf{F}^n)'$ we get $Q^{\perp}$ in F^n; from $Q^{\perp}$ in F^n we get $(Q^{\perp})^{\perp}$—also written $Q^{\perp\perp}$—in $(\mathsf{F}^n)'$. Similarly, we go from V in F^n to $V^{\perp\perp}$ in F^n.

7.4. THEOREM. *(the "duality relation"). $V = V^{\perp\perp}$ and $Q = Q^{\perp\perp}$.*

The second relation, for example, says that if we take all solutions of a set of linear equations and then *all* linear equations that those solutions satisfy, we get the span of the original equations. The first one tells us that *any* subspace V of F^n can be described as the solution space of some sutiably chosen set of equations (namely of any set $A_1, \ldots, A_m$ that spans $V^{\perp}$). This is the *second* description of a subspace (the first one describes V as span of some vectors) mentioned in Chapter 2, Section 1.

PROOF OF THEOREM 7.4. First, it is almost a tautology that V is contained in $V^{\perp\perp}$. Any v in V satisfies all the linear equations corresponding to row vectors in $V^{\perp}$, since $V^{\perp}$ by definition consists of those equations that are satisfied by *every* vector in V. Furthermore, from Theorems 7.3 and 7.3′ one sees that $\dim V = \dim V^{\perp\perp}$. But then V and $V^{\perp\perp}$ are equal (Chapter 2, Section 2.4)). A similar approach is used for Q. ■

Example

In R^5: $A_1 = (1,2,0,-1,3)$ and $A_2 = (1,3,-1,1,2)$; $Q = ((A_1, A_2))$; equations for $Q^{\perp}$ are $A_1 \cdot X = 0$ and $A_2 \cdot X = 0$, where row-echelon form is

$$\begin{pmatrix} 1 & 2 & 0 & -1 & 3 \\ 0 & 1 & -1 & 2 & -1 \end{pmatrix}.$$

There are three independent solutions $X_1 = [-2, 1, 1, 0, 0]$, $X_2 = [5, -2, 0, 1, 0]$, and $X_3 = [-5, 1, 0, 0, 1]$ and $Q^{\perp} = ((X_1, X_2, X_3))$. Equations for $Q^{\perp\perp}$ (equations for the "equations") are $A \cdot X_1 = 0$, $A \cdot X_2 = 0$, and $A \cdot X_3 = 0$. Matrix with X_i *as rows*, in row-echelon form (except for scalar factors)

$$\begin{pmatrix} 2 & -1 & -1 & 0 & 0 \\ 0 & 1 & 5 & 2 & 0 \\ 0 & 0 & 5 & 3 & 1 \end{pmatrix};$$

that is, equations

$$2a_1 - a_2 - a_3 = 0,$$

$$a_2 + 5a_3 + 2a_4 = 0$$

$$5a_3 + 3a_4 + a_5 = 0,$$

a_4, and a_5 are free, the solution space has dimension two, a basis is $(\frac{1}{5}, 1, -\frac{3}{5}, 1, 0), (\frac{2}{5}, 1, -\frac{1}{5}, 0, 1)$, check that these two vectors form another basis for Q.

PROBLEMS

1. Find, separately, row rank and column rank of the matrix

$$\begin{pmatrix} 1 & 2 & 3 & 4 \\ 3 & 2 & 2 & 3 \\ -1 & 2 & 4 & 5 \end{pmatrix}.$$

2. Solve Problem 3 of Section 5 by the following shheme: First, find (a basis for) $V^{\perp}$, then decide whether Z is in $V^{\perp\perp}$.

3. With the data from Problem 4 of Section 5, justify and carry out the following procedure to find $V_1 \cap V_2$
a. Find $Q_1 = V_1^{\perp}$ and $Q_2 = V_2^{\perp}$;
b. Find (a basis or a spanning set for) $Q_1 + Q_2$;
c. Find $(Q_1 + Q_2)^{\perp}$.
This is it.

4. Let V and W be subspaces of F^n, with $V \subset W$. Show that $W^{\perp} \subset V^{\perp} (\subset (\mathsf{F}^n)')$; note reversal of order.

5. Let V and W be subspaces of F^n. Prove the relations $(V+W)^\perp = V^\perp \cap W^\perp$ and $(V \cap W)^\perp = V^\perp + W^\perp$; describe in words what they mean.

6. For $V = (([1,2,1,0],[2,1,0,1],[3,0,-1,2]))$ find first $V^\perp$, then $V^{\perp\perp}$, and check that V equals $V^{\perp\perp}$.

7. Find the (inhomogeneous) system of equations whose solution set is the flat $[1,1,1,1] + (([-1,-2,1,2],[2,-2,1,-1]))$.

8. Let M be a matrix in row-echelon form. Describe what kind of a matrix one gets by bringing M to column-echelon form.

9. If the rows of a matrix are independent, so are the columns. True or false?

5

LINEAR EQUATIONS, ABSTRACTLY

The duality between equations and solutions is treated on a more conceptual basis, through the concept of **dual** of a vector space. **Dual bases**, which appear in many theoretical and practical applications, are treated conceptually and computationally. We describe how all this is handled in an abstract vector space with the help of components relative to a basis.

1. THE DUAL SPACE; DUAL BASES

In our development above it has turned out fairly useful that row vectors (or the corresponding linear functions) form a vector space (namely $(\mathsf{F}^n)'$), so that we can add them and multiply by scalars. This idea makes sense and is useful also for the abstract case: Let U be a vector space. Then there is the set or collection of all linear functionals on U; we denote this set by U'. An element of U' is a linear functional on U—a rather abstract, but nevertheless well-defined object. Be sure to distinguish a linear functional φ, an element of U', from any of its *values* $\varphi(u)$ for y in U, which are elements of F. Admittedly there is some ambiguity in writing AX for the linear functional on F^n corresponding to a given row vector A: One should think of X *not* as numerical (then AX would be a scalar and not a function), but as a symbol for which (or rather, for whose components) we *could* substitute numerical values.

Now there is a quite obvious way (analogous to what we did with row vectors in Chapter 4, Section 3) of adding two linear functionals φ and ψ: We define $\varphi+\psi$ by saying that its value at any u in U is $\varphi(u)+\psi(u)$; in brief, $(\varphi+\psi)(u)=\varphi(u)+\psi(u)$. Similarly, we define the scalar multiple $r\varphi$ so that $(r\varphi)(u)=r\cdot\varphi(u)$; the value of $r\varphi$ at u is r times the value of φ at u.

(These are the only sensible definitions for $\varphi+\psi$ and $r\varphi$.) One must check, of course, that $\varphi+\psi$ and $r\varphi$, as defined, are indeed linear functionals; for example,

$$(r\varphi)(u+v)=r\cdot\varphi(u+v)=r\cdot(\varphi(u)+\varphi(v))$$
$$=r\cdot\varphi(u)+r\cdot\varphi(v)=(r\varphi)(u)+(r\varphi)(v).$$

It should not come as a surprise that U', with this definition of + and $\cdot$, is a vector space; that is, that axioms VS_1–VS_9 of Chapter 1, Section 1 are satisfied. Checking them is very simple; one just has to be careful about definitions. Thus, every vector space U brings a "friend" along, the vector space U'. One calls U' the **dual** of U. In particular and as an example, $(\mathsf{F}^n)'$ is the dual of F^n.

We establish a symmetry ("duality") between U and U' similar to that between F^n and $(\mathsf{F}^n)'$.

1.1. THEOREM. *U' is finitely generated, and* $\dim U'=\dim U$.

PROOF. Let $\beta=\{u_1,\ldots,u_n\}$ be a basis for U. For each i from 1 to n we get an element φ_i of U' setting $\varphi_i(u)=x_i=i$th entry of the column vector X representing u relative to β, for any u in U. That these φ_i are indeed linear functionals should be clear. We have $\varphi_i(u_i)=1$ for all i, and $\varphi_i(u_j)=0$ for $i\neq j$, since u_i is represented by E^i. (Putting $\delta_{ij}=1$ if $i=j$, and 0 if $i\neq j$, the "Kronecker delta," one writes in short: $\varphi_i(u_j)=\delta_{ij}$.)

We claim that the φ_i are a *basis* of U' (called the **dual basis,** β', **to** β). To show this, we take any n scalars $a_1,\ldots,a_n$ and form the element $\psi=\Sigma a_i\varphi_i$ of U'. We have the crucial equation $\psi(u_i)=a_i$, from the 1- and 0-values of $\varphi_i(u_j)$; for example,

$$\psi(u_1)=a_1\cdot\varphi_1(u_1)+a_2\cdot\varphi_2(u_1)+\cdots=a_1\cdot 1+a_2\cdot 0+\cdots=a_1.$$

Proposition 1.1 of Chapter 4 tells us now that the every linear functional occurs as such a ψ for suitable and moreover uniquely determined a_i, but that just means that the φ_i are a basis for U'. ■

We consider a simple (almost too simple) example: The standard basis $E^1,\ldots,E^n$ of F^n has a dual the standard basis $E_1,\ldots,E_n$ of $(\mathsf{F}^n)'$. (Check!)

Next we need the converse of Chapter 4, Section 1.1; we are going "backwards" now, which is why it is harder; the result would not even be true if we did not have finite dimension; the concrete case was easier because $(\mathsf{F}^n)'$ is so much like F^n.

1.2. PROPOSITION. *Let $\varphi_1,\ldots,\varphi_n$ be a basis of U', and let $b_1,\ldots,b_n$ be given scalars. There exists a (unique) vector u_0 in U such that $\varphi_i(u_0)=b_i$, $i=1,\ldots,n$.*

PROOF. This is the most abstract argument we shall encounter. One has to keep firmly in mind: U is an (abstract) vector space. On U one has linear functionals; they can be added, and multiplied by scalars, and form the *dual* vector space U'. Thus a vector of U' is a linear functional *on* U.

Now we have to go a step further. U' is a vector space in its own right, even though a rather abstract one, and therefore one can consider linear functionals defined *on* U' (not on U!). Some such linear functionals are easy to manufacture. Every vector u of U generates a linear functional, say l_u, *on* U' by the formula $l_u(\varphi)=\varphi(u)$ for any φ in U'. In other words, we hold u fixed in the expression $\varphi(u)$ and let φ range over U'. This is a *function* on U'; it assigns to the vector φ the scalar value $\varphi(u)$. It is a *linear* functional by the definition of $+$ and $\cdot$ in U': $l_u(\varphi+\psi)=(\varphi+\psi)(u)$ $=\varphi(u)+\psi(u)=l_u(\varphi)+l_u(\psi)$, and similarly for $\cdot$.

We show now that we get all linear functionals on U' that way. We verify

$$l_{u+v}=l_u+l_v\colon\ l_{u+v}(\varphi)=\varphi(u+v)=\varphi(u)+\varphi(v)=l_u(\varphi)+l_v(\varphi)=(l_u+l_v)(\varphi)$$

for any φ; similarly $l_{ru}=r\cdot l_u$. Generally, if $u=\Sigma a_i u_i$, the $l_u=\Sigma a_i l_{u_i}$.

Suppose $u_1,\dots,u_k$ are independent, then so are the l_{u_i}. If $\Sigma a_i l_{u_i}=0$, then, with $u=\Sigma a_i u_i, l_u=0$. This means $l_u(\varphi)=\varphi(u)=0$ *for all* φ, and so $u=0$, by Chapter 4, Section 1, Problem 4. But the relation $\Sigma a_i u_i=u=0$ implies that all a_i are 0, since the u_i are independent.

Let now $u_1,\dots,u_n$ be a basis for U; then the l_{u_i} are n independent linear functionals on U'. Since $\dim U'=n$, the l_{u_i} are a basis for the dual space of U' by Theorem 1.1 applied to U' instead of U. This shows that *any* linear functional on U' is of the form $\Sigma a_i l_{u_i}$ or l_u with $u=\Sigma a_i u_i$. (In fact, the u is unique here.)

Now we consider Proposition 1.2. By Proposition 1.1 of Chapter 4, applied to U', and its basis $\varphi_1,\dots,\varphi_n$, there is a linear functional on U' that takes the value b_i at φ_i. As we just saw, the linear functional is of form l_{u_0} for some (in fact unique) u_0 in U. And the equations $l_{u_0}(\varphi_i)=b_i$ simply mean $\varphi_i(u_0)=b_i$. ■

An immediate consequence is that any basis $\varphi_1,\dots,\varphi_n$ of U' is dual basis of *some* basis $u_1,\dots,u_n$ of U. Choose the b_i so that one is 1 and all others 0, and find the corresponding vector in U. Thus we can form dual bases "in either direction," from U to U' or from U' to U.

We come now to the abstract version of the notion of "homogeneous system of linear equations." It simply amounts to the idea of annihilator $Q^{\perp}$ of a subspace Q of U', or "dually" of annihilator $V^{\perp}$ of a subspace V of U, just as in Definition and Theorem 4.7.3′ and 4.7.3. We immediately prove the analogs of these theorems; the proofs are very short. Let $u_1,\dots,u_k$ be a basis for V; extend to a basis $u_1,\dots,u_n$ of U; let $\varphi_1,\dots,\varphi_n$ be the dual basis. $V^\perp$ is then precisely $((\varphi_{k+1},\ldots,\varphi_n))$ (and thus of the dual dimension, $\dim V^\perp = n - \dim V$). For a φ of the form $\Sigma_1^n a_i\varphi_i$ to vanish at u_j means precisely that the coefficient a_j is 0; thus, to vanish on V, we must have $a_1 = \cdots = a_k = 0$. Symbolically,

$$U\colon \overbrace{u_1,\ldots,u_k}^{V}, u_{k+1},\ldots,u_n.$$

$$U'\colon \varphi_1,\ldots,\varphi_k,\underbrace{\varphi_{k+1},\ldots,\varphi_n}_{V^\perp}.$$

The procedure is similar for $Q^\perp$. This proves the analogs of Theorems 4.7.3 and 4.7.3′. The argument for Theorem 4.7.4 carries over to the abstract situation without change. ■

We look at inhomogeneous systems in the abstract case. Given $\varphi_1,\ldots,\varphi_k$ in U' and constants $b_1,\ldots,b_k$, the solutions of $\varphi_i(u) = b_i$, if there are any at all, form a translate of $\ker(\varphi_1,\ldots,\varphi_n)$. The condition for existence of a solution is that whenever $\Sigma a_i\varphi_i = 0$, then also $\Sigma a_i b_i = 0$ ("rank M = rank M', any relation between the rows of M also holds for the augmented matrix")—not a very helpful condition. We will get a better way to look at this later. (To understand this, one should see that a relation $\Sigma a_i\varphi_i = 0$ between linear functions means that $\Sigma a_i\varphi_i(u) = 0$ for *all* vectors u, for *each* u the *numbers* $\varphi_1(u), \varphi_2(u),\ldots,\varphi_k(u)$ satisfy $\Sigma a_i x_i = 0$.)

PROBLEMS

1. Show that $V \subset W$ ($\subset U$) implies $W^\perp \subset V^\perp$ ($\subset U'$).
2. Show, with the notation above, that if $l_u = l_v$, then $u = v$.
3. Show in detail that u_0 in Proposition 1.2 is unique.

2. THE DUAL BASIS, CONCRETELY

We repeat the definition: Let $A_1,\ldots,A_n$ be a basis for $(\mathsf{F}^n)'$; the *dual basis* of F^n consists of vectors $X_1,\ldots,X_n$ satisfying the relation $A_i\cdot X_j = \delta_{ij}$ (= 1, if $i = j$, and = 0, if $i \neq j$; the "Kronecker delta"). We will see that the X_i are well determined by these relations and form indeed a basis for F^n. How does one find the dual basis? This amounts, of course, simply to solving systems of linear equations; for instance, for X_1:

$$A_1\cdot X_1 = 1, A_2\cdot X_1 = \cdots = A_n\cdot X_1 = 0.$$

Here is a simplified scheme: Let M be the matrix $[A_1,\ldots,A_n]$ with the A_i as rows, and let M' be the *big augmented* $n\times 2n$ matrix $(M, E^1,\ldots,E^n)$

obtained by adding the columns $E^1,\ldots,E^n$ on the right. Now use row operations on M' to bring it into row-echelon form (remember that the rows have length $2n$). The part M of M' will change thereby to a matrix with 1's on the main diagonal (at the $(1,1)$-, $(2,2)$-, $\ldots,(n,n)$ position), and 0's below it (because of the independence of the A_i there are no jumps). By further row operations on the big matrix change the elements *above* the main diagonal also to zero (use the 1 at $(2,2)$ to make the $(1,2)$-element 0; then the 1 at $(3,3)$ to make the $(2,3)$- and $(1,3)$-entries 0, etc.), getting a new matrix, say N'. The first n columns of N' are $E^1,\ldots,E^n$. And, surprise, the second n columns (# $n+1,\ldots,2n$) are the dual basis $X_1,\ldots,X_n$.

This follows since the matrix (M,E^j) for each j is the augmented matrix for the inhomogeneous system of equations for X_j; using the big augmented matrix means that we treated the n problems together. Furthermore, the left side of the *new* equations are extremely simple. For the ith equation it is simply x_i; thus the new equations for the jth problem (the problem to find X_j) read $x_i = i$th entry of the $(n+j)$th column of N' (i.e., of the jth column in the second batch of n columns). All this is really a recapitulation of what we saw in Chapter 4, Section 5.4. As noted there, the $n\times n$ matrix $(E^1,\ldots,E^n)$, with 1's on the main diagonal and 0's elsewhere, is called the $n\times n$ *identity* matrix or *unit* matrix, denoted by I_n or just plain I. Thus, our scheme is to start with $M'=(M,E^1,\ldots,E^n)$ $=(M,I)$, and by row operations to bring it into the form (I,N) with the initial block equal to I. Here N is a square matrix that evolves in the process; and its columns are the vectors $X_1,\ldots,X_n$ that form the dual basis to the A_i; in brief: $N=(X_1,\ldots,X_n)$.

Example

$$M=\begin{pmatrix}1 & 2\\ 1 & 3\end{pmatrix}\cdot\begin{pmatrix}1 & 2 & 1 & 0\\ 1 & 3 & 0 & 1\end{pmatrix}\to\begin{pmatrix}1 & 2 & 1 & 0\\ 0 & 1 & -1 & 1\end{pmatrix}$$

$$\to\begin{pmatrix}1 & 0 & 3 & -2\\ 0 & 1 & 1 & 1\end{pmatrix}.\quad N=\begin{pmatrix}3 & -2\\ 1 & 1\end{pmatrix}.$$

By reversing the row operations we can change from (I,N) back to (M,I). It follows that the X_j are independent, since any linear relation—or, equivalently, solution of $NY=0$—would persist under row operations, and we can change N to I, whose columns are independent. Conversely we can start with independent X_j. The argument just given reverses, and we can reduce N to I, since no 0-row can appear. This yields the "dual" definition and fact: Given n independent column vectors $X_1,\ldots,X_n$ (a basis for F^n), we find the *dual basis* $A_1,\ldots,A_n$ for $(\mathsf{F}^n)'$ by writing $N=(X_1,\ldots,X_n)$, applying row operations to the $n\times 2n$ matrix (N,I) to bring it into the form (I,M), and reading off the *rows* of M (writing $M=[A_1,\ldots,A_n]$). They satisfy $A_i\cdot X_j=\delta_{ij}$.

PROBLEMS

1. $A_1=(2,0,3), A_2=(0,-1,1)$, and $A_3=(1,2,0)$ form a basis for $(\mathsf{R}^3)'$. Find the dual basis.

2. $[4,3,2,1]$, $[5,4,3,1]$, $[2,2,1,1]$, and $[9,4,3,2]$ form a basis for F^4. Find the dual basis.

3. A COMMENT ON ABSTRACT VECTOR SPACES

All specific problems and numerical computations so far have taken place in R^n or C^n (or their duals) with column and row vectors. How about specific problems (as opposed to general theorems and such) in abstract vector spaces?

The answer is that these are almost always reduced to the F^n case by the device of taking components relative to a basis (see Chapter 3, Section 4). In the vector space U at hand we find some basis $\{u_1,\ldots,u_n\}$ or β in short; such a basis is either given or constructed or in the worst case assumed. (Frequently there are obvious bases around, for example, for the vector space P^n of polynomials of degree $\leqslant n$ the polynomials $1,x,x^2,\ldots,x^n$.) We recall that any vector u of U has its expansion $u=\sum x_i u_i$ with respect to the basis β, with unique scalars $x_1,\ldots,x_n$. The column vector $X=[x_1,\ldots,x_n]$ represents u with respect to β; we write $u\underset{\beta}{\leftrightarrow}X$. ($u$ and X determine each other; furthermore, $r\cdot u\underset{\beta}{\leftrightarrow}r\cdot X$, and if also $v\underset{\beta}{\leftrightarrow}Y$, then $u+v\underset{\beta}{\leftrightarrow}X+Y$.) The point is now that we perform all computations on the representing column vectors $X, Y,\ldots$. If linear functionals enter the problem, we imagine the dual basis $\{\varphi_1,\ldots,\varphi_n\}=\beta'$ chosen in U'; any φ then appears as $\varphi=\sum a_i\varphi_i$. We represent φ by the row vector $A=(a_1,\ldots,a_n)$ and write $\varphi\underset{\beta}{\leftrightarrow}A$. It is then true that $\varphi(u)=A\cdot X$. Questions about kernel of linear functionals and the like are now translated into the language of linear equations. (In fact, one does not even need the dual basis β' to construct A from φ. The a_i are simply the values $\varphi(u_i)$ that φ takes at the u_i. Proof. From $u=\sum x_i u_i$ we get $\varphi(u)=\varphi(\sum x_i u_i)=\sum x_i\varphi(u_i)=\sum x_i a_i=A\cdot X$.)

Example

1. $U=P^3$, and basis $\beta=\{1,x,x^2,x^3\}$. What is the dimension of the space

spanned by $(1-x)^3$, $x-2x^2+x^3$, and $1-x-x^2+x^3$? The representing vectors are $[1,-3,3,-1]$, $[0,1,-2,1]$, and $[1,-1,-1,1]$. Column-echelon form gives

$$\begin{pmatrix} 1 & 0 & 1 \\ -3 & 1 & -1 \\ 3 & -2 & -1 \\ -1 & 1 & 1 \end{pmatrix} \to \begin{pmatrix} 1 & 0 & 0 \\ -3 & 1 & 2 \\ 2 & -2 & -4 \\ -1 & 1 & 2 \end{pmatrix} \to \begin{pmatrix} 1 & 0 & 0 \\ -3 & 1 & 0 \\ 3 & -2 & 0 \\ 1 & 1 & 0 \end{pmatrix}.$$

The dimension is two, and the first two vectors form a basis.

2. $U=P^3$. Let φ be the linear functional defined by $\varphi(p(x))=p(1)-p'(1)$ (value at 1 minus value of the derivative at 1). Problem: find $\ker\varphi$.

We choose the following basis (suggested by the fact that $x=1$ is used in the definition of φ): 1, $x-1$, $(x-1)^2$, and $(x-1)^3$. Any p can be written $p(x)=a+b(x-1)+c(x-1)^2+d(x-1)^3$; the representing column is $[a,b,c,d]$. (The remarks above suggest naming the coefficients x_1,x_2,x_3,x_4; we shall not do that.) Then $p'(x)=a+2b(x-1)+3c(x-1)^2$. Thus, $p(1)=a$, $p'(1)=b$, and $\varphi(p(x))=a-b$. $\ker\varphi$ appears now as the solution space (in $\mathbf{R}^4$) of the equation $a-b=0$; b, c, and d are free; the dimension is three, a basis is $[1,1,0,0]$, $[0,0,1,0]$, and $[0,0,0,1]$. The corresponding polynomials are $p_1(x)=1+x-1=x$, $p_2(x)=(x-1)^2=1-2x+x^2$, and $p_3(x)=(x-1)^3=-1+3x-3x^2+x^3$. They form a basis for the subspace $\ker\varphi$ of P^3. Incidentally, the form $a-b$ of the equation tells us that the row vector representing φ with respect to the dual basis (whatever that is; no need to compute it!) is $(1,-1,0,0)$.

PROBLEMS

1. In the space P^4 consider the vector $p_0(x)=x^4+x^2+1$, and the linear functional η defined by $\eta(p(x))=\int_0^1 p(x)\,dx$ for any polynomial $p(x)$ in P^4. With respect to the basis $\beta=\{1,x-1,(x-1)^2,(x-1)^3,(x-1)^4\}$,

a. find the column vector X_0 representing $p_0(x)$;

b. find the row vector A for η;

c. verify $\eta(p_0(x))=A\cdot X_0$.

6

MATRICES

Matrices, which appeared earlier as devices for computing, are now studied in greater detail. The notions of **matrix product** and of **transposed** matrix are introduced. The important notions of **invertible** matrix and the **inverse** of such a matrix appear, in connection with the problem of solving n independent equations in n unknowns. Matrix language is used to describe what happens to the components of a vector if one changes the basis; **transition matrix** is the relevant concept.

1. MATRIX OPERATIONS

Matrices have been discussed several times, as describing either a set of column vectors $M=(Z_1,\ldots,Z_n)$ or a set of row vectors $M=[A_1,\ldots,A_m]$. They will appear again soon with several other interpretations and uses; in this chapter we consider operations on matrices that will be useful for those new interpretations. The components or entries of our matrices are in F ($=\mathsf{R}$ or C); they can also be symbols (a's, x's, etc.) that stand for unspecified scalars.

To begin with, there are the two operations of addition and multiplication-by-scalars:

$+$: Matrices *of the same shape* $m\times n$ can be *added* by adding corresponding entries: $A=(a_{ij})$ and $B=(b_{ij})$ lead to $A+B$ with (i,j)-entry $a_{ij}+b_{ij}$; here $1\leqslant i\leqslant m$ and $1\leqslant j\leqslant n$.

$\cdot$: A matrix is *multiplied* by a scalar entrywise: If $A=a_{ij})$, then $r\cdot A$ has (i,j)-entry $r\cdot a_{ij}$.

This is very similar to what happens to vectors in F^n or $(\mathsf{F}^n)'$. In fact, by arranging the elements of A in a single row (by putting each row of A to the right of the preceding row), we could think of A as a vector in $(\mathsf{F}^{m\cdot n})'$, a row with $m\cdot n$ elements; and $+$ and $\cdot$ would then be just as for such vectors. It should be clear from this that matrices of a fixed size $m\times n$ form a *vector space* over F, which we denote by $\mathsf{F}^{m,n}$; axioms VS_1–VS_9 (Chapter 1, Section 1) hold. The matrix with all entries 0 is the 0 of this space, and is, again, called 0. The "standard basis" consists of the matrices E_i^j with 1 as (i,j)-entry and all other entries 0. Note $\mathsf{F}^{n,1}=\mathsf{F}^n$ and $\mathsf{F}^{1,n}=(\mathsf{F}^n)'$; $\dim \mathsf{F}^{m,n}=m\cdot n$.

We come to something new, the **product** of an $m\times n$ matrix A and an $n\times p$ matrix B (note that the number of columns of A must equal that of rows of B). Write A as $[A_1,\ldots,A_m]$ (a column of m rows) and B as $(B^1,\ldots,B^p)$ (a row of p columns); then $A\cdot B$, or AB in short, is the $m\times p$ matrix whose (i,j)-entry is $A_i\cdot B^j$. Here $A_i\cdot B^j$ is the value of the row A_i (of n terms) on the column B^j (of also n terms) introduced earlier (Chapter 4, Section 2). Writing this out, we have $A_i=(a_{i1},\ldots,a_{in})$, $B^j=[b_{1j},b_{2j},\ldots,b_{nj}]$, and $A_i\cdot B^j=\sum_{k=1}^n a_{ik}b_{kj}$. Note that i and j are fixed and that the sum goes over k; "we multiply corresponding elements of the ith row of A and the jth column of B, and add." It should be clear that the actual computation of the elements of $A\cdot B$ involves a fairly large number of multiplications and additions.

Note 1. In order for *both* products $A\cdot B$ and $B\cdot A$ to be defined, A and B must be of shape $m\times n$ and $n\times m$; then $A\cdot B$ and $B\cdot A$ are "square" matrices of shape $m\times m$ and $n\times n$. If both A and B are $n\times n$ matrices (square and of the same size), then so are both products $A\cdot B$ and $B\cdot A$.

Note 2. Our old concept $A\cdot X=\sum a_i x_i$ for the product of a row vector A ($1\times n$ matrix) and a column vector X ($n\times 1$ matrix) agrees with the one defined now (which yields a 1×1 matrix), by the device of identifying any *scalar* r with the 1×1 matrix (r).

Examples

1.

$$\begin{pmatrix}1 & 0\\ 2 & -1\end{pmatrix}\cdot\begin{pmatrix}2 & 1 & -1\\ 1 & 2 & 3\end{pmatrix}=\begin{pmatrix}1\cdot 2+0\cdot 1 & 1\cdot 1+0\cdot 2 & 1\cdot -1+0\cdot 3\\ 2\cdot 2-1\cdot 1 & 2\cdot 1-1\cdot 2 & 2\cdot -1-1\cdot 3\end{pmatrix}$$

$$=\begin{pmatrix}2 & 1 & -1\\ 3 & 0 & -5\end{pmatrix}.$$

2.
$$\begin{pmatrix}1 & 2\\ 2 & 4\end{pmatrix}\cdot\begin{pmatrix}2 & 4\\ -1 & -2\end{pmatrix}=\cdots=\begin{pmatrix}0 & 0\\ 0 & 0\end{pmatrix}.$$

3. $$\begin{pmatrix} 1 & 2 \\ 2 & 3 \end{pmatrix} \cdot \begin{pmatrix} 2 & 1 \\ 2 & -1 \end{pmatrix} = \begin{pmatrix} 6 & -1 \\ 10 & -1 \end{pmatrix}.$$

4. $$\begin{pmatrix} 2 & 1 \\ 2 & -1 \end{pmatrix} \cdot \begin{pmatrix} 1 & 2 \\ 2 & 3 \end{pmatrix} = \begin{pmatrix} 4 & 7 \\ 0 & 1 \end{pmatrix}.$$

5. $$\begin{pmatrix} 2 & 1 & -1 \\ 1 & 0 & 3 \end{pmatrix} \cdot \begin{pmatrix} a & d \\ b & e \\ c & f \end{pmatrix} = \begin{pmatrix} 2a+b-c & 2d+e-f \\ a+3c & d+3f \end{pmatrix}.$$

Example 1 is straightforward; Example 2 shows that the product of two nonzero matrices can be 0; and Examples 3 and 4 show that $A \cdot B$ and $B \cdot A$ need not be equal, that is, that matrix multiplication is *not commutative* (but, of course, occasionally AB will equal BA, for instance, if $A = B$!). Example 5 illustrates a case where the entries involve variables.

Right away we mention the use of matrix multiplication for linear equations: Let A be an $m \times n$ matrix, with row vectors $A_1, \ldots, A_m$; let $X = [x_1, \ldots, x_n]$ be an arbitrary vector in F^n. Then the *system* of equations $A_1 \cdot X = 0, \ldots, A_m \cdot X = 0$ can be written compactly as $A \cdot X = 0$ (where 0 means the zero vector of F^m), since $A \cdot X$ is the $m \times 1$ matrix (= column) with entries $A_1 \cdot X, \ldots, A_m \cdot X$. (Or, equivalently, the column vector AX equals the linear combination of the columns of A with coefficients x_i; compare Chapter 4, Section 5.1.) More generally, with any $C = [c_1, \ldots, c_m]$, the nonhomogeneous equations $A_1 \cdot X = c_1, \ldots, A_m \cdot X = c_m$ can be written simply as $A \cdot X = C$; for example, the equations of Chapter 4, Section 4 read

$$\begin{pmatrix} 1 & 1 & 2 & 0 \\ 0 & 2 & 1 & 1 \\ 1 & 2 & 1 & 2 \\ 1 & 1 & 1 & 1 \end{pmatrix} \cdot \begin{pmatrix} x_1 \\ x_2 \\ x_3 \\ x_4 \end{pmatrix} = \begin{pmatrix} 8 \\ 6 \\ 2 \\ 2 \end{pmatrix},$$

since the product on the left multiplies out to the column

$$\begin{pmatrix} x_1 & + \; x_2 & +2x_3 & \\ & 2x_2 & + \; x_3 & + \; x_4 \\ x_1 & +2x_2 & + \; x_3 & +2x_4 \\ x_1 & + \; x_2 & + \; x_3 & + \; x_4 \end{pmatrix}.$$

PROBLEMS

1. Multiply out

$$\begin{pmatrix} 2 & 5 \\ 1 & 2 \\ -6 & 1 \\ 0 & 3 \end{pmatrix} \cdot \begin{pmatrix} 1 & 2 & 1 & 3 & 1 \\ 1 & 1 & 2 & 1 & 3 \end{pmatrix}.$$

2. Multiply out

$$\begin{pmatrix} 1+i & -1 \\ i & 1 \end{pmatrix} \cdot \begin{pmatrix} 2 & 1-i \\ 1+i & 2-1 \end{pmatrix}.$$

3. Multiply out

$$\begin{pmatrix} 1 & 2 & 3 \\ 3 & -1 & 0 \\ 2 & 2 & 1 \end{pmatrix} \cdot \begin{pmatrix} 1 \\ 0 \\ 0 \end{pmatrix} \text{ and } \begin{pmatrix} 1 & 2 & 3 \\ 3 & -1 & 0 \\ 2 & 2 & 1 \end{pmatrix} \cdot \begin{pmatrix} 0 \\ 1 \\ 0 \end{pmatrix}.$$

4. Extract from Problem 3 a general law about $A \cdot E^i$.

5. Multiply out

$$\begin{pmatrix} a & b \\ c & d \end{pmatrix} \cdot \begin{pmatrix} d & -b \\ -c & a \end{pmatrix}.$$

6. Multiply out

$$\begin{pmatrix} 1 & 2 \\ 3 & 4 \end{pmatrix} \cdot \begin{pmatrix} a & 0 \\ 0 & b \end{pmatrix} \text{ and } \begin{pmatrix} a & 0 \\ 0 & b \end{pmatrix} \cdot \begin{pmatrix} 1 & 2 \\ 3 & 4 \end{pmatrix}.$$

2. SOME RULES; THE TRANSPOSE; SOME SPECIAL MATRICES

First, as a convention, whenever we write $A+B$, we tacitly assume that A and B have the same shape; similarly, if we write $A \cdot B$, we assume that the number of columns of A equals the number of rows of B (so that $A+B$ or $A \cdot B$ make sense).

Now for a few simple and obvious rules that hold for matrix products:

a. $A \cdot (B \cdot C) = (A \cdot B) \cdot C$; associative law.

b. $A \cdot (B+C) = AB + AC$ and $(B+C) \cdot A = BA + CA$; distributive laws.

c. $r \cdot (A \cdot B) = (rA) \cdot B = A \cdot (rB)$; r is a scalar.

d. $0 \cdot A = 0$; $A \cdot 0 = 0$; 0 stands for 0-matrices.

e. $A \cdot I_n = A$, $I_m \cdot A = A$; I_n and I_m are identity matrices (of appropriate size; i.e., A is $m \times n$).

Rule (a) is an exercise in rearranging double sums; for example, the (i,j)-entry of $A \cdot (B \cdot C)$ is $\Sigma_k (a_{ik} \cdot \Sigma_l b_{kl} c_{lj})$. Rule (b) is trivial consequence of the distributive law for scalars: $a_{ik}(b_{kj}+c_{kj})=a_{ik}\cdot b_{kj}+a_{ik}c_{kj}$; Rule (c) is similar. Rule (d) is obvious, while Rule (e) uses the fact that for a row vector $R=(r_1,\dots,r_n)$ the product $R\cdot E^i$ is just r_i, and the corresponding fact $E_i\cdot C=c_i$. This law justifies the name "identity" or "unit" matrix, since it says that I acts like 1 under multiplication.

Using Rule (a), we define the *powers* A^r, for $r=2,3,\dots$, of a square matrix A by $A^2=A\cdot A, A^3=A\cdot A\cdot A$, etc.; the first power A^1 means A itself. The symbol A^0 (zeroth power) means I_n. We have, of course, $A^r\cdot A^s=A^{r+s}$. This leads to *polynomials* in A, of the form $a_0A^k+a_1A^{k-1}+\cdots+a_{k-1}A+a_kI$.

We must consider one more important operation: If A is an $m\times n$ matrix (a_{ij}), we define the **transpose** A^t of A to be the $n\times m$ matrix (note reversal of order) whose (i,j)-entry is a_{ji}. In slightly dangerous notation we write $A^t=(a^t_{ij})$ and $a^t_{ij}=a_{ji}$. To get A^t we turn A 180° with the main diagonal as axis.

Example

$$A=\begin{pmatrix}2 & 1 & 3\\ 0 & -1 & 4\end{pmatrix},\qquad A^t=\begin{pmatrix}2 & 0\\ 1 & -1\\ 3 & 4\end{pmatrix}.$$

We verify the following rules.

f. $(A^t)^t=A$

g. $(rA)^t=r\cdot A^t$

h. $(A+B)^t=A^t+B^t$

i. $(A\cdot B)^t=B^t\cdot A^t$

Rules (f), (g), and (h) are quite obvious; Rule (f) says that transposition is "of order 2." The important rule is Rule (i), the fact that transposition reverses the order of the factors in a product.

PROOF FOR RULE (i). The (i,j)-entry of $(A\cdot B)^t$ is the (j,i)-entry of $A\cdot B$, that is, $\Sigma_k a_{jk}b_{ki}$; with $b_{ki}=b^t_{ik}$ and $a_{jk}=a^t_{kj}$ this becomes $\Sigma_k b^t_{ik}\cdot a^t_{kj}$, the (i,j)-entry of $B^t\cdot A^t$. ■

For a matrix $A=(a_{ij})$ over C (complex numbers) we have the notion of complex conjugate $\bar{A}=(\bar{a}_{ij})$; we take the comp lex conjugates of all entries. There are some obvious rules:

$$\overline{(A+B)}=\bar{A}+\bar{B};\ \overline{AB}=\bar{A}\cdot\bar{B};\ \overline{(cA)}=\bar{c}\cdot\bar{A};\ (\bar{A})^t=\overline{A^t}.$$

The last expression $\overline{A^t}$ is called the **adjoint** of A and denoted by A^*.

Some special classes of matrices (all of them square) will play a role later.

1. A **scalar** matrix is one of the form $r \cdot I_n$; it has the same entry r all along the main diagonal, and 0's everywhere else.

2. A **diagonal** matrix has all entries *not* on the main diagonal equal to 0; for such a matrix whose diagonal entries are $a_1, \ldots, a_n$ (moving downwards) we write $\operatorname{diag}(a_1, \ldots, a_n)$.

3. An **upper triangular** matrix is one with 0's below the main diagonal: $a_{ij} = 0$ if $i > j$; similarly for **lower triangular**.

4. An **idempotent** matrix or **projection** (name to be justified later) is one that is equal to its own square, $A = A^2$; for example, $A = \begin{pmatrix} 4 & -6 \\ 2 & -3 \end{pmatrix}$. Check $A = A^2$.

5. A **nilpotent** matrix is one some power of which is 0; for example,

$$B = \begin{pmatrix} 1 & -1 & 0 \\ 2 & -3 & 1 \\ 3 & -5 & 2 \end{pmatrix}.$$

Check $B^2 \neq 0, B^3 = 0$.

6. A **symmetric** matrix is one that equals its transpose, $A = A^t$, so that $a_{ij} = a_{ji}$; for example,

$$\begin{pmatrix} 1 & 3 & 4 \\ 3 & 2 & 1 \\ 4 & 1 & -1 \end{pmatrix}.$$

7. A **skew** matrix is one that is the negative of its transpose, $A = -A^t$ so that $a_{ij} = -a_{ji}$; for example

$$\begin{pmatrix} 0 & 1 & -2 \\ -1 & 0 & 3 \\ 2 & -3 & 0 \end{pmatrix}.$$

8. A **Hermitean** matrix is one that equals its adjoint, $A = \overline{A}^t = A^*$, so that $a_{ij} = \overline{a_{ji}}$; for example,

$$\begin{pmatrix} 0 & i & 2 \\ -i & 1 & 1+i \\ 2 & 1-i & -1 \end{pmatrix};$$

these are also called **self-adjoint**. (For real A this the same as symmetric.)

It is easy to see that each of the Classes 1, 2, and 3 is closed under addition, multiplication by a scalar, and also under matrix product. The last one is not quite so obvious, particularly for Class 3, and should be worked out. A skew matrix has diagonal terms 0; for a Hermitean matrix the diagonal terms are real.

A notion that comes up at times is **direct sum** $A \oplus B$ of two matrices A, B. It is the matrix $\begin{pmatrix} A & 0 \\ 0 & B \end{pmatrix}$ with 0-matrices of appropriate shape; if A is $m \times n$ and B is $p \times q$, then $A \oplus B$ is $(m+p)\times(n+q)$. This generalizes to more than two summands.

As a special case of interest, if $M_1,\ldots,M_k$ are square matrices of shape $n_1 \times n_1,\ldots,n_k \times n_k$, then $M_1 \oplus \cdots \oplus M_k$ is also written $\operatorname{diag}(M_1,\ldots,M_k)$. It is square of shape $n \times n$, where $n = n_1 + \cdots + n_k$, with the M_i "along the main diagonal" and 0 everywhere else.

PROBLEMS

1. Verify associativity (rule (a) above) for the three matrices $\begin{pmatrix} 1 & 2 \\ 3 & -1 \end{pmatrix}$, $\begin{pmatrix} 2 & -1 \\ 1 & 2 \end{pmatrix}$, and $\begin{pmatrix} 0 & 3 \\ 2 & 3 \end{pmatrix}$.

2. Verify Rule (i) for the two matrices in Problem 1 of Section 1.

3. With $A = (a_{ij})$ and $B = \operatorname{diag}(b_1,\ldots,b_n)$, what do AB and BA look like?

4. For $A = \begin{pmatrix} 1 & 0 & 0 \\ 0 & 1 & 0 \\ 0 & 0 & 0 \end{pmatrix}$ compute A^2 and A^3.

5. For $B = \begin{pmatrix} 0 & 0 & 0 \\ 1 & 0 & 0 \\ 0 & 1 & 0 \end{pmatrix}$ compute B^2 and B^3.

6. Prove that the product of two upper triangular matrices is again upper triangular.

7. Is the product of two symmetric matrices symmetric? Give proof or counterexample.

8. Prove that the diagonal entries of a skew matrix are 0 and that those of a Hermitean matrix are real.

9. What can one say about the product of two diagonal matrices?

10. Show: If A is any square matrix, then $A + A'$ is symmetric and $A - A'$ is skew.

11. Show that any square matrix can be written as sum of a symmetric and a skew matrix, in one and only one way. Illustrate this with an example. (*Hint.* If A were equal to $A_1 + A_2$ with A_1 symmetric and A_2 skew, what would A^t be?)

12. Let $A = \begin{pmatrix} 1 & 2 & 3 \\ 2 & 1 & -1 \\ 3 & 0 & 1 \\ 1 & 2 & -3 \end{pmatrix}$, $M_1 = \begin{pmatrix} 0 & 1 & 0 \\ 1 & 0 & 0 \\ 0 & 0 & 1 \end{pmatrix}$, $M_2 = \begin{pmatrix} 1 & 0 & 0 \\ 0 & 2 & 0 \\ 0 & 0 & 1 \end{pmatrix}$, and $M_3 = \begin{pmatrix} 1 & 0 & 0 \\ 0 & 1 & 1 \\ 0 & 0 & 1 \end{pmatrix}$.
Compute $A \cdot M_1$, $A \cdot M_2$, and $A \cdot M_3$. Compute the sum of the second and third column vector of A. Describe what you notice.

13. (Generalization of Problem 12.) One defines three kinds of "elementary" matrices ($n \times n$), corresponding to the elementary operations (Chapter 2, Section 4). They are constructed by *modifying* the identity matrix I as follows: For Operation 1 (interchanging two vectors) we choose two indices i and j, make the (i,i)- and (j,j)-entry 0 and the (i,j)- and (j,i)-entry 1. For Operation 2 (multiplying a vector by a nonzero scalar c) we choose an i and make the (i,i)-entry c. For Operation 3 (adding a multiple of vector j to vector i) we choose i and j and make either the (j,i)-entry (for columns) or the (i,j)-entry (for rows) equal to some chosen scalar r. Show that multiplying a matrix M on the right by an elementary matrix amounts to the corresponding operation on the columns of M; show that the same is true for multiplication on the left and rows.

14. Show, using Problem 13, that to any matrix M there exist matrices A and B such that AM is M-brought-to-row-echelon-form and MB is M-brought-to-column-echelon-form.

15. For the matrix $\begin{pmatrix} 3 & 1 \\ 2 & 1 \end{pmatrix} = A$, compute $A^2 + 3A + 3I$.

3. RANK INEQUALITY; INVERSE MATRIX

We prove a fact about rank (Chapter 4, Section 7).

3.1. THEOREM. *The rank of AB is not larger than the rank of A or that of B; with min (for minimum) meaning "the smallest of", this reads*

$$\rho_{AB} \leqslant \min(\rho_A, \rho_B).$$

PROOF. Close inspection of a matrix product reveals that the jth column of AB is a linear combination of the columns of A with coefficients equal to the entries of the jth column of B. Therefore, the column space of AB is a subspace of that of A and $\rho_{AB} \leqslant \rho_A$ follows. The method for finding the inequality for the second factor is similar; the row space of AB is contained in that of B. (For the second factor we could have used the result for the first factor plus transposition: Clearly, for any matrix M we have $\rho_M = \rho_{M'}$. And now, using Rule (i) of Section 2 and $\rho_{AB} \leqslant \rho_A$, we find $\rho_{AB} = \rho_{(AB)'} = \rho_{B'\cdot A'} \leqslant \rho_{B'} = \rho_B$.)

We come now to the important notion "inverse of a nonsingular matrix." Let us consider the system $A \cdot X = C$ for the case of a *square* $n \times n$ A with *independent* rows (or columns), that is, of rank $\rho_A = n$. By Chapter 4, Section 5.1 we know that this has a (unique) solution for *any* C. We plan to develop what amounts to a **formula** for X in terms of C. Let us, purely formally, replace the arbitrary constants c_i by indeterminates y_i, putting $Y = [y_1, \ldots, y_n]$. Applying the row-echelon process to the augmented matrix (A, Y), we reduce it (see Chapter 4, Section 5.4) to the new form (I_n, Y'). Here the entries y_i' of Y' are linear combinations of the variables y_j with certain numerical coefficients, as obtained during the process; we can write y_i' as $B_i \cdot Y$ with B_i the row vector formed by these coefficients. The equations corresponding to the new matrix are very simple (see Chapter 4, Section 5.4, again); namely, $x_i = y_i'$ or $x_i = B_i \cdot Y$. Rewriting this as $X = B \cdot Y$ with the matrix $B = [B_1, \ldots, B_n]$, we can say that the original equations $A \cdot X = Y$ have the solution $X = B \cdot Y$; we have "solved for the x's in term of the y's." Example:

$$\begin{matrix} x_1 + 2x_2 = y_1 \\ x_1 + 3x_2 = y_2 \end{matrix} \quad A = \begin{pmatrix} 1 & 2 \\ 1 & 3 \end{pmatrix}; \quad \begin{pmatrix} 1 & 2 & y_1 \\ 1 & 3 & y_2 \end{pmatrix} \to \begin{pmatrix} 1 & 2 & y_1 \\ 0 & 1 & -y_1 + y_2 \end{pmatrix}$$

$$\to \begin{pmatrix} 1 & 0 & 3y_1 - 2y_2 \\ 0 & 1 & -y_1 + y_2 \end{pmatrix}.$$

New equations

$$\begin{matrix} x_1 = 3y_1 - 2y_2 \\ x_2 = -y_1 + y_2 \end{matrix} \quad B = \begin{pmatrix} 3 & -2 \\ -1 & 1 \end{pmatrix}; \quad X = B \cdot Y.$$

We could let Y take any scalar values; then BY would be the scalar solution vector X satisfying $AX = Y$. Conversely, if we take any scalar values for X, we get scalar values for $Y = AX$; because of the reversible nature of the row operations the relation $X = BY$ will hold. Substituting one relation into the other, we get the two equations $ABY = Y$ *and* $BAX = X$, holding for *any* scalar column vector X or Y. This implies, of course, that both AB and BA equal the identity matrix I_n (proof as exercise). This is a very important relation: Two $n \times n$ matrices A, B such that $AB = I_n = BA$ are called **inverse** to each other; we write $A = B^{-1}$ and $B = A^{-1}$. We will see below that the inverse is unique (if it exists!). A matrix that has an inverse is called "invertible." Recalling the notion "nonsingular" (rank n), we have proved the direct part of the following.

3.2. THEOREM. *Any nonsingular (square) matrix is invertible, and conversely.*

The converse follows from Theorem 3.1, which yields $n=\rho_I=\rho_{AB}\leqslant\rho_A$; we also know $\rho_A\leqslant n$, of course. Thus, $\rho_A=n$.

We saw that if A is nonsingular, then the equations $AX=Y$ are solved by $X=A^{-1}\cdot Y$. ■

Next we consider a remark that may help to understand the concept "inverse" and that shows how to *compute* the inverse of a matrix (if it exists): The entries of AB are the $A_i\cdot B^j$ (row·column). The equation $AB=I$ means, therefore, that the rows of A and the columns of B are *dual bases* of $(\mathsf{F}^n)'$ and F^n. Thus, we can use the scheme of Chapter 5, Section 2, to compute A^{-1} $(=B)$: Start with (A,I_n), use row operations to get (I_n,A^{-1}). (If A is singular, this will show up automatically; we *cannot* produce I_n by row operations.)

Of the two relations $AB=I$ and $BA=I$ that define the inverse, only one is really needed.

3.3. THEOREM. *If to a given $n\times n$ matrix A there exists an $n\times n$ matrix C, such that either $AC=I$ or $CA=I$ holds, then A is nonsingular and C is inverse to A. Furthermore, the inverse of A is uniquely determined.*

PROOF. That A and C are nonsingular follows, as above, from the rank inequality (Theorem 3.1). Suppose we have $AC=I$. Write A^{-1} for any inverse of A (such exist by Theorem 3.2), an nultiply the relation $AC=I$ from the left by A^{-1}. We get $A^{-1}\cdot A\cdot C=A^{-1}$. Because of $A^{-1}\cdot A=I$, the left side is C. We see $C=A^{-1}$. This shows that C is inverse of A and also that there is only one possibility for the inverse, namely C. Similarly in case $CA=I$. ■

To finish up, we re-prove formally that the equations $AX=Y$ are solved by $X=A^{-1}\cdot Y$ if A is nonsingular. We substitute $A^{-1}\cdot Y$ for X in AX; we obtain $A\cdot(A^{-1}\cdot Y)$. Using the associative law (to state, for once, every gory detail), this becomes $(A\cdot A^{-1})\cdot Y=I\cdot Y=Y$. This says that $A^{-1}Y$ is a solution of $AX=Y$. Furthermore, if X is any solution of $AX=Y$, we multiply from the left by A^{-1} and get $A^{-1}AX=A^{-1}Y$—that is, $X=A^{-1}Y$—so that $A^{-1}Y$ is also the *only* possibility for a solution. ■

We sketch briefly another method for finding the inverse of a matrix, "pivoting." Consider the system $\Sigma a_{ij}x_j=y_i$. Take a nonzero a_{ij} (the pivot) and solve the ith equation for x_j; substitute the result for x_j into the other equations. The new system has x_j instead of y_i on the right-hand side. Solve some other equation for some other x, and proceed as before. After n steps all the x's are on the right; the equations read $\Sigma b_{ij}y_j=x_i$; the b_{ij} form the inverse matrix. (If the inverse does not exist, the process grinds to a halt earlier.)

PROBLEMS

1. Prove the fact (used in the text for $M=AB$): if $MY=Y$ for all vectors Y, then $M=I$.

2. Find the inverse of the matrix

$$\begin{pmatrix} 2 & -1 & 1 \\ 1 & 2 & -1 \\ 1 & -1 & 1 \end{pmatrix}.$$

3. Find the inverse of the matrix

$$\begin{pmatrix} 2 & -1 & 1 & 0 \\ 7 & -2 & 0 & 1 \\ -6 & 3 & 0 & 1 \\ 6 & -2 & 2 & 1 \end{pmatrix}.$$

4. Let A and B be invertible $n\times n$ matrices, then one of $A^{-1}B^{-1}$ and $B^{-1}A^{-1}$ is the inverse of AB. Decide which one is the inverse and explain why or how.

5. Suppose a square matrix A satisfies the relation $A^2-A+I=0$. Show that then A is invertible, and that in fact $I-A$ is the inverse of A. Verify for $A=\begin{pmatrix} 3 & -7 \\ 1 & -2 \end{pmatrix}$.

6. Construct two matrices A and B so that ρ_{AB} is less than $\min(\rho_A, \rho_B)$. (Theorem 3.1 only says "less than or equal to.")

7. Show that if B is invertible, then $\rho_{AB}=\rho_A$.

8. Solve the system

$$\begin{aligned} 2x_1 \qquad - x_3 &= y_1 \\ -x_1+3x_2-2x_3 &= y_2 \\ x_1-2x_2+ x_3 &= y_3 \end{aligned}$$

for the x's in terms of the y's by the procedure described in the text.

9. Show that the inverse of an invertible upper triangular matrix is again upper triangular. (Interpret in terms of solving equations.)

10. Let A be a row vector and X a column vector of length n. The matrix $M=X\cdot A$ (not $A\cdot X$) is $n\times n$. Find the rank of M.

11. Suppose A is invertible. What is the inverse of A^{-1}? In fact, is A^{-1} invertible? Be careful with your arguments.

12. Show that the elementary matrices (see Section 4, Problem 14) are invertible and describe the inverses.

13. Show that a square nonsingular matrix can be written as a product of elementary matrices.

14. Work out Problem 13 for the matrix of Problem 2.

15. Find the inverse of the (complex) matrix $\begin{pmatrix} 1 & -2i \\ -1 & i \end{pmatrix}$.

16. Construct a 2×3 matrix P and a 3×2 matrix Q such that $PQ = I_2$. (P is "left inverse" to Q, and Q is "right inverse" to P.) Can one find P and Q, with shapes as given, such that $QP = I_3$?

17. Show: If the rows of the matrix M and the columns of the matrix N are dual bases of $(\mathsf{F}^n)'$ and F^n, then the same holds with M and N interchanged.

4. TRANSITION MATRIX

Earlier we noted the scheme for concrete problems in abstract vector spaces: Choose any basis $\beta = \{u_1, \ldots, u_n\}$ and represent each vector u by a column vector $X = [x_1, \ldots, x_n]$ with $u = \Sigma_i x_i u_i$, then work with these column vectors. This brings up the question: What happens if somebody else wants to use *another* basis, say $\gamma = \{v_1, \ldots, v_n\}$, for U? Any vector u also has components with respect to γ, via $u = \Sigma_j x'_j v_j$, giving rise to a *different* column X'; we have $u \underset{\beta}{\leftrightarrow} X$ and $u \underset{\gamma}{\leftrightarrow} X'$. What is the relation between X and X' (the vectors representing the *same* u with respect to two different bases)?

Example

In P^3 we can take $\beta = \{1, x, x^2, x^3\}$ and $\gamma = \{1, x-1, (x-1)^2, (x-1)^3\}$. Take, for instance, $f(x) = 2 - 2x - x^2 + x^3$; it can also be written as $-1(x-1) + 2(x-1)^2 + (x-1)^3$ (check!); therefore, $f(x) \underset{\beta}{\leftrightarrow} [2, -2, -1, 1]$ and $f(x) \underset{\gamma}{\leftrightarrow} [0, -1, 2, 1]$.

The answer goes as follows. Each element v_j of the "new" basis (like *any* vector in U) can be expended in terms of the "old" basis, so that there are equations $v_j = \Sigma_{i=1}^n p_{ij} u_i$, for $j = 1, \ldots, n$. We write the coefficients with two indices; p_{ij} is the factor with which u_i enters the expansion of v_j (note the *order* of the indices). The p_{ij} form an $n \times n$ matrix P, called the **transition matrix** (*from* β *to* γ). Note that the components of v_j with respect to β appear as the jth column P^j of P; that is, $v_j \underset{\beta}{\leftrightarrow} P^j$. We feed the expressions for the v_j into $u = \Sigma x_i u_i = \Sigma_j x'_j v_j$, getting $\Sigma_j x'_j (\Sigma_i p_{ij} u_i)$ for the latter term. The sum can be rearranged to $\Sigma_i (\Sigma_j p_{ij} x'_j) u_i$, with $\Sigma_j p_{ij} x'_j$ as the factors of u_i. This still equals $\Sigma x_i u_i$; since the u_i are independent, corresponding coefficients must be equal, and we get $x_i = \Sigma_j p_{ij} x'_j$ as the relation between the "old" and the "new" components of u. In matrix notation this reads:

(T) $$X = P \cdot X'.$$

This is the *basic formula* for the change or transition of components (coordinates) under change of basis. We recall that P is the $n \times n$ matrix, whose jth column represents v_j with respect to $\beta = \{u_1, \ldots, u_n\}$ and that X and X' are the column vectors representing a vector u with respect to β and γ. We emphasize that to obtain P, we expand the *new* basis vectors in terms of the *old* ones. Once we have P, it describes the *old* coordinates of any vector in terms of the *new* ones. This reversal of order cannot be avoided (choosing a different notation will only push it to another place in the theory). If we want the new coordinates in terms of the old ones, then, unfortunately, we have to solve the system (T) for X' in terms of X. (We will see below that P is invertible.)

If we want to be more explicit about the bases used, we write T_γ^β instead of P. Finding T_γ^β can involve considerable numerical work.

Example (continuing the example above)

We have the equations $1 = 1$, $x - 1 = -1 + x$, $(x-1)^2 = 1 - 2x + x^2$, and $(x-1)^3 = -1 + 3x - 3x^2 + x^3$. We read off

$$P = \begin{pmatrix} 1 & -1 & 1 & -1 \\ 0 & 1 & -2 & 3 \\ 0 & 0 & 1 & -3 \\ 0 & 0 & 0 & 1 \end{pmatrix}.$$

(Once again, the expansions of 1, $x-1$, $(x-1)^2$, and $(x-1)^3$, with respect to 1, x, x^2, and x^3, give the *columns* of P.) The statement about the two vectors representing the polynomial $f(x)$ above, combined with formula (T), means that we must have the relation

$$\begin{pmatrix} 2 \\ -2 \\ -1 \\ 1 \end{pmatrix} = \begin{pmatrix} 1 & -1 & 1 & -1 \\ 0 & 1 & -2 & 3 \\ 0 & 0 & 1 & -3 \\ 0 & 0 & 0 & 1 \end{pmatrix} \cdot \begin{pmatrix} 0 \\ -1 \\ 2 \\ 1 \end{pmatrix}.$$

(Check!)

Example

$U = R^2$, first basis $= \sigma = \{E^1, E^2\}$, second basis $= \gamma = \{X_1 = [2, 1],\ X_2 = [-1, 1]\}$. Since $X_1 = 2E^1 + E^2$ and $X_2 = -E^1 + E^2$, the transition matrix P

is $\begin{pmatrix} 2 & -1 \\ 1 & 1 \end{pmatrix}$. An arbitrary vector $X=[x_1,x_2]$ has components x_1, x_2 relative to $\sigma(X=x_1E^1+x_1E^2)$ and components x_1', x_2' relative to γ $(X=x_1'X_1+x_2'X_2)$. They are related by the equation $X=PX'$; in detail this reads $x_1=2x_1'-x_2'$, $x_2=x_1'+x_2'$. This is a case of the (familiar) change of coordinates or introduction of new axes in the plane (and our whole development simply generalizes this). The two equations are easily solved: $x_1'=1/3(x_1+x_2)$, $x_2'=1/3(-x_1+2x_2)$. The x_1'-axis is the line with equation $x_1-2x_2=0$ (i.e., $x_2'=0$), and the x_2'-axis is given by $x_1+x_2=0$. (See Figure 12.)

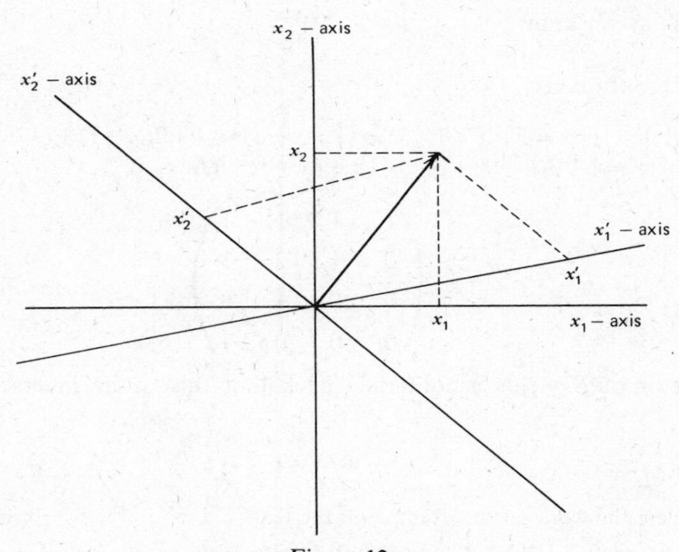

Figure 12.

The symbols X and X' in formula (T) can be numerical or general. We can use (T) to "substitute for X in terms of X'."

Suppose we take a third basis $\delta=\{w_1,\ldots,w_n\}$. There will be a transition metric Q (from γ to σ) $=(q_{ij})$ defined by $w_j=\sum_i q_{ij}v_i$; with $u=\sum_k x_k''w_k$ we have the analog of (T), namely $X'=Q\cdot X''$. There is also a transition matrix R from β to σ via $w_j=\sum_i r_{ij}u_i$, and $X=R\cdot X''$. There is a relation between P, Q, and R, which we get by substituting $X'=Q\cdot X''$ into $X=P\cdot X'$, which gets $X=P\cdot Q\cdot X''$ (associativity of matrix product). Comparing this to $X=R\cdot X''$ yields $R=P\cdot Q$, since X'' is arbitrary.

Indicating the bases in question, we can express the relation as

(Pr) $$T_\delta^\beta=T_\gamma^\beta\cdot T_\delta^\gamma$$

The transition matrix from β to δ is the product of that from β to γ and that from γ to δ.

Application. We take δ (the third basis) equal to β (the first one); that is, we are looking for the transition matrix T_β^γ in the opposite direction, from γ to β. We get $T_\gamma^\beta \cdot T_\beta^\gamma = T_\beta^\beta$. But T_β^β is, of course, the identity matrix: The equations expressing the u_i in terms of the u_i are $u_1 = 1 \cdot u_1 + 0 \cdot u_2 + \cdots + 0 \cdot u_n$, $u_2 = 0 \cdot u_1 + 1 \cdot u_2 + 0 \cdot u_3 + \cdots$, etc. It follows from Theorem 3.3 that T_γ^β $(= P)$ is invertible and that T_β^γ is the inverse of T_γ^β; the two transition matrices between two bases β and γ (one from β to γ, the other from γ to β) are inverse to each other. This checks with what we get from (T), once we know that P is invertible; $X = P \cdot X'$ can be "solved", giving $X' = P^{-1} \cdot X$, as we know.

Example (continued)

We have $1 = 1$; $x = 1 + x - 1$; $x^2 = (1 + x - 1)^2 = 1 + 2(x-1) + (x-1)^2$; $x^3 = (1 + x - 1)^3 = 1 + 3(x-1) + 3(x-1)^2 + (x-1)^3$. Thus

$$T_\beta^\gamma = \begin{pmatrix} 1 & 1 & 1 & 1 \\ 0 & 1 & 2 & 3 \\ 0 & 0 & 1 & 3 \\ 0 & 0 & 0 & 1 \end{pmatrix}.$$

Each vector of β yields a column. Check that this *is* the inverse of the earlier P.

PROBLEMS

1. Complete the work on the example in the text.

2. The vectors [2,2,1]. [2,1,2], and [3,2,2] form a basis, say β, for $\mathbf{R}^3$.
a. Find the transition matrix from the standard basis to the basis β.
b. For any X in $\mathbf{R}^3$ let X' be the column that represents X with respect to β. Write out the equations expressing X in terms of X', and also the equations expressing X' in terms of X. What matrix operation is involved here?
c. Find X' for the vector $X = [3,3,1]$. (You should realize that this is nothing more than the old problem of expressing the vector X in terms of the three vectors given in the problem.)

3. The vectors $X_1 = [2,1,1]$ and $X_2 = [1,2,-1]$ span the same subspace V of $\mathbf{F}^3$ as $Y_1 = [1,5,-4]$ and $Y_2 = [0,3,-3]$. Find the transition matrix from the basis X_1, X_2 to the basis Y_1, Y_2 of V. Find the coefficients needed to express the vector $3Y_1 - 4Y_2$ as linear combination of X_1 and X_2. (*Comment*. The transition matrix P can be defined by $(Y_1, Y_2) = (X_1, X_2) \cdot P$. Using only the first two components of all vectors, this becomes $M = NP$ with invertible N; that is one way to find P.)

7

DETERMINANTS

The concept **determinant** of a (square) matrix is developed from its geometric root, which is the idea of volume of the figure spanned by n vectors in R^n, generalizing the parallelogram. We derive a number of the most important properties, among them the **det-criterion** for independence of n vectors in F^n, the "formula" for the inverse of an invertible matrix and Cramer's rule, the Laplace expansion along a row or column, and the product rule. In an appendix we discuss the idea of **orientation** of R^n (or any vector space over R).

1. DEFINITION; BASIC PROPERTIES

We approach determinants through the notion of **volume** in n dimensions, which generalizes the notion of length, area, volume in line, plane, space.

For R^1 it is natural to say that the vector $[a]$ (the vector "from 0 to a") has length a; the vector from 0 to 1 has length 1; vectors "to the left," with negative a, have negative length—this indicates their direction.

In R^2 let us take two vectors $X_1=[a,b]$ and $X_2=[c,d]$ and form the parallelogram Π with vertices $0, X_1, X_2, X_1+X_2$; we get all points on or within Π as $a_1X_1+a_2X_2$, with a_1 and a_2 ranging from 0 to 1. Elementary considerations, based on Figure 13, show that the area of Π is $ad-bc$; here we think of the axes in R^2 as the usual rectangular coordinate system, and use $1/2 \cdot$base$\cdot$altitude as area of a triangle; for example, $\triangle OY_2X_1$ has area $1/2ab$. The formula $ad-bc$ is now to be applied with *any* X_1 and X_2, not just in the first quadrant as shown. One verifies that the value is *negative*, if the sense of rotation from X_1 to X_2 is opposite to that from the x_1-axis to the x_2-axis (from E^1 to E^2); for example, $X_1=[1,2]$ and $X_2=[4,1]$, $ad-bc=-7$. (See Figure 14.) In space, take three vectors $X_1=[a_1,a_2,a_3]$, $X_2=[b_1,b_2,b_3]$, and $X_3=[c_1,c_2,c_3]$ and form the parallelepiped Π determined by X_1, X_2, and X_3, consisting of all points $r_1X_1+r_2X_2+r_3X_3$

with $0 \leqslant r_i \leqslant 1$. By a similar argument for the plane case, although a good deal more complicated (involving a lot of planes and tetrahedra), one gets for the volume of Π the value $a_1b_2c_3 + a_2b_3c_1 + a_3b_1c_2 - a_3b_1c_2 - a_3b_2c_1 - a_1b_3c_2 - a_2b_1c_3$. Again, this comes out positive if X_1, X_2, and X_3 are situated like E^1, E^2, and E^3 (rotation from X_1 to X_2 together with a shift in the X_3 direction produces a screw motion of the same sense as for E^1, E^2, and E^3 —"right-hand screw"), and negative if they are oriented like E^1, E^2, and $-E^3$. Also, the volume is 0 if the three vectors lie in a plane, that is, are dependent; similarly, the area in the plane is 0 for two dependent vectors, and the length in R^1 is 0 for a dependent vector (i.e., for 0).

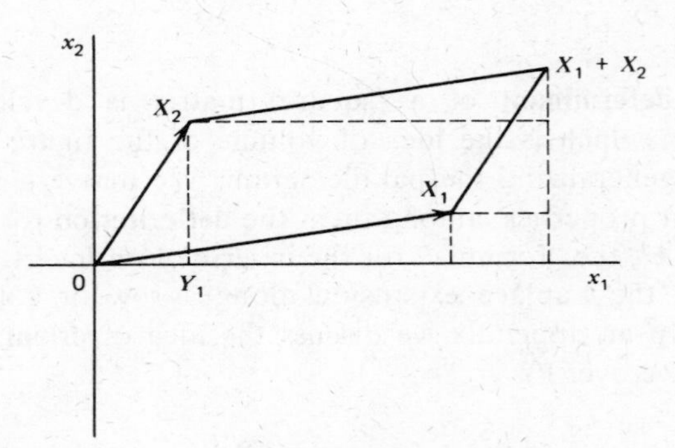

Figure 13.

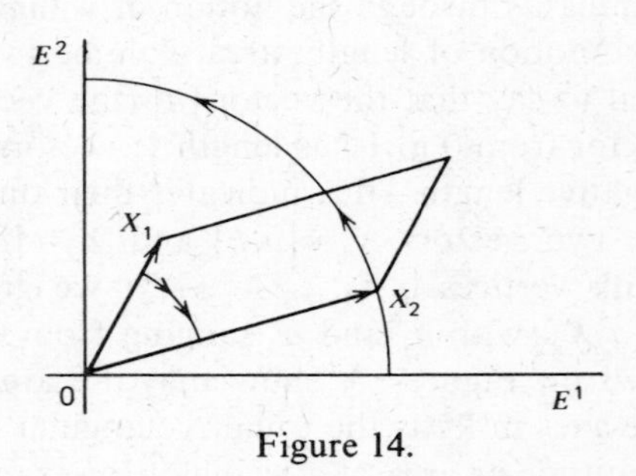

Figure 14.

Now we generalize to R^n: Any n vectors $X_1, \ldots, X_n$ determine a *parallelotope*, the set of all points $r_1X_1 + r_2X_2 + \cdots + r_nX_n$ with $0 \leqslant r_i \leqslant 1$ (generalizing segment, parallelogram, and parallelepiped); we want to define the volume of that figure so that it has properties similar to the cases $n = 1, 2, 3$. We may replace a family of n vectors $X_1, \ldots, X_n$ by the matrix $A = (X_1, \ldots, X_n)$ with the X_i as columns, and conversely. What we are trying to construct then is a function, traditionally called determinant (instead of volume) and denoted in this book by det, that assigns to any n vectors $X_1,\ldots,X_n$ of $\mathbb{R}^n$, or again to any $n\times n$ matrix M, a number $\det(X_1,\ldots,X_n)$ or $\det M$. We require a few properties (suggested by properties of area, etc.) to hold:

1. $\det(X_1,\ldots,r\cdot X_j,\ldots,X_n)=r\cdot\det(X_1,\ldots,X_n)$; multiplying one vector by r changes det by a factor r.
2. $\det(X_1,\ldots,X_j+X_k,\ldots,X_n)=\det(X_1,\ldots,X_n)$; that is, det does not change it for some j, k with $j\neq k$, one replaces X_j by X_j+X_k. We will see that Properties 1 and 2 imply

2′. $\det(X_1,\ldots,X_j+rX_k,\ldots,X_n)=\det(X_1,\ldots,X_n)$; that is, we "may" add any *multiple* of X_k to X_j.

3. $\det(E^1,\ldots,E^n)=1$; "normalization."

Here (1) means that if one expands one side of a parallelotope by a factor r, the volume changes also by r (sign change included). For the nontrivial property (2) or (2′) we have Figure 15: $BB'=CC'=r\cdot OA$; area $OABC$ $=$area $OAB'C'$; volume is invariant under "shear", a special case of "Cavalieri's principle". Property 3 just fixes the unit of volume: the unit-segment, -square, and -cube have length, area, and volume 1.

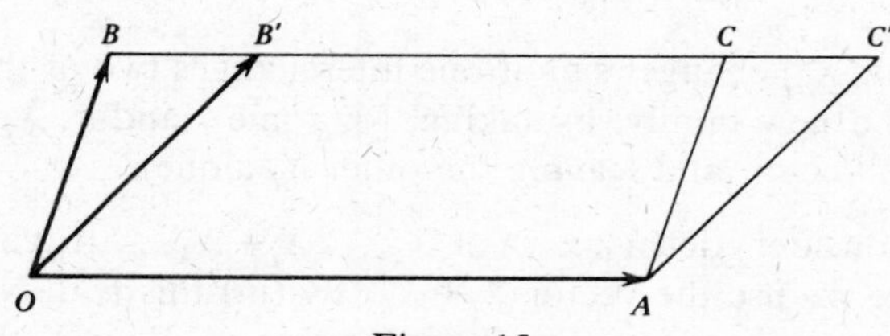

Figure 15.

We draw consequences that give important properties and also lead to a formula ((D) in Section 2) generalizing the formulae for $n=1,2,3$, which will show that there actually *exists* such a function det as we want, and in fact only one.

i. $\det(X_1,\ldots,X_n)=0$ if one of the X_j is 0: Use (1), with $r=0$.

ii. Proof of property 2′:

$$\begin{aligned} r\cdot\det(X_1,\ldots,X_n) &\overset{(1)}{=} \det(X_1,\ldots,rX_k,\ldots,X_n)\\ &\overset{(2)}{=} \det(X_1,\ldots,X_j+rX_k,\ldots,rX_k,\ldots,X_n)\\ &\overset{(1)}{=} r\cdot\det(X_1,\ldots,X_j+rX_k,\ldots,X_n). \end{aligned}$$

Now cancel r.

iii. $\det(X_1,\ldots,X_n)=0$ if the X_j are dependent. By (2′) we can reduce one of the X_j to 0 by adding multiples of the others without changing the determinant; now use (i). There is an important special case: $\det M=0$, if two columns are equal.

iv. $\det(X_1,\ldots,X_j'+X_j'',\ldots,X_n)=\det(X_1,\ldots,X_j',\ldots,X_n)+\det(X_1,\ldots,X_j'',\ldots,X_n)$. "Additivity" in each vector is variable. Together with (1) this says that det is a linear function in each variable, when all other X_k are held fixed.

PROOF OF (iv). For simplicity take $j=n$. We may assume that $X_1,\ldots,X_{n-1}$ are independent; otherwise by (iii) everything is 0. If X_n' and X_n'' are dependent on $X_1,\ldots,X_{n-1}$, then again everything is 0. Say X_n' is not so dependent, then $X_1,\ldots,X_{n-1}, X_n'$ form a basis on F^n; we can write $X_n''=aX_n'+Y$ with Y in $((X_1,\ldots,X_{n-1}))$. And now

$$\begin{aligned}\det(X_1,\ldots,X_{n-1},X_n'+X_n'')&=\det(X_1,\ldots,X_{n-1},(1+a)X_n'+Y)\\ &\overset{(2')}{=}\det(X_1,\ldots,X_{n-1},(1+a)X_n')\overset{(1)}{=}(1+a)\det(X_1,\ldots,X_n')\\ &\overset{(1)}{=}\det(X_1,\ldots,X_{n-1},X_n')+\det(X_1,\ldots,X_{n-1},aX_n')\\ &\overset{(2')}{=}\det(X_1,\ldots,X_{n-1},X_n')+\det(X_1,\ldots,X_{n-1},aX_n'+Y).\end{aligned}$$ ■

v. $\det(X_1,\ldots,X_n)$ changes sign if one interchanges two of the vectors (i.e., if one makes a new family, by taking, for some j and k, X_j as kth vector and X_k as jth vector, and leaving the other X_l alone).

PROOF. Consider $\det(X_1,\ldots,X_j+X_k,\ldots,X_j+X_k,\ldots,X_n)$, where at jth and kth place we use the vector X_j+X_k. By (iii) this is 0. Now "expand", using (iv) twice; we get four terms, with either X_j,X_j or X_j,X_k or X_k,X_j or X_k,X_k at jth and kth place. The first and last are 0 by (iii); the remaining two give what we want, since they add up to 0. ■

PROBLEMS

1. Complete the argument for the formula that gives the area of a parallelogram in R^2.

2. Look up "Cavalieri's Principle."

3. In R^2 find X in the first quadrant and Y and Z in the third quadrant such that $\det(X,Y)$ is positive and $\det(X,Z)$ is negative.

4. Show: If $ad-bc=0$, then the two vectors $[a,b]$ and $[c,d]$ of F^2 are dependent. (If you want an honest argument here, you should not divide by any one of a, b, c, or d without assuming—or at least without realizing that tacitly you are assuming—that the quantity in question is not 0.)

5. Consider $\det(X_1,\ldots,X_{17})$, in F^{17}. If we interchange X_1 and X_{17}, we change the sign of det by v). Instead, we could interchange first X_1 and X_2 (getting $\det(X_2,X_1,\ldots,X_{17})$), then X_1 and X_3 (getting $\det(X_2,X_3,X_1,\ldots,X_{17})$), etc. After X_1 has arrived at the back, we bring X_{17} to the front in the same stepwise way. At each step the sign changes. Do we end up with a minus sign? Does it depend on 17 being odd? What would happen if we took 18 vectors in F^{18} and did the analogous things?

2. THE MAIN FORMULA FOR THE DETERMINANT

We develop a formula to which our requirements (1), (2), and (3) of Section 1 on the function det (presumed to exist) lead. First, as an example, consider the case $n=2$. $X_1=aE^1+bE^2$ and $X_2=cE^1+dE^2$ give, using (1) and (iv) several times,

$$\det(X_1,X_2)=ac\cdot\det\mathit{det}(E^1,E^1)+ad\cdot\det(E^1,E^2)$$
$$+bc\cdot\det(E^2,E^1)+bd\cdot\det(E^2,E^2).$$

Now $\det(E^1,E^1)$ and $\det(E^2,E^2)$ are 0 by (iii), $\det(E^1,E^2)$ is 1 by (3), and $\det(E^1,E^2)$ is -1 by (v) and (3). Thus $\det\begin{pmatrix} a & c \\ b & d\end{pmatrix}=ad-bc$, agreeing with the old formula. Similarly, without details, for $n=3$, $X_1=a_1E^1+a_2E^2+a_3E^3$, etc. Using (1) and (iv) repeatedly, one gets $\det(X_1,X_2,X_3)$ as a sum of 27 terms like $a_1b_1c_3\det(E^1,E^1,E^3)$. Twenty-one of these are 0, since they have two or three equal to E^i. For the remaining six the det-factor is ± 1 (using (3) and (v)), and we get our old formula. We note the following rule for the formula: Write the 3×5 matrix (X_1,X_2,X_3,X_1,X_2), with X_1 and X_2 as fourth and fifth columns. Then the six terms correspond to six diagonals with signs as shown in Figure 16. (*Warning*. This is *only* for 3×3 matrices.)

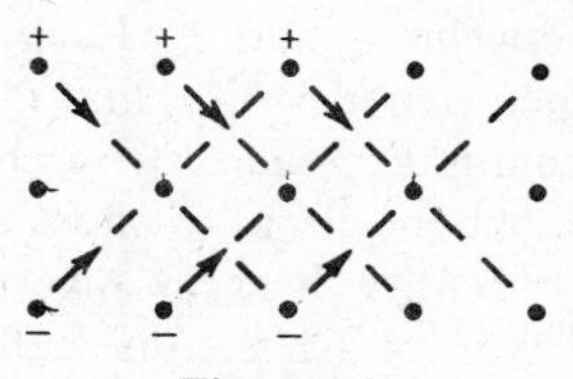

Figure 16.

Now we consider the general case: We write

$$A=(a_{ij}),\quad \text{or}\quad X_j=[a_{1j},a_{2j},\ldots,a_{nj}]=a_{1j}E^1+a_{2j}E^2+\cdots+a_{nj}E^n.$$

We substitute this expression for each X_j in $\det(X_1,X_2,\ldots,X_n)$ and expand by repeated (many times!) use of (1) and (iv). First, we work on X_1:

$$\det(X_1,\dots,X_n)=a_{11}\cdot\det(E^1,X_2,\dots,X_n)$$
$$+a_{21}\cdot\det(E^2,X_2,\dots,X_n)+\cdots+a_{n1}\cdot\det(E^n,X_n).$$

Then we work on X_2 in each term, and so on. Altogether we get n^n terms of the form $a_{i_11}\cdot a_{i_22}\cdot\cdots\cdot a_{i_nn}\cdot\det(E^{i_1},E^{i_2},\dots,E^{i_n})$; the second indices of the a's go from 1 to n, each a_{i_jj} coming from X_j. The indices $i_1,\dots,i_n$ run independently from 1 to n. The factor $\det(E^{i_1},\dots,E^{i_n})$ is, of course, 0 if two of the indices are equal. There remain the sequences $i_1,\dots,i_n$, which consist precisely of the numbers $1,2,\dots,n$ in some order. Such a sequence is called a **permutation** of $1,\dots,n$, and denoted by a symbol such as σ or τ; we often write $\sigma(r)$ instead of i_r. We take it as known from elementary combinatorics that there are $n!=1\cdot2\cdot3\cdot\cdots\cdot n$ possible permutations of $\{1,\dots,n\}$. (One can choose i_1 in n ways, then i_2 in $n-1$ ways, and so on.) The value of $\det(E^{i_1},\dots,E^{i_n})$ is determined by the permutation $\sigma=\{i_1,\dots,i_n\}$ as follows: Let w be the number of "inversions" of σ, that is, the number of pairs r, s with $r<s$, but $\sigma(r)>\sigma(s)$; then the value is $(-1)^w$. We need some combinatorics to show this. σ can be reduced to the "identity" permutation $\sigma_1=\{1,\dots,n\}$ (where the numbers stand in their natural order) by a number of "transpositions"; here a transposition interchanges two numbers and leaves all others alone. Suppose we need t such transpositions (t is not unique; there are many ways of reducing σ to the identity) then $\det(E^{i_1},\dots,E^{i_n})=(-1)^t$, by use of (v) and (3).

Now we claim that $(-1)^w=(-1)^t$, that is, w and t differ by an *even* number (possibly 0). The latter relation is based on the fact that each transposition changes the number of inversions by exactly 1 (up or down). Suppose we interchange i_r and i_s; depending on whether i_r is greater or less than i_s, we either lose or gain an inversion. We next look at those i_p which have $r<p<s$ *and* i_p between i_r and i_s. Each one of those contributes exactly one inversion when paired with i_r and i_s, both before and after the interchange. For the remaining i_q there is no change at all. The total is a loss or gain of 1. The statement about w and t follows now: we are trying to reduce w to 0 (the identity has no inversions); each transposition changes w by ±1; each gain of 1 has to be canceled by a loss of 1. Thus the number of steps is $w+$even number. We write $\epsilon(\sigma)$ for $(-1)^w$ and call it the **parity** of σ; we have, finally, the formula $\det(E^{i_1},\dots,E^{i_n})=\epsilon(\sigma)$. ■

The last equation, together with what we found above, gives us the main formula

(D) $$\det A=\sum_\sigma \epsilon(\sigma)a_{\sigma(1)1}\cdot a_{\sigma(2)2}\cdot\cdots\cdot a_{\sigma(n)n}$$

(the sum extends over all permutations σ of $1,\dots,n$).

This is a formidable formula. It is a sum of $n!$ terms, one term for each σ; each term is a product of certain entries a_{ij}, with a $\pm$ sign $\epsilon(\sigma)$, where $\epsilon(\sigma)$ is the parity of σ. The "general" term displayed in the formula is $\pm a_{i_1 1} \cdot a_{i_2 2} \cdot \cdots \cdot a_{i_n n}$, with $\sigma = \{i_1, \ldots, i_n\}$; each such term is obtained by picking one entry from each column of the matrix, never using the same row twice, multiplying these entries together, and prefixing the sign $\epsilon(\sigma)$. For example, the "first" term, corresponding to the "identity" permutation $\sigma_1 = \{1, 2, \ldots, n\}$ (which has $\epsilon(\sigma_1) = +1$), is $a_{11} \cdot a_{22} \cdot a_{33} \cdot \cdots \cdot a_{nn}$, the product of the diagonal entries—sometimes called the principal term of the sum. Incidentally, exactly half the terms have $\epsilon = +1$.

We write out the sum (D) for $n = 2$; there are two permutations, $\sigma_1 = \{1, 2\}$ (the identity), and $\sigma_2 = \{2, 1\}$; $\epsilon(\sigma_1) = 1$, $\epsilon(\sigma_2) = -1$ (one inversion); $\det\begin{pmatrix} a_{11} & a_{12} \\ a_{21} & a_{22} \end{pmatrix} = 1 \cdot a_{11} a_{22} + (-1) \cdot a_{21} a_{12}$, which is the familiar $ad - bc$.

Next, for $n = 3$, there are six permutations of $\{1, 2, 3\}$, by $3! = 6$. They are $\{1, 2, 3\}$, $\{3, 1, 2\}$, $\{2, 3, 1\}$, $\{3, 2, 1\}$, $\{1, 3, 2\}$, and $\{2, 1, 3\}$. The first three have parity $+1$, the last three have -1; for example, there are two inversions in $\{3, 1, 2\}$, namely 3, 1 and 3, 2. Therefore, $\epsilon = (-1)^2 = +1$. The sum (D) gives

$$\det A = a_{11} \cdot a_{22} \cdot a_{33} + a_{31} a_{12} a_{23} + a_{21} a_{32} a_{13} - a_{31} a_{22} a_{13} - a_{11} a_{32} a_{23} - a_{21} a_{12} a_{33};$$

this is, with a slight change of notation, the formula for the volume of a three-dimensional parallelepiped quoted in the beginning, and worked out, more or less, above before (D).

Formula (D) at this point means *uniqueness*: *If* there is a det-function (satisfying requirements (1), (2), and (3) of Section 1), its value for any matrix A has to be the well-determined one given by the right side of (D). (Never mind that the formula is complicated; we will try never to use it for actual computation. The point is that it *is* an *explicit* formula.)

Now we consider the existence of the function det. We turn the use of (D) around and *define*, for each A, the value $\det A$ as the sum on the right. This is at any rate a well-determined *function* that to each matrix A assigns a number. We have to check Properties 1, 2, and 3. Here (3) is the easiest. For $A = I$ the only term in the sum that is not 0 is the principal term—the one with σ equal to σ_1—which equals $+1 \cdot 1 \cdot \cdots \cdot 1. = 1$. All other terms vanish, since $a_{ij} = 0$ if $i \neq j$. Thus, $\det I = 1$.

Property 1 is also simple. If we replace X_j by rX_j, each term $\pm a_{i_1 1} \times a_{i_2 2} \cdots a_{i_n n}$ in formula (D) acquires a factor r, since the jth a-factor gets multiplied by r. And then the number r factors out from the whole sum.

Property 2 is a good deal more complicated. What we get from formula (D) on the left side is a sum (over σ), where in each term the factor $a_{i_j j}$ has been replaced by $a_{i_j j}+a_{i_j k}$. Multiplying out we get two terms: the first one is exactly the original term of (D); for the second one the factor $a_{i_j j}$ has been replaced by $a_{i_j k}$. Summing over σ, the first terms give precisely the sum of (D); therefore, to prove (2), we have to show that the sum of the second terms is 0. Now this second sum looks almost like that in (D); the only difference is that we are not using the originally given X_j as the jth vector, but that we have replaced it by X_k. In other words, we are talking about the value of the sum (D) in case that two of the n vectors are equal, and we have to show that in this case the value is 0. We proceed to do this. To simplify, let us assume that the first and second vector are equal (the general case goes by the same argument, but has somewhat more messy notation). Thus we have $a_{i1}=a_{i2}$ for all i. The product $a_{i_1 1}\cdot a_{i_2 2}\cdot a_{i_3 3}\cdot\cdots\cdot a_{i_n n}$, corresponding to a permutation $\sigma=\{i_1,i_2,\ldots,i_n\}$, therefore equals (because of $a_{i_1 1}\cdot a_{i_2 2}=a_{i_1 2}\cdot a_{i_2 1}=a_{i_2 1}\cdot a_{i_1 2}$) the product $a_{i_2 1}\cdot a_{i_1 2}\cdot a_{i_3 3}\cdot\cdots\cdot a_{i_n n}$, which corresponds to the permutation $\sigma'=\{i_2,i_1,i_3,\ldots,i_n\}$. Here σ' is obtained from σ by interchanging the first two entries. Consequently, the parities $\epsilon(\sigma)$ and $\epsilon(\sigma')$ are *opposite*. In the sum the two terms for σ and σ' cancel each other; in other words, in the sum the terms cancel each other in pairs, and the whole sum is 0.

Existence and uniqueness of the determinant, with Properties 1, 2, and 3 of Section 1, are now established. We emphasize that A is a (square) matrix, then $\det A$ is a certain numerical value assigned to A. It can be computed by Formula (D); for example,

$$\det\begin{pmatrix} 3 & 2 & 2\\ 1 & 4 & 1\\ -2 & -4 & -1\end{pmatrix}=3\cdot 4\cdot -1+\cdots=?\,\cdot$$

(Work this out!) Do not confuse matrix and determinant. One cannot reconstruct the matrix from the single number $\det A$. For the example just given, there are many other 3×3 matrices with the same det-value. The computation involved in (D) is quite long, if n is more than 3; for $n=4$ there are 24 terms, for $n=5$ there are 120 terms, and so on. We will develop better ways to compute $\det A$ below, and shall avoid Formula (D) as much as possible.

Note that det is only defined for square matrices.

PROBLEMS

1. Determine from Properties v and 3 of Section 1 whether $\det(E^3,E^1,E^5,E^2,E^4)$ is $+1$ or -1.

2. For the permutation $\{3,2,5,1,6,4\}$
a. find the number w of inversions and
b. find a number t of transpositions that reduce it to the identity permutation. Check that w and t differ by an even number.

3. For the general 4×4 matrix A write out all terms of formula (D), including the sign.

4. Use Problem 3 to compute the following:

a.
$$\det\begin{pmatrix} 1 & 2 & 3 & 4 \\ 2 & 1 & 0 & -1 \\ 2 & 0 & 4 & 2 \\ 7 & 3 & 1 & -1 \end{pmatrix}.$$

b.
$$\det\begin{pmatrix} 3 & 2 & 2 \\ 1 & 4 & 1 \\ -2 & -4 & -1 \end{pmatrix}.$$

c.
$$\det\begin{pmatrix} 1 & 0 & -2 \\ -2 & 3 & 0 \\ 1 & -2 & 1 \end{pmatrix} \quad\text{and}\quad \det\begin{pmatrix} 0 & 1 & -2 \\ 3 & -2 & 0 \\ -2 & 1 & 1 \end{pmatrix}.$$

d.
$$\det\begin{pmatrix} 1 & 2 & 3 \\ 2 & 1 & 1 \\ -1 & 4 & 7 \end{pmatrix}.$$

5. For an $n\times n$ matrix A find the k for which the equation $\det(rA)=r^k\cdot\det A$ holds.

3. FURTHER PROPERTIES OF DETERMINANTS

a. If A is triangular, then $\det A = a_{11}\cdot a_{22}\cdot\cdots\cdot a_{nn}$.

Of the sum in (D) only the term for σ_1, where σ is the identity, occurs; every other term must have a factor a_{ij} with $i>j$, as well as one with $i<j$, but then the term is 0. This holds, in particular, for diagonal matrices.

b. $\det A$ is 0 exactly if the columns of A are dependent.

The elementary column operations change A by a *nonzero* factor at worst. Interchange of columns gives a factor -1, multiplying a column by $r(\neq 0)$ gives r as factor; the third operation does not change $\det A$ at all by (1) and (iv) of Section 1. And for A in column-echelon form the statement follows from Property (a) above: A matrix A in column-echelon form is triangular and is nonsingular exactly if the leading 1's are all on the diagonal.

This is a property of det that fits well with the volume interpretation of det (see Section 1). The volume of the parallelotope spanned by $X_1, \ldots, X_n$ is 0 if the parallelotope "collapses"; that is, if the X_i lie in a subspace of $\mathbf{R}^n$ of dimension $<n$ (or, equivalently, are linearly dependent). We even have the reverse: Vanishing of the volume implies dependence. (One could make this more quantitative and say that small volume means *nearly* dependent: Either some vectors are very small, or they all lie *nearly* in a space of lower dimension.)

We have here what sometimes is called the **det-criterion**: A is singular if and only if $\det A = 0$.

c. $\det A = \det A^t$; a matrix and its transpose have the same determinant.

PROOF BY FORMULA (D). With $a_{ij}^t = a_{ji}$, each term of the sum for $\det A^t$ appears as $\epsilon(\sigma) a_{1i_1} \cdot a_{2i_2} \cdot \cdots \cdot a_{ni_n}$; here $\sigma = \{i_1, \ldots, i_n\}$. This looks like the σ-term for $\det A$, except the indices of the a's are reversed. By interchanging the factors, we can write this as $\epsilon(\sigma) a_{r_1 1} \cdot a_{r_2 2} \cdot \cdots \cdot a_{r_n n}$, where $r_1, \ldots, r_n$ is a certain permutation σ^{-1} of $1, \ldots, n$ called the "inverse" of σ (specifically, r_k is the number of the *place where* one finds the entry k in σ). For example, $a_{12}a_{23}a_{31}$ can be rearranged to $a_{31}a_{12}a_{23}$; thus for $\sigma = \{2,3,1\}$ we have $\sigma^{-1} = \{3,1,2\}$. One checks that σ and σ^{-1} have the same parity: Their inversions correspond to each other (any pair $i_j = p$ and $i_k = q$ for σ gives rise to a pair $r_p = j$ and $r_q = k$ for σ^{-1}; either both pairs are inversions or neither is). Thus $\epsilon(\sigma) = \epsilon(\sigma^{-1})$. The above term of $\det A^t$ is now recognized as $\epsilon(\tau) a_{\tau(1)1} \cdot a_{\tau(2)2} \cdot \cdots \cdot a_{\tau(n)n}$ for a certain τ, namely σ^{-1}. As σ runs over *all* permutations, so does σ^{-1} (they determine each other; in fact, $(\sigma^{-1})^{-1} = \sigma$); therefore, summing over σ we get precisely $\det A$. ■

An important consequence is that everything that was done for columns ((1, (2), (i), and (ii) of Section 1) also holds for the rows of the matrix.

d. We introduce the process "expansion of $\det A$ along a column". Choose j and write X_j as $a_{1j}E^1 + \cdots + a_{nj}E^n$. Expand $\det A$ by (1) and (iv) of Section 1 to get $\det A = a_{1j}T_1 + a_{2j}T_2 + \cdots + a_{nj}T_n$, where T_i means $\det(X_1, \ldots, E^i, \ldots, X_n)$ with E^i at jth place. Since we may subtract E^i from the other X_k without changing $\det A$ (by (2′)), we see that T_i is det of the matrix obtained from A by changing the entry a_{ij} to 1, and all other entries of ith row and jth column to 0. We show that, up to a sign $(-1)^{i+j}$, this equals $\det A_{ij}$, where A_{ij} is the $(n-1) \times (n-1)$ matrix obtained from A by *removing* the ith row and jth column. For the proof we could use Formula (D), but we do it by a different "trick": To each $(n-1) \times (n-1)$ matrix B form the $n \times n$ matrix B' with 1 as (i,j)-entry, 0's otherwise in ith row and jth column, and the remaining elements chosen so that $B'_{ij} = B$. (This is "B

blown up from (i,j).") We define a function f on the set of $(n-1)\times(n-1)$ matrices by $f(B)=\det(B')$. It is quite obvious that f satisfies Properties (1) and (2) for det.

As for (3), we note that the matrix I'_{n-1} is not quite I_n (unless $i=j$) but differs from I_n by a permutation of the columns. The trouble is that E^i stands at jth place. We move it to ith place by $|j-i|$ interchanges with its neighbors, getting the value $(-1)^{|j-i|}$, which equals $(-1)^{i+j}$. (Check!) The result is that the function $(-1)^{i+j}\cdot f$ has all three properties of det (for $(n-1)\times(n-1)$ matrices), and is, therefore, by *uniqueness*, equal to it, or $\det B'=(-1)^{i+j}\det B$ for all B. In particular, T_i above is $(-1)^{i+j}\det A_{ij}$. This gives the formula for "expansion along the jth column," also called "Laplace expansion":

$$(L^j)\quad \det A=\sum_i(-1)^{i+j}a_{ij}\det A_{ij}=(-1)^{1+j}a_{1j}\det A_{1j}+(-1)^{2+j}a_{2j}\det A_{24}$$

$$+\cdots+(-1)^{n+j}a_{nj}\det A_{nj}\qquad \text{(for any fixed } j).$$

The term $(-1)^{i+j}\det A_{ij}$ is often called the **cofactor** of a_{ij} in A. There is, of course, also expansion along a row (see Property c). Complicated as it looks, this gives a good way of computing determinants (it reduces an $n\times n$ det to n $(n-1)\times(n-1)$ det's). It is particularly good if the chosen column X_j has many 0's in it (i.e., many of the values $a_{1j}, a_{2j},\ldots$ are 0), and one can often arrange that by the third row or column operation. The "best" case is that where X_j has only *one* nonzero entry. Note that the signs $(-1)^{i+j}$ form a chess board pattern of $+1$'s and -1's, with $+1$ at the $(1,1)$-position.

Example

$$\det\begin{pmatrix}1&2&3&4\\2&1&0&-1\\2&0&4&2\\7&3&1&-1\end{pmatrix}=?$$

Using X_2 we change A (without changing $\det A$, (by Section 1, (2)) to

$$\begin{pmatrix}-3&2&3&6\\0&1&0&0\\2&0&4&2\\1&3&1&2\end{pmatrix}.$$

Expanding along the second row gives $\det A = +1\cdot\det A_{22}$; because of the 0's the expansion reduces to one term. We can now use the 3×3 formula or operate again (by rows, say)

$$\begin{pmatrix} -3 & 3 & 6 \\ 2 & 4 & 2 \\ 1 & 1 & 2 \end{pmatrix} \rightarrow \begin{pmatrix} 0 & 6 & 12 \\ 0 & 2 & -2 \\ 1 & 1 & 2 \end{pmatrix}.$$

Thus, expanding along the first column, $\det A_{22} = +1\cdot\det\begin{pmatrix} 6 & 12 \\ 2 & -2 \end{pmatrix}$ $= -12-24 = -36$.

This surely beats computing 24 ($=4!$) terms and summing.

e. Formula (L^j) has an unexpected consequence. The right side can be looked at as the value (in our usual sense) of the row vector $((-1)^{1+j}\det A_{1j}, (-1)^{2+j}\det A_{2j},\ldots,(-1)^{n+j}\det A_{nj})$ on the column vector $[a_{1j},\ldots,a_{nj}]$. If instead we use the column vector $[a_{1k},\ldots,a_{nk}]$ for some $k\neq j$, we get, by that very formula, the det not of A but of A with its jth column replaced by the kth (the A_{ij}, $1\leqslant i\leqslant n$, do not change!); but that det is 0 since now two columns are equal. Thus, $\Sigma_i(-1)^{i+j}a_{ik}\det A_{ij} = \det A$ if $j=k$, and $=0$ if $j\neq k$. This leads us to define a new matrix $\tilde{A}$, associated to A, the "adjunct" of A, whose (i,j)-entry $\tilde{a}_{ij}$ is $(-1)^{i+j}\det A_{ji}$. (Note the reversal of indices.) What we just said can then be stated very briefly: The product $\tilde{A}\cdot A$ is the *scalar* matrix $\det A\cdot I$ (indeed, the (j,k)-entry of $\tilde{A}\cdot A$ is $\Sigma_i\tilde{a}_{ji}a_{ik} = \Sigma_i(-1)^{i+j}A_{ij}a_{ik}$; these are precisely the sums we just discussed, which are equal to $\det A\cdot\delta_{jk}$). Similarly, using row expansion, we get $A\cdot\tilde{A} = \det A\cdot I_n$.

Finally, suppose $\det A\neq 0$ (this holds exactly if A is of rank n, by Property b), then we can divide by $\det A$ and get

(I) $$A^{-1} = \frac{1}{\det A}\cdot\tilde{A}.$$

Thus, we have arrived at a *formula* for the inverse of a (nonsingular) matrix. This means that one can compute the inverse in a completely mechanical way (the entries of $\tilde{A}$ are computed as $\pm$ determinants of certain submatrices of A). However, while it is of great theoretical interest that there is such a formula, in practice one almost never uses it. The number of steps is too large; we use the scheme of Chapter 7, Section 3 instead.

f. *Product rule.* If A and B are $n\times n$ matrices, then $\det(A\cdot B) = \det A\cdot\det B$. For a formal proof, one should and could take formula (D) for A and B, "multiply out and rearrange the terms" and show that one gets $\det A\cdot B$. We proceed differently (compare Property d): If A is singular (of rank $<n$), so is $A\cdot B$ (Theorem 4.3.1), and then $\det A$ and $\det A\cdot B$ are 0 by (b). Suppose A is nonsingular. We let B vary over *all* $n\times n$ matrices and consider the function f that assigns to B the value $f(B)=\det(A\cdot B)$. Now we know that the jth column of $A\cdot B$ is a linear combination of the columns of A, with coefficients the jth column of B. It follows easily that f satisfies Laws (1) and (2) of Section 1 for $\det B$. As for (3), we have $f(I)=\det(A\cdot I)$ $=\det A$ $(\neq 0)$. Therefore the function $(1/\det A)\cdot f$ satisfies all three laws for det, and by uniqueness *equals* det; that is $(1/\det A)\cdot\det(A\cdot B)=\det B$ for any B. This proves our point.

g. Generalization. So far, the entries of our matrices have been real numbers. There is absolutely no change in the definition of $\det A$ and its properties, if we allow complex numbers (only the interpretation of det as volume of a parallelotope gets lost). In fact, we can allow even more general entries, as long as addition and multiplication (commutative!) with the usual properties are present. For instance, the entries could be *polynomials* in a variable x (or y or ...). In particular the det of a matrix whose entries are polynomials in x is again a polynomial in x. For example, if

$$A=\begin{pmatrix}1-x & x & 2+x^2\\ 2x & 1+x & -3\\ -1 & x^2 & x\end{pmatrix},$$

then $\det A$, the sum of the usual six terms, equals $2x^5-x^3+4x^2+6x+2$.

Strictly speaking, in the general case envisaged here we should use (2′) of Section 1 instead of (2); the proof of (2′) from (2) requires the law "$a\cdot b=0$ implies $a=0$ or $b=0$ (absence of zero-divisors)".

h. Cramer's rule. We know that if A is $n\times n$ nonsingular and if C is a vector in F^n, then the system of n linear equations $A\cdot X=C$ has the solution $X=A^{-1}\cdot C$. Furthermore we have a "formula" for A^{-1}, namely (I) of (e). One can put this a bit differently, and also prove at least part of it in a simpler way: If $Z_1,\ldots,Z_n$ are the columns of A, then the equations $A\cdot X=C$ can be rewritten as $\Sigma x_iZ_i=C$. Nonsingularity of A implies that the Z_i are a basis for F^n, and, therefore, the above equations are guaranteed to have a (unique) solution $X=[x_1,\ldots,x_n]$, whatever C. We choose an index j and compute $\det(Z_1,\ldots,\overset{j}{C},\ldots,Z_n)$, where C has replaced Z_j. Using linearity of det we get, by expanding $C=\Sigma x_iZ_i$, a sum of n determinants. The jth of these is $x_j\det(Z_1,\ldots,Z_n)$. All others have x_rZ_r at jth place, for some $r\neq j$; thus jth and rth columns are dependent, and the term is 0. Since

$\det(Z_1,\ldots,Z_n)=\det A$ is not 0 (by Property b), we can write the result as

(Cr) $$x_j=\frac{1}{\det A}\cdot\det\Big(Z_1,\ldots,\overset{j}{C},\ldots,Z_n\Big).$$

This is Cramer's rule, an explicit formula for the solution of $A\cdot X=C$; x_j is the quotient of the det of A-with-jth-column-replaced-by-C by $\det A$. Note that all x_j have the same denominator $\det A$. This is a very explicit formula for the solution, but it is not often used for computation since it involves calculating $n+1$ different determinants.

Actually (Cr) is just the relation $X=A^{-1}\cdot Y$ in disguise: Expanding the numerator along the jth column, we get

$$x_j=\frac{1}{\det A}\cdot\Big(c_1\cdot(-1)^{1+j}\det A_{1j}+c_2\cdot(-1)^{2+j}\det A_{2j}+\cdots\Big)$$

$$=\frac{1}{\det A}\cdot\sum_i \tilde{a}_{ji}c_i.$$

By the formula for A^{-1} (see (e)) this is the jth entry of $A^{-1}\cdot C$.

i. We have defined det for a square matrix, or for n vectors in F^n. One might wonder whether one can, in the *abstract* situation, define a det of n vectors in a vector space of dimension n. The answer is that one can do something, but it is not so immediate. We indicate two directions: (1) One can define the det of n vectors $v_1,\ldots,v_n$ *with respect to a basis* $\beta=\{u_1,\ldots,u_n\}$ of U; namely, one defines a matrix $A=(a_{ij})$ by the expansions $v_j=\sum_i a_{ij}u_i$ and takes $\det A$ as det. But this value definitely depends on the basis (find out how it changes with the basis!). (2) There is an abstract equivalent of det, constructed with the help of the so-called exterior powers of U ("skew tensors"). We shall not enter into this.

PROBLEMS

1. Compute

$$\det\begin{pmatrix}1&2&3&4&5\\0&2&3&4&5\\0&0&3&4&5\\0&0&0&4&5\\0&0&0&0&5\end{pmatrix}.$$

2. Determine via det whether the three vectors $[2,1,1]$, $[1,2,-1]$, and $[1,5,-4]$ of F^3 are dependent.

3. Compute

$$\det\begin{pmatrix} 1 & 2 & 3 & 1 \\ -1 & 0 & 1 & 1 \\ 1 & 3 & 2 & -1 \\ -1 & -2 & -2 & 1 \end{pmatrix}.$$

(Compare the example in (d).)

4. Find the inverse of the permutation $\{3,2,5,1,6,4\}$, and check that both permutations have the same number of inversions.

5. For the following matrices find the det, the adjunct matrix and the inverse (if it exists, i.e., if the det is not 0):

a.

$$\begin{pmatrix} 2 & 2 & 1 \\ 1 & 2 & 1 \\ 1 & 1 & 1 \end{pmatrix}$$

b.

$$\begin{pmatrix} 1 & 2 \\ 1 & 4 \end{pmatrix}.$$

c. $\begin{pmatrix} a & b \\ c & d \end{pmatrix}$, the "general" matrix. (Of course here you will have to *assume* that det is not 0 for the a,b,c,d under consideration.)

6. Let M be a matrix of the form $\begin{pmatrix} A & 0 \\ 0 & B \end{pmatrix}$, with A $m\times m$, B $n\times n$, and 0 standing for appropriate zero matrices. Prove that $\det M = \det A \cdot \det B$. (A possible approach: M factors as $\begin{pmatrix} A & 0 \\ 0 & I_n \end{pmatrix}\cdot\begin{pmatrix} I_m & 0 \\ 0 & B \end{pmatrix}$.)

7. Extend Problem **6** to the case of an M of the form $\begin{pmatrix} A & D \\ 0 & B \end{pmatrix}$.

8. Compute

$$\det\begin{pmatrix} 1-x & 1 & 1 & 1 \\ 1 & 1-x & 1 & 1 \\ 1 & 1 & 1-x & 1 \\ 1 & 1 & 1 & 1-x \end{pmatrix}.$$

Find the values of x for which this is 0.

9. Solve by Cramer's rule: $x+2y=4$ and $x+3y=5$.

10. Suppose that $X_1,\ldots,X_{n-1}$ are given vectors in $\mathbf{R}^n$. With a variable vector $X=[x_1,\ldots,x_n]$ we can form the determinant $\det(X_1,\ldots,X_{n-1},X)$. By one of the main properties of det this is a *linear* function of X (see (iv) of Section 1). *Question*: What is the relation of the nullspace of this linear function (a certain subspace of $\mathbf{R}^n$ of codimension 1 or 0) to the given vectors $X_1,\ldots,X_{n-1}$?

11. Apply Problem **10** to the following:
a. $[1,2,3]$ and $[2,3,4]$ in $\mathbf{R}^3$.
b. $[-1,1,0,5]$, $[-1,0,2,3]$, and $[1,-2,1,-3]$ in $\mathbf{R}^4$.

4. ORIENTATION

We indicated in Section 1 that the sign of det has a geometric meaning in $\mathbf{R}^2$ and $\mathbf{R}^3$: In the plane, if $\det(X,Y)>0$, then the sense of rotation from X to Y is the same as that from E^1 to E^2; in space, $\det(X,Y,Z)>0$ means that X, Y, and Z form a "right-handed screw" just like E^1, E^2, and E^3. We generalize to $\mathbf{R}^n$ simply by saying that a basis $X_1,\ldots,X_n$ is **positively oriented**, if $\det(X_1,\ldots,X_n)$ is positive (and negatively oriented, if det <0). Note that the *order* of the vectors is important for this; if we exchange two vectors, we *change* the sign of the det. In particular, the standard basis is positively oriented.

The notion of orientation turns out to be important when one studies the idea of gradual or continuous change of a basis: Imagine that during some time interval the vectors $X_1,\ldots,X_n$ move around gradually, without any sudden jumps, in such a way that at every moment they form a basis. The question arises naturally: Given two bases β and γ, can we get from β to γ by such a gradual change?

To formulate this precisely, we need the concept "continuous vector function": This is a continuous function $X(t)=[x_1(t),\ldots,x_n(t)]$ that to each t in some given interval $[a,b]$ on $\mathbf{R}$ (sorry, here $[a,b]$ is not a vector) assigns a vector $X(t)$ in $\mathbf{R}^n$; continuity simply means that the n ordinary real-valued functions $x_i(t)$ are continuous in the usual sense. For example, $[a,b]=[0,1]$, $n=3$, $X(t)=[1-t^2,e^t,\sin t]$; thus $X(0)=[1,1,0]$, $X(1/2)=[3/4,\sqrt{e},\sin 1/2]$, and so on.

Now, to *connect* two bases β and γ means to have n continuous vector function $X_1(1),\ldots,X_n(t)$, defined on some $[a,b]$, so that (1) $\beta=\{X_1(a),\ldots,X_n(a)\}$, (2) $\gamma=\{X_1(b),\ldots,X_n(b)\}$, and (3) for each t in $[a,b]$ the $X_i(t)$ are independent. Property (3) means simply that the determinant $\det(X_1(t),\ldots,X_n(t))$ ($=d(t)$ in short) is never 0 on $[a,b]$. It is clear from the determinant expansion formula (D) of Chapter 7, Section 3.d that $d(t)$ is also a continuous function. One of the principal properties of continuous functions implies then that $d(t)$ cannot change sign on $[a,b]$ (otherwise it would be 0 at some t). In particular, the determinants of the matrices representing β and γ must have the same sign. In other words, β and γ must have the same orientation.

It turns out that the condition is also sufficient; we sketch a proof using column operations. Let then $\beta=\{Y_1,\ldots,Y_n\}$ and $\gamma=\{Z_1,\ldots,Z_n\}$ be two bases with sign $\det M = \operatorname{sign}\det N$, where M and N are the matrices formed with the Y's and Z's.

We know that we can reduce M and N to I by the elementary column operations. Now some of these operations can be done in a *continuous* fashion. For operation (3), adding rX_k to X_j, we add $t \cdot rX_k$ to X_j and let t go from 0 to 1. Similarly for *part* of (2), multiplying X_j by a *positive* r, we multiply by t and let t go from 1 to r (then t is never 0, and so the det is never 0). Now it is not hard to see that with these two operations alone one can change M and N to the form $\operatorname{diag}(\pm 1, \pm 1, \ldots, \pm 1)$. (One can no more *interchange* two vectors; however, one can replace, in several steps, X_j and X_k by X_k and $-X_j$.) The det never changes sign, of course. Finally, as substitute for multiplying a vector by a *negative* r, we can multiply *two* vectors by -1. Suppose we work on the first two vectors; we multiply our matrix by $N(t)=\operatorname{diag}(R(t),1,\ldots,1)$, where $R(t)$ is the 2×2 matrix

$$\begin{pmatrix} \cos t & -\sin t \\ \sin t & \cos t \end{pmatrix},$$

and let t go from 0 to π. Note that $\det R(t)$ is constant $(=1)$. Thus, we can reduce any *two* -1's in the diagonal form above to $+1$'s; it should be clear that we can, in this fashion, get the *same* diagonal matrix from M or N if, and only if, $\det M$ and $\det N$ have the same sign. In the case of equal sign we connect β and γ by putting all the individual steps together in the obvious way.

It is easy to extend all this to abstract vector spaces. The only difficulty is that a basis now does not have a determinant (see Section 3.i), and so there is no natural way to decide whether a basis is positively or negatively oriented. At first sight this seems to spoil the whole idea of orientation. However, two bases β and γ of our vector space U (over $\mathbf{R}$) have a transition matrix T_γ^β. We shall say that β and γ have the *same* orientation, if $\det T_\gamma^\beta$ is positive, and the *opposite* orientation otherwise. From the product formula for det and the product relation 6.4.(P) for transition matrices, one sees that the collection of *all* bases of U falls into precisely *two* classes such that within each class any two bases have the same orientation, but two bases not from the same class have opposite orientation. To *orient* U means to pick one of the two classes (quite arbitrarily) and to call those bases positive, and the others negative. Our construction above shows that two bases β and γ of the same orientation can be connected by a *continuous* family of bases—a family $\beta(t)$ of bases, depending on a variable or parameter t, beginning with β and ending with γ, and continuous in the sense that the transition matrix from β to $\beta(t)$ is a continuous function of t (i.e., all entries of the matrix are continuous functions). Clearly, all we have to do is to take T_γ^β, whose det is positive by assumption, and to connect it to I by a continuous family of nonsingular matrices, but that is just what we did above.

PROBLEMS

1. Connect the two bases $\sigma=\{E^1,E^2\}$ and $\beta=\{[3,1],[2,1]\}$ of R^2 by a continuous family of bases.

2. Connect the two bases $\sigma=\{E^1,E^2,E^3\}$ and $\gamma=\{[2,3,1],[1,2,1],[1,1,1]\}$.

3. Construct negatively oriented bases in R^n for $n=1,2,3$.

4. What happens to the idea of orientation when we replace R by C?

5. Work out a proof for the remark in the text on the replacement of X_j and X_k by X_k and $-X_j$.

8 LINEAR TRANSFORMATIONS

The notion **linear transformation**—a map, from one vector space to another, that "preserves" linear relations—touched upon in Chapter 1, is now taken up in earnest. This is the "dynamic" (and bigger) half of linear algebra, as opposed to the "static" description of a single vector space. Even a single vector space cannot really be understood without using **operators** (linear transformations of the space to itself). We define the concept, "represent" linear transformations by matrices, interpret earlier results and notions from the new perspective, and study linear transformations geometrically and computationally. In several appendices we discuss some ideas of interest, which however will not be used seriously in the remainder of the book: **transpose** of a linear transformation (the abstract version of transposed matrix), **affine** transformations (a natural generalization of linear transformations), **quotient spaces**, and **exactness**.

1. BASIC DEFINITIONS

To repeat the definition, a **linear transformation** T (also $F, G, \ldots, R, \ldots$) from a vector space U to a vector space V (both over the same field F), denoted by $T\colon U \to V$, is a *function* from U to V, that is, a rule or law that to each vector u in U assigns unambiguously a vector $T(u)$ (or Tu) in V (called the T-image of u, or the value of T at u), subject to the *linearity* requirements

L_1: $T(u+u') = T(u) + T(u')$, for all u, u' in U

L_2: $T(ru) = r \cdot T(u)$, for all u in U and r and F (We write $u \underset{T}{\mapsto} v$, or only $u \mapsto v$, if v is T-image of u.)

We recall that a linear transformation from a vector space U to *itself* ($V=U$) is called an **operator** on U.

We also recall that $T(0)=0$ *always* (take $r=0$ in L_2). We note that L_1 and L_2 imply, by repeated application, the formula $T(r_1u_1+r_2u_2)=r_1\cdot T(u_1)+r_2\cdot T(u_2)$. This extends to $T(\Sigma a_iu_i)=\Sigma a_iTu_i$; one says that T *preserves* linear combinations. In particular, if $u_1,\ldots,u_k$ are dependent (in U), then $Tu_1,\ldots,Tu_k$ are dependent (in V); namely, $\Sigma a_iu_i=0$ implies $\Sigma a_iTu_i=T(\Sigma a_iu_i)=T0=0$.

Note. For two linear transformations S and T to be identical they (a) must go from the same U to the same V and (b) must satisfy $S(u)=T(u)$ for *every* u in U. There could be (and are) many different T's from a given U to a given V.

For some examples, see Chapter 1, Section 3 and Chapter 4, Section 1 (linear functionals!, $V=\mathsf{F}$!); we add a few more.

1. We consider another geometric example, a modification of an earlier one: Let U=ordinary 3-space (with origin) and V a plane in space (through O). Choose a fixed direction d not parallel to V; map U into itself as follows: Given any vector OA, draw the line parallel to d through A, let it intersect V at A', put $T(OA)=OA'$. (See Figure 17.) Convince yourself that this T is linear; it is called the "projection of U onto V along d."

2. $U=P^n$ (polynomials); $V=\mathsf{F}^k$. We define a T as follows: Let $x_1,\ldots,x_k$ be k values of the variable x, in F, arbitrarily chosen. Put $T(p(x))=[p(x_1),\ldots,p(x_k)]$; to the polynomial $p(x)$ assign the vector in F^k whose components are the values $p(x_1),\ldots,p(x_k)$ of $p(x)$ at $x_1,\ldots,x_k$. This is *linear*, because of the definition of + and $\cdot$ in P^n and F^k. (Check!) For example, $n=3$; $k=2$, $x_1=0$, $x_2=1$. Then $T(1+2x+x^2)=[1+2\cdot 0+0^2, 1+2\cdot 1+1^2]=[1,4]$, $T(1-x)^3=[1,0],\cdots$.

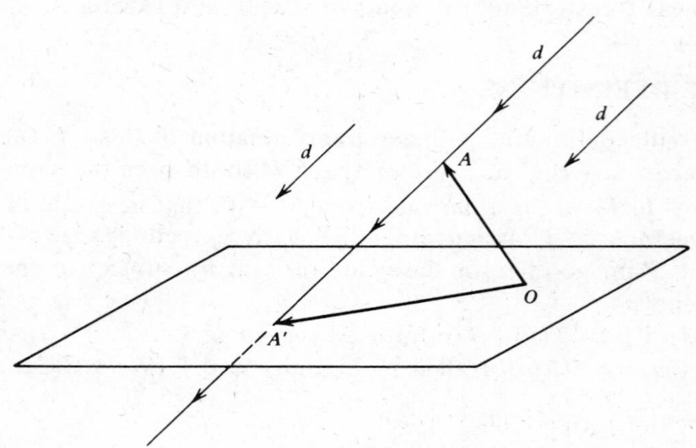

Figure 17.

3. Let U be the vector space of solutions $f(x)$ of a linear differential equation $y^{(n)}+p_1(x)y^{(n-1)}+\cdots+p_n(x)y=0$; $V=\mathsf{R}^k$. Define T as follows: $T(f(x))=[f(0),f'(0),\ldots,f^{(k-1)}(0)]$; that is, we assign to the function $f(x)$ the k-vector whose components are the values of f and its derivatives at 0. Again linearity is pretty obvious.

4. The following is called the "standard example." Let A be an $m\times n$ matrix. We define T_A: $\mathsf{F}^n\to\mathsf{F}^m$ by: T_A assigns to any $X\in\mathsf{F}^n$ the vector $A\cdot X$ in F^m; that is, $T_A(X)=A\cdot X$. (The example of Chapter 1, Section 3, is of this form, with $A=\begin{pmatrix}2 & -1 & 0\\ 1 & 1 & -1\end{pmatrix}$.) That such a T_A is linear follows from Chapter 6, Section 2.b and 2.c). Thus, we have as many linear transformations here as there are matrices. To be precise, if A and B are two different matrices, then the linear maps T_A and T_B are different. If A and B are of different shape, then T_A and T_B do not go between the same vector spaces and therefore clearly are different. If they are of the same shape, but differ say in the jth column, then the T_A- and T_B-values at E^j are different (they are just the jth columns of A or B). That is enough to make $T_A\neq T_B$.

5. For any U and V we define the zero-map 0: $U\to V$ assigning to any u in U the vector 0 in V; $0(u)=0$ for all u. This is trivially linear; compare with the linear functional 0.

6. For any U we have the identity operator 1_U, or just 1, defined by $1(u)=u$; that is, the 1-image of u is u itself. "Every vector stays put"; this is again trivially linear. Generalizing, we have for any scalar s the "scalar operator" $s\cdot 1$ (sometimes written just s), with $s\cdot 1(u)=s\cdot u$; for example, $-1(u)=-u$. (Linearity of this depends on commutativity of the product in F!)

7. We end up with two "intuitive" examples: Rotations of plane and space. Visualize the plane as a sheet of paper, with a thumbtack at the origin. Rotate the paper through some angle. The shift from the initial to the final position amounts to a linear transformation: every point moves to a new position; thus any vector OA is sent to a new vector OA'. Sum goes to sum, since a parallelogram, with O as one vertex, goes to another such parallelogram. Similarly, $r\cdot OA$ turns to $r\cdot OA'$. Now consider space; visualize it as the earth, with the center of the earth as origin. Rotate the earth by waiting six hours. Every point in and on (and outside!) the earth moves from its initial to its final position (we disregard the fact that the center also moves, and act as if it were standing still). Again this amounts to a linear transformation: Parallelograms, with the center as vertex, rotate to parallelograms (so that "sum goes to sum"), and similarly for real multiples. (Both examples have very special properties. For instance, the *distance* of any two points is unaffected by the map; it equals that of the image points. This does not hold for most of the linear transformations

that we consider; in fact, so far we have not even defined the concept "distance" for vector spaces different from line, plane, or space.) ■

One can understand linear transformations by considering their affect on a basis of U (compare Proposition 4.1.1).

1.1. PROPOSITION. (*construction principle*). *Let $u_1,\ldots,u_n$ be a basis for U. A linear transformation T: $U \to V$ is completely determined by the images $T(u_1),\ldots,T(u_n)$ of the basis vectors ("uniqueness"). Furthermore, given any vectors $w_1,\ldots,w_n$ of V, there is a (unique) linear transformations T: $U \to V$ with $T(u_1)=w_1,\ldots,T(u_n)=w_n$ ("existence").*

We note right away that while the u_i are a basis for U, the w_i are quite arbitrary; they do not have to be independent or span V.

PROOF. Any u in U can be written uniquely as $\Sigma x_i u_i$, with scalars x_i. By repeated use of L_1 and L_2 we get $T(u)=\Sigma x_i T(u_i)(=x_i \cdot T(u_1)+x_2 \cdot T(u_2)+\cdots+x_n \cdot T(u_n))$; this shows that $T(u)$ is *determined* once one knows the vectors $T(u_1),\ldots,T(u_n)$; thus uniqueness. Now we look at the second part, with $w_1,\ldots,w_n$ chosen arbitrarily in V. For any u in U, $=\Sigma x_i u_i$, we *define* $T(u)=\Sigma x_i w_i$; this is a well-determined vector in V, and so we have at any rate a *function* from U to V. And linearity is proved just as for linear functionals: Take u', $=\Sigma x_i' u_i$, in U, then the expansion of $u+u'$ is, of course, $\Sigma(x_i+x_i')u_i$. We get $T(u')=\Sigma x_i' w_i$, $T(u+u')=\Sigma(x_i+x_i')w_i$ by the recipe for T, and we see $T(u)+T(u')=T(u+u')$. The procedure is similar for L_2. ■

Using the basis $\{\varphi_1,\ldots,\varphi_n\}$ of U' dual to $\{u_1,\ldots,u_n\}$, as defined in 5.1, we can condense the construction of T into the equation $T(u)=\Sigma\varphi_i(u)\cdot w_i$; indeed, $\varphi_i(u)=x_i$. We note a quite special case: Take φ in U' and w in V; they define a linear transformation $T_{\varphi,w}$ or $\varphi \cdot w$ by $u \mapsto \varphi(u)\cdot w$. Such a T is called a rank-one-map, or sometimes a "dyad"; see Section 4 for rank.

We see from Proposition 1.1 that there are as many linear transformations from U to V as there are sequences $\{w_1,\ldots,w_n\}$ of vectors in V. Clearly (?), two different sequences yield different transformations. But we emphasize that the basis $u_1,\ldots,u_n$ that we used to describe T has nothing intrinsic to do with T; any other basis would serve just as well.

Example

$U=\mathsf{R}^2$, $V=\mathsf{R}^3$. Basis for U: E^1 and E^2. We define a T by $E^1 \mapsto [1,-1,2]$ and $E^2 \mapsto [2,0,-1]$. Then, for example, $T[1,2]=T(E^1+2E^2)=[1,-1,2]+2\cdot[2,0,-1]=[5,-1,0]$. Let us take another basis for U, say $\gamma=\{X_1,X_2\}$ with $X_1=[1,1]$, $X_2=[1,-1]$. The *same* T is described by $X_1 \mapsto [3,-1,1]$, $X_2 \mapsto [-1,-1,3]$ (namely, $TX_1=T(E^1+E^2)=TE^1+TE^2$; TX_2 is obtained similarly). Now $[1,2]$ can be written as $3/2X_1-1/2X_2$, and $T[1,2]=T(3/2X_1-1/2X_2)=3/2TX_1-1/2TX_2$

checks with the earlier result for $T[1,2]$. (Check this!) An analogous result holds for any $[x,y]$ instead of $[1,2]$.

We note a fact about the transformations T_A of (iv) above associated to matrices A: *Any* linear transformation T: $\mathsf{F}^n \to \mathsf{F}^m$ is a T_A, with a suitable (and uniquely determined) matrix A. "Proof": see Problem 4 below.

We consider one more concept, a way to construct linear transformations from others: Let T: $U \to V$ be a linear transformation, and let W be a subspace of U. We define a linear transformation from W to V, denoted by $T|W$ (and called **restriction of T to W**) by $w \to T(w)$ for any w in W. In other words, we use the *given* T, but only for vectors in W; the vectors of U that are not in W are simply forgotten.

Finally, the linear functionals φ: $U \to \mathsf{F}$, which we studied in Chapter 4, are in fact linear transformations, from U to the one-dimensional vector space F $(=\mathsf{F}^1)$!

PROBLEMS

1. Find vector (= polynomials) whose T-images are $[1,-1]$ and $[0,3]$ for Example 2 above. In fact, find two such vectors in each case.

2. $X_1=[1,2,1]$, $X_2=[1,1,1]$, $X_3=[3,1,2]$ form a basis for F^3. We define T: $\mathsf{F}^3 \to \mathsf{F}^2$ by $X_1 \mapsto Y_1=[1,2]$, $X_2 \mapsto Y_2=[2,1]$, and $X_3 \mapsto Y_3=[-1,1]$. For $Z=[1,-1,1]$ find $T(Z)$; also find TE^1, TE^2, and TE^3. (We have several procedures to perform the necessary task of expressing Z and the E^i as linear combinations of the X_i.)

3. For the T_A with $A=\begin{pmatrix} 1 & 0 & 2 & 1 \\ 2 & -1 & 0 & 3 \\ 1 & -2 & 3 & 1 \end{pmatrix}$ find the images of the vectors $[1,1,-1,1]$, $[2,-1,3,0]$, $[1,0,0,0]$, and $[0,1,0,0]$.

4. Give a proof for the "important fact" in the text above. (*Hint*. For a matrix A, what is $A \cdot E^i$?)

5. For Problem 2 find the matrix M such that $T=T_M$.

2. DESCRIPTION OF LINEAR TRANSFORMATIONS BY MATRICES

Let U, V be two vector spaces; suppose $\beta=\{u_1,\ldots,u_n\}$ is a basis for U, and $\gamma=\{v_1,\ldots,v_m\}$, a basis for V. We recall that any u in U has then a coordinate vector X with respect to β (from $u=\sum x_j u_j$); similarly for v in V we have, say, Y (from $v=\sum y_i v_i$).

Let now T: $U \to V$ be a linear transformation. We come to an important construction: We assign to T an $m \times n$ matrix, say $A=(a_{ij})$, by

$$Tu_j = \sum_i a_{ij} v_i, \qquad j=1,\ldots,n;$$

the jth *column* A^j of A is formed by the γ-coordinates of the T-image of u_j, or $Tu_j \underset{\gamma}{\leftrightarrow} A^j$. Note carefully the position of the indices i, j. We say that A **represents** T with respect to the bases β and γ and write $T \underset{\beta,\gamma}{\leftrightarrow} A$. In the case of an *operator* ($U = V$) one usually takes γ *equal* to β.

Example

1. $U = V = P^3$ and $\beta = \gamma = \{1, x, x^2, x^3\}$; $T = D$ is differentiation. From $D1 = 0$, $Dx = 1$, $Dx^2 = 2x$, and $Dx^3 = 3x^2$, we read off the matrix

$$\begin{pmatrix} 0 & 1 & 0 & 0 \\ \cdot & 0 & 2 & 0 \\ \cdot & \cdot & 0 & 3 \\ 0 & \cdot & \cdot & 0 \end{pmatrix}.$$

What is the matrix if we take $\beta = \gamma = \{x^3, x^2, x, 1\}$?

2. U = vector space of solutions of $y'' + y = 0$, that is, the two-dimensional space of functions of the form $f(x) = a \cdot \sin x + b \cdot \cos x$. Define T: $U \to \mathsf{R}^2$ by $f(x) \mapsto [f(0), f(\pi)]$. Basis for U is $\sin x$ and $\cos x$; basis for R^2 is standard. From $T(\sin x) = [0, 0]$ and $T(\cos x) = [1, -1]$ we read off the matrix

$$\begin{pmatrix} 0 & 1 \\ 0 & -1 \end{pmatrix}.$$

3. The following is a special case of the next example: T: $\mathsf{R}^3 \to \mathsf{R}^2$ by $E^1 \mapsto [1, 2]$, $E^2 = [2, 1]$, and $E^3 \mapsto [1, 1]$; $\beta = \{E^1, E^2, E^3\}$ and $\gamma = \{E^1, E^2\}$. (The E's should really have subscripts 3 and 2, respectively.)

$$A = \begin{pmatrix} 1 & 2 & 1 \\ 2 & 1 & 1 \end{pmatrix}$$

4. The following is called the standard example: Let T_A: $\mathsf{F}^n \to \mathsf{F}^m$ be the standard operator defined by an $m \times n$ matrix A (see 1.iv). The matrix representing T_A with respect to the standard bases of F^n and F^m is precisely A!; indeed, $T_A(E^j_{(n)}) = A \cdot E^j_{(n)}$ is the jth column of $A = \Sigma a_{ij} E^i_{(m)}$; we read off that the jth column of the representing matrix equals the jth column of A.

5. Any 0-map is always (with respect to any bases) represented by the corresponding 0-matrix.

6. The identity operator $\mathbb{1}$ is always (with respect to any basis) represented by the identity matrix I.

Incidentally, note that any given $m \times n$ matrix A can appear as the matrix representing a suitable T: $U \to V$. All one has to do is *define* T by $Tu_j = \Sigma_i a_{ij} v_i$ (see Proposition 1.1).

We go back to the situation at the beginning of this section and describe what the matrix A is really meant for, what one means by saying that A

represents T: Take any u in U, and let X be the representing coordinate vector (so that $u \underset{\beta}{\leftrightarrow} X$). We abbreviate Tu to v and let Y be the representing coordinate vector for v (i.e., $v \underset{\gamma}{\leftrightarrow} Y$). We claim that the relation $v = Tu$ translates into the formula

$$\text{(R)} \qquad Y = A \cdot X$$

or, written out,

$$\begin{aligned} y_1 &= a_{11}x_1 + a_{12}x_2 + \cdots + a_{1n}x_n \\ y_2 &= a_{21}x_1 + a_{22}x_2 + \cdots + a_{2n}x_n \\ &\vdots \\ y_m &= a_{m1}x_1 + a_{m2}x_2 + \cdots + a_{mn}x_n. \end{aligned}$$

PROOF. From $u = \Sigma_j x_j u_j$ we get $Tu = \Sigma_j x_j Tu_j$. Substituting for the Tu_j, we get $Tu = \Sigma_j x_j (\Sigma_i a_{ij} v_i) = \Sigma_i (\Sigma_j a_{ij} x_j) v_i$ (by rearranging the sum); this equals $v = \Sigma_i y_i v_i$. Therefore, $y_i = \Sigma_j a_{ij} x_j$; in matrix language this says $Y = A \cdot X$. ■

This is then the sense in which A represents T: The matrix enables us to *compute* the components of the image vector (Tu) from those of the vector (u), by the simple formula (R) above.

We could say that once bases have been chosen (and thus coordinates introduced), *any* linear transformation "looks like" the standard example T_A: $\mathsf{F}^n \to \mathsf{F}^m$, with suitable A. We symbolize this with the diagram

$$\begin{array}{ccc} u & \mapsto Tu = & v \\ \beta \updownarrow & & \updownarrow \gamma \\ X & \mapsto AX = & Y \end{array}.$$

Example

1. See Example 1 above. The relation $D(2 - 3x + x^2 - 4x^3) = -3 + 2x - 12x^2$ translates into $[-3, 2, -12, 0] = A \cdot [2, -3, 1, -4]$ with the A above. (Check!)
2. See example 2 above. The relation $T(\cos x) = [1, -1]$ translates into

$$\begin{pmatrix} 1 \\ -1 \end{pmatrix} = \begin{pmatrix} 0 & 1 \\ 0 & -1 \end{pmatrix} \cdot \begin{pmatrix} 0 \\ 1 \end{pmatrix}.$$

Note that the bases β and γ, used to construct A, have no direct relation to T; they were chosen quite arbitrarily in advance. What happens if one chooses different bases? The matrix naturally changes. Suppose $\beta' = \{u'_1,\ldots,u'_n\}$ and $\gamma' = \{v'_1,\ldots,v'_m\}$ are two other bases for U and V. We have transition matrices $T^{\beta}_{\beta'}$ and $T^{\gamma}_{\gamma'}$ (see Chapter 6, Section 4). The vector u has X' as β'-coordinate vector; similarly, v has Y' as γ'-vector; and we have $X = T^{\beta}_{\beta'}\cdot X'$ and $Y = T^{\gamma}_{\gamma'}\cdot Y'$. We substitute this into our relation $Y = A\cdot X$, getting $T^{\gamma}_{\gamma'}\cdot Y' = A\cdot T^{\beta}_{\beta'}\cdot X'$ or $Y' = (T^{\gamma}_{\gamma'})^{-1}\cdot A\cdot T^{\beta}_{\beta'}\cdot X'$.

This makes clear that our new matrix A' for T is given by

(CL) $$A' = (T^{\gamma}_{\gamma'})^{-1}\cdot A\cdot T^{\beta}_{\beta'}.$$

(Change of matrix for linear transformation under change of bases.) We recall that the first factor on the right could be written as $T^{\gamma'}_{\gamma}$. Abbreviating $T^{\beta}_{\beta'}$ and $T^{\gamma}_{\gamma'}$ to P and Q, we also write $A' = Q^{-1}\cdot A\cdot P$.

In the case of an operator ($U = V$) with $\beta = \gamma$ and $\beta' = \gamma'$, we have then $A' = (T^{\beta}_{\beta'})^{-1}\cdot A\cdot T^{\beta}_{\beta'} = P^{-1}\cdot A\cdot P$. There is a name for this last relation: Two matrices M and N are **similar** if there is an *invertible* P such that

(S) $$N = P^{-1}\cdot M\cdot P.$$

We can say that the matrices that represent an operator with respect to different bases are similar *via* the transition matrix.

We note an important fact: Two similar matrices, A and $P^{-1}\cdot A\cdot P$, have the same determinant, by the product rule for det. Therefore, an operator T also has a well-defined determinant $\det T$, namely $\det A$ for any representing matrix.

PROBLEMS

1. Let T: $P^3 \to P^3$ be defined by $Tp(x) = p(x+1)$ (i.e., substitute $x+1$ for x). Find the matrix for T with respect to the basis $\{x^3, x^2, x, 1\}$ (in this order!). Find $\det T$.

2. Let T: $\mathsf{F}^2 \to \mathsf{F}^3$ be defined by $T[x_1,x_2] = [x_1, x_1+x_2, -4x_1+3x_2]$.
a. Find the matrix A for T with respect to the standard bases.
b. Find the matrix A' for T with respect to the following bases: $[2,3]$ and $[1,2]$ for F^2, and $[2,5,1]$, $[1,3,2]$, and $[0,0,1]$ for F^3. (Find the two transition matrices.) Interpret the result.

3. Consider Example 3 in the text. We define a subspace W of R^3 by the equation $x_1+2x_2-x_3=0$ and define S: $W\to\mathsf{R}^2$ by $S = T|W$ (restriction of T to W). *Choose* a basis β for W, and find the matrix B for S with respect to β and the standard basis of R^2.

4. Suppose T: $U\to U$ is represented by a matrix A with respect to a basis $\beta = \{u_1,\ldots,u_n\}$. We introduce the new basis $\gamma = \{u_2,u_1,\ldots,u_n\}$ (u_1 and u_2 are interchanged). Find the transition matrix and describe how the matrix A' for T with respect to γ differs from A.

5. Let T: $U \to V$ be a linear transformation. Suppose $\beta = \{u_1, \ldots, u_n\}$ is a basis for U, and the images $Tu_i = v_i$ happen to form a basis, say γ, for V. What is the matrix of T with respect to β and γ?

6. Let T: $\mathbf{C}^n \to \mathbf{C}^m$ be a linear transformation, with matrix A relative to the standard bases. Consider $\mathbf{C}^n$ and $\mathbf{C}^m$ as vectors spaces over $\mathbf{R}$ (see Chapter 3, Section 1, Problem 6). Now what is the matrix of T, relative to the bases E^1, iE^1, E^2, $iE^2, \ldots$?

3. IMAGE; KERNEL

We introduce the two most important characteristics of a linear transformation T: $U \to V$. (Actually, we shall see in Comment vii that they are old friends in disguise.)

3.1. DEFINITION. *The image of T, denoted by* $\operatorname{im} T$, *or also by* $T(U)$, *is the subspace of V consisting of all vectors that are T-images of vectors in U. The kernel or null space of T, denoted by* $\ker T$, *or also by* $T^{-1}(0)$, *is the subspace of U consisting of all vectors with T-image* 0. *This can be stated in symbols*:

$$\operatorname{im} T = \{v \in V: \ v = T(u) \text{ for some } u \text{ in } U\}, \qquad \ker T = \{u \in U: \ T(u) = 0\}.$$

Comments

i. We check that ker and im are actually subspaces; by now that should be routine. For example, if v comes up as $T(u)$ and v' as $T(u')$, then $v + v' = Tu + Tu' = T(u + u')$; that is, $\operatorname{im} T$ is closed under $+$. If $T(u) = 0$ and $T(u') = 0$, then $T(u + u') = Tu + Tu' = 0$, that is, ker is closed under $+$. im and ker are never empty; they always contain 0. This all depends on the *linearity* of T. We will see that frequently $\operatorname{im} T$ and $\ker T$ are nontrivial subspaces of V and U.

ii. The notation $T(U)$ can be substituted for $\operatorname{im} T$; more generally, if A is any sub*set* of U, one denotes by $T(A)$ and calls (direct) T-image of A the set of all v that come up as $T(u)$ for some u *in* A. (Naturally, any such $T(A)$ is *contained* in $T(U) = \operatorname{im} T$.) Note, however, that for a given v there may be several vectors $u, u', \ldots$ in A with $v = T(u) = T(u') = \cdots$; it is, therefore, a bit misleading to write the elements of $\operatorname{im} T$ (or, more generally, of any $T(A)$) as $T(u)$—it is not the specific u that counts, only that there *is at least* one u with $v = T(u)$.

iii. The notation $T^{-1}(0)$ is also used for $\ker T$. If B is any subset of V, one denotes by $T^{-1}(B)$ and calls the inverse image of B under T the set of *all those* u in U that have their image $T(u)$ in B. This is not meant to imply that there is a (linear) transformation called T^{-1} (there is only a very special case where that is so, see "isomorphism" below); thus, although the symbol T by itself makes sense (it denotes a linear transformation), T^{-1} by itself is meaningless for us. Only $T^{-1}(B)$ makes sense, for any part B of V; it denotes a certain part of U (possibly the empty subset, namely if there is *no* u with $T(u)$ in B, which can happen.)

iv. If A is a sub*space* of U, then $T(A)$ is a sub*space* of V. The proof is the same as that for $\operatorname{im} T$ $(=T(U))$. Similarly, if B is a sub*space* of V, then $T^{-1}(B)$ is one of U; the proof is a minor modification of that for $T^{-1}(0)$ (if $T(u)$ and $T(u')$ belong to B, then $T(u+u')=Tu+Tu'$ also does, and so with u, u' we have $u+u'$ in $T^{-1}(B)$. The procedure is similar for $\cdot$).

v. We discuss the nature of $T^{-1}(v)$ for an indivual vector v of V. There are two cases: (1) If v is not in the subspace $\operatorname{im} T$, then there is no u with $T(u)=v$; that is, $T^{-1}(v)$ is the empty set $\varnothing$. (2) If v is in $\operatorname{im} T$, then $T^{-1}(v)$ is a linear variety (see Chapter 4, Section 4) of type $\ker T$. If u and u' are two vectors with $T(u)=T(u')=v$, then, of course, $T(u-u')=v-v=0$; that is, $u-u' \in \ker T$. On the other hand, if $T(u)=v$ and w is *any* vector in $\ker T$, then $T(u+w)=Tu+Tw=v+0=v$. This shows that if u_0 is a vector with $T(u_0)=v$, then $T^{-1}(v)$ (the set of *all* u with $T(u)=v$) is $u_0+\ker T$.

This, then, is how one should "understand" the nature of the general T: The space U is divided into all the linear varieties of type $\ker T$; each such linear variety is sent ty T to a single point in V (lying necessarily in the subspace $\operatorname{im} T$). Different linear varieties go to different points, and one obtains precisely all the points in $\operatorname{im} T$ this way.

Note the geometric example 1 of Section 1: Here the image is the whole plane V; the kernel is the line through O in direction d; and for any vector OA' in V the inverse image (the vectors OA that project to OA') is the whole line through A' of direction d (i.e., all vectors OA with A on that line).

vi. Let $T: U \to V$ be described by a basis $\{u_1,\ldots,u_n\}$ of U and the T-images v_i of the u_i as above. Then the image $T(U)$ is precisely the subspace $((v_1,\ldots,v_n))$ of V, spanned by the v_i. Any u in U is $\Sigma x_i u_i$, and so $T(u)$ is $\Sigma x_i T(u_i)$ or $\Sigma x_i v_i$; the x_i can be taken arbitrarily, thus it is precisely the linear combinations of the v_i that appear as T-images. This shows that, given V and a subspace W, one can construct a linear transformation T from some U to V with $\operatorname{im} T$ exactly equal to W: Take any spanning set $v_1,\ldots,v_n$ for W, take a vector space of dimension n, with a basis $u_1,\ldots,u_n$ (e.g., F^n, $\{E^1,\ldots,E^n\}$) and define T by $T(u_i)=v_i$.

The moral here is that finding the image space $\operatorname{im} T$ of a linear transformation amounts to constructing the span of some vectors. (Note that in the above, instead of the T-image of a basis for U, we can take the T-image of any spanning set for U.)

vii. How does all this look "in coordinates"? We chose bases β and γ in U and V and represent T by the matrix A, as in Section 2.

A vector u of U is in $\ker T$, if the corresponding column X satisfies $AX=0$ (since the vector $v=0$ of V is represented by the column $Y=0$). Therefore, $\ker T$ is simply the solution space of the homogeneous system $AX=0$ (or better, it corresponds to it; for each solution $X=[x_1,\ldots,x_n]$ we take the vector $u=\Sigma x_i u_i$). Finding $\ker T$ amounts to our standard process of finding the solutions of the system $AX=0$.

Similarly, a vector v of V is in $\operatorname{im} T$, if the representing column Y is of the form AX for *some* X. Now the vector AX is nothing but the linear combination of the columns of A, with the x_i as coefficients, and, therefore, the space of all possible AX is precisely the column space of A. Thus, $\operatorname{im} T$ is the column space of A (or, again, better: It corresponds to it; to each $Y=[y_1,\ldots,y_m]$ in the column space we are supposed to take the represented vector $v=\Sigma y_j v_j$ of V). To find a basis for the column space, we bring A into *column*-echelon form.

Example

Let T_A: $\mathbf{R}^4 \to \mathbf{R}^3$ be defined by

$$A=\begin{pmatrix} 1 & 2 & 0 & -1 \\ -1 & 1 & 3 & -5 \\ 2 & 1 & -3 & 4 \end{pmatrix};$$

find im and ker. For im we reduce A to column-echelon form;

$$A\to\cdots\to\begin{pmatrix} 1 & 0 & 0 & 0 \\ -1 & 1 & 0 & 0 \\ 2 & -1 & 0 & 0 \end{pmatrix};$$

thus $[1,-1,2]$ and $[0,1,-1]$ are a basis for $\operatorname{im} T_A = T_A(\mathbf{R}^4)$. For ker we bring A (the original A, not its column-echelon form) to row-echelon form;

$$A\to\cdots\to\begin{pmatrix} 1 & 2 & 0 & -1 \\ 0 & 1 & 1 & -2 \\ 0 & 0 & 0 & 0 \end{pmatrix}.$$

For the resulting two equations x_3 and x_4 are free; we get $[2,-1,1,0]$ and $[-3,2,0,1]$ as basis for the solution space $=\ker T_A$.

viii. Let T be an operator on U. We describe an important concept that fits into the present context: A subspace W of U is called T-invariant (or just plain invariant if it is clear what T one is talking about) if $T(W)\subset W$; that is, if for any w in W the image Tw is also in W.

Examples

1. $U=P^{17}$. $T=xD$; that is, $T(p(x))=x\cdot p'(x)$. Then $W=P^{16}$ is a T-invariant subspace; if $p(x)$ is of degree $\leqslant 16$, so is $x\cdot p'(x)$. Similarly, $P^{15},P^{14},\ldots$ are all T-invariant.

2. $U=\mathbb{R}^3$. T defined by $T[x_1,x_2,x_3]=[2x_1+x_2,x_1+2x_2,x_1+x_2+x_3]$. (Exercise. What is the matrix of this T?) Then the subspace W defined by the equation $x_1=x_2$ is T-invariant; if $x_1=x_2$, then the first two coordinates of TX are equal.

3. There are two trivial examples: The whole space U and the subspace 0 are always invariant (under any T).

4. Now we look at a less trivial example: For any T the two subspaces $\ker T$ and $\operatorname{im} T$ are T-invariant. We have $T(\ker T)=0\subset\ker T$, and $T(\operatorname{im} T)\subset T(U)$ since $T(A)\subset T(U)=\operatorname{im} T$ for any subset A of U.

If W is T-invariant, one can consider T (or better: $T|W$) as an operator *on* W (since $T|W$ maps W into itself). This is called the *induced* operator.

ix. An even more special, but nevertheless important, situation is that of the vector space U being the direct sum $V\oplus W$ of two T-invariant subspaces. A simple case of this occurs when one rotates ordinary space (around some axis, through some angle). The axis itself is a one-dimensional invariant subspace; the plane perpendicular to the axis is a two-dimensional invariant subspace; and line and plane have space as their direct sum.

How does this look in coordinates? Naturally, we take a basis for U that is adapted to the direct sum relation, namely one made up of a basis $\{u_1,\ldots,u_r\}$ for V *together* with a basis $\{v_1,\ldots,v_{n-r}\}$ for W. The T-invariance of the two subspaces implies that each Tu_j involves only the u_i, and similarly each Tv_q involves only the v_p. The matrix A for T, relative to this basis, clearly has the form $\begin{pmatrix} B & 0 \\ 0 & C \end{pmatrix}$, where B is the $r\times r$ matrix for T on V, C is the $(n-r)\times(n-r)$ matrix for T on W, and the O's mean O-matrices of appropriate shapes. In short, $A=B\oplus C$, or also $A=\operatorname{diag}(B,C)$.

Conversely, if the matrix A for an operator T relative to a basis $\{u_1,\ldots,u_n\}$ has the form $\operatorname{diag}(B,C)$, then the two subspaces V and W, spanned by $u_1,\ldots,u_r$ and $u_{r+1},\ldots,u_n$, are T-invariant and have U as direct sum.

PROBLEMS

1. Check the computation in Comment vii.

2. Construct an example of an operator T on some vector space U with $\ker T = \operatorname{im} T$.

3. Find $\ker D$ and $\operatorname{im} D$ for Examples 1–3 in Section 2.

4. A linear transformation T: $\mathbf{R}^3 \to \mathbf{R}^4$ is defined by $X_1 = [2,0,-1] \mapsto [1,2,1,1]$, $X_2 = [0,3,-2] \mapsto [2,2,3,0]$, and $X_3 = [1,-2,1] \mapsto [1,0,2,-1]$. (The X_i form a basis of $\mathbf{R}^3$.) Find $\operatorname{im} T$ and $\ker T$.

5. Let T: $U \to V$ be given and let W be a subspace of U. Recall the concept $T|W$ (see Section 1). Make sense of and prove the relation $\ker T|W = W \cap \ker T$.

6. With the data of Problem 5, show that if $\operatorname{im} T|W = \operatorname{im} T$ (i.e., if $T(W) = T(U)$), then $W + \ker T = U$, and conversely.

7. Let T: $U \to U$ be an operator; suppose W is an invariant subspace. Extend a basis for W to a basis β for U. Show that the matrix A for T relative to β has the form $\begin{pmatrix} B & C \\ 0 & D \end{pmatrix}$, where 0 means an 0-matrix and B is the (square) matrix representing the "induced" operator $T|W$ on W.

8. Construct examples for Problem 7 in $\mathbf{R}^2$ and $\mathbf{R}^3$.

9. Prove the statement in the last sentence of Section 3.

4. RANK–NULLITY

For any T: $U \to V$ the dimension of $\operatorname{im} T$ is called the rank ρ_T of T, and the dimension of $\ker T$ is called the nullity ν_T of T.

From what we said above, it is clear that these definitions agree with the earlier ones for any representing matrix A; that is, $\rho_T = \rho_A$ and $\nu_T = \nu_A$. They are related by one fundamental relation, the rank–nullity law.

4.1. THEOREM. *Rank ρ_T and nullity ν_T of a linear transformation T: $U \to V$ are related by $\rho_T + \nu_T = \dim U$.*

Note that $\dim V$ does *not* enter this law.

PROOF. This theorem is simply an abstract form of Theorem 4.7.2. We give the abstract proof; it is very important and should be clearly understood. The proof uses a careful choice of bases, as befits a theorem about dimensions: Let $u_1, \ldots, u_r$ be a basis for the vector space $\ker T$ (thus $r = \nu_T = \dim \ker T$); extend this to a basis $u_1, \ldots, u_r, \ldots, u_n$ of the whole space U. The images $Tu_1, \ldots, Tu_r$ of the first r vectors are all 0, since those u_i are in $\ker T$. Therefore, the vectors $v_{r+1} \overset{\text{def}}{=} T(u_{r+1})$, $v_{r+2} = T(u_{r+2}), \ldots, v_n = T(u_n)$ span $\operatorname{im} T$, by (vi.) in Section 1. We show—and that is the crux of the argument—that they are independent: Let $a_{r+1}v_{r+1} + \cdots + a_n v_n = 0$ be a relation between them. This means that the T-image of the vector $u_0 = a_{r+1}u_{r+1} + \cdots + a_n u_n$ is 0, implying that u_0 is in $\ker T$. Then u_0 is a linear combination $a_1u_1 + \cdots + a_ru_r$ of the basis $u_1,\ldots,u_r$ for $\ker T$. But the relation $a_1u_1 + \cdots + a_ru_r = a_{r+1}u_{r+1} + \cdots + a_nu_n$ can hold only if all the a_i are 0, since $u_1,\ldots,u_n$ ar independent. In particular, $a_{r+1} = \cdots = a_n = 0$; our claim is proved. The vectors $v_{r+1},\ldots,v_n$ form now a *basis* for $\operatorname{im} T$; thus $\rho_T = \dim \operatorname{im} T = n - r$. Together with $r = \nu_T$ this proves the theorem. Note that the vectors $u_{r+1},\ldots,u_n$ span a complement in U to $\ker T$. ■

Example

In the example of (vii) in Section 3, we have $n=4$, $\rho_T = 2$, and $\nu_T = 2$.

The rank–nullity law is the only restriction on a linear transformation in the following sense: Let U and V be two vector spaces, and let $U' \subset U$ and $V' \subset V$ be subspaces such that the relation $\dim U' + \dim V' = \dim U$ holds, then there is a linear transformation T: $U \to V$ (in fact there are many such) with $\ker T = U'$ and $\operatorname{im} T = V'$. For its construction (suggested by the proof of the rank–nullity law), let $u_1,\ldots,u_n$ be a basis for U such that the first r vectors $u_1,\ldots,u_r$ form a basis for U'.Note that $n-r$ is then the dimension of V', by our hypothesis. Let $v_{r+1},\ldots,v_n$ (there are $n-r$ of these!) be a basis for V'. Now define T by $Tu_1 = \cdots = Tu_r = 0$ for the first r u's, and $Tu_{r+1} = v_{r+1},\ldots,Tu_n = v_n$ for the last $n-r$ u's using the construction principle—Proposition 1.1.

The image clearly is the span $((v_{r+1},\ldots,v_n)) = V'$ (we can get those v's and their linear combinations, but nothing else, as T-image). The kernel, again clearly, *contains* the span $((u_1,\ldots,u_r)) = U'$, since each such u_i goes to 0 under T. It is not quite so clear that $\ker T$ is *equal* to U'. There are two arguments for this: By dimension, we know $\dim\ker T = n - \dim\operatorname{im} T = n - (n-r) = r$, by the rank–nullity law, and so $\dim\ker T = \dim U'$. Together with $U' \subset \ker T$ this shows $U' = \ker T$ (see Chapter 3, Section 2.4). For a direct proof, let $u = a_1u_1 + \cdots + a_nu_n$ be any vector. Then $Tu = a_{r+1}v_{r+1} + \cdots + a_nv_n$ (since $Tu_i = 0$ for $1 \leqslant i \leqslant r$ and $Tu_i = v_i$ for $r+1 \leqslant i \leqslant n$). If u is in $\ker T$, then Tu is 0, but the $v_{r+1},\ldots,v_n$ are independent. Thus $a_{r+1} = \cdots = a_n = 0$, and so $u = a_1u_1 + \cdots + a_ru_r$ is in U'. ■

We see from the construction, choosing the u's in some fixed way, that there are as many such linear transformations as there are bases for V' (the different choices of $v_{r+1},\ldots,v_n$). Here U could equal V, that is, T could be an operator; U' and V' are then any two subspaces of U whose dimensions add up to $\dim U$. (Nothing is said about their mutual position, e.g., their intersection, there is no need for the intersection to be 0.)

PROBLEMS

1. Construct T: $\mathsf{F}^5 \to \mathsf{F}^6$ with $\ker T = ((E^1 + E^2, E^1 + E^3, E^1 + E^4))$ and $\operatorname{im} T$

$=((E^1+E^5, E^1+E^6, E^5-E^6))$—"construct" means to describe the images of the E^i or give the matrix A with $T=T_A$.

2. Construct operators T_0, T_1, and T_2 on R^4 such that
a. always $\dim \ker T_i = \dim \operatorname{im} T_i = 2$ and
b. $\dim \ker T_i \cap \operatorname{im} T_i = i$.

3. Can one construct an operator T: $\mathsf{R}^{17} \to \mathsf{R}^{17}$ with $\ker T = \operatorname{im} T$?

5. OPERATIONS ON LINEAR TRANSFORMATIONS

The linear transformations from a given U to a given V can be added and multiplied by scalars: If S and T are two such maps from U to V, the sum $S+T$ assigns to any u in U the vector $S(u)+T(u)$ in V. In brief, $(S+T)(u)=S(u)+T(u)$. And, rT is defined by $(rT)(u)=r\cdot T(u)$. All this is similar to what we did with linear functionals (Chapter 4, Section 1), and, as there, it is quite trivial to verify that with these definitions of + and · the set of *all* linear transformations from U to V becomes a vector space, that is, the axioms VS_1–VS_9 of Chapter 1, Section 1, are satisfied. We call this space $\mathcal{L}(U,V)$; we will see soon that is is finitely generated, and we will find its dimension in terms of those of U and V. If $V=U$, that is, if we have to do with operators, we write $\mathcal{L}(U)$ for $\mathcal{L}(U,U)$.

There is another way of combining linear transformations that is more important,—**composition**. Let U, V, and W be three vector spaces (over the same F, of course); let $T\colon U\to V$ and $S\colon V\to W$ be linear transformations. We get a new linear transformation $S\circ T\colon U\to W$, the "composition" of T and S (also written as plain ST), by the definition $S\circ T(u)=S(T(u))$; we first move u into V by T, getting $T(u)$, and then move *that* vector in W, getting $S(T(u))$. The symbolic picture is $U\overset{T}{\to}V\overset{S}{\to}W$; "from U to W via V." Note that in $S\circ T$ one first applies T (to u), then S (to $T(u)$). One has to check that $S\circ T$ is again linear; that is pretty clear. $u+u'$ goes first to $Tu+Tu'$, and this goes under S to $S(Tu)+S(Tu')$, which equals $S\circ T(u)+S\circ T(u')$. A similar procedure works for ·. A few fairly obvious rules about composition and the operations + and · above: (i) $S\circ(T+T')=S\circ T+S\circ T'$, (ii) $(S+S')\circ T=S\circ T+S'\circ T$, and (iii) $S\circ(rT)=(rS)\circ T=r\cdot S\circ T$. This is verified by applying the maps to any u in U and tracing the result carefully; for example, $(S\circ(T+T'))(u)=S((T+T')(u))=S(Tu+T'u)=S(Tu)+S(T'u)=S\circ T(u)+S\circ T'(u)$, etc. If W' is a fourth space (I am running out of letters) and $R\colon W\to W'$ a linear transformation, we have $R\circ(S\circ T)$ and $(R\circ S)\circ T$, linear transformations from U to W'. It is true and easy to see that they are equal: $R\circ(S\circ T)=(R\circ S)\circ T$, associative law for composition; we write both sides as $R\circ S\circ T$. (PROOF. $(R\circ(S\circ T))(u)=R(S\circ T(u))=R(S(T(u)))$; $((R\circ S)\circ T)(u)=(R\circ S)(T(u))=R(S(T(u)))$; one has to be careful about the defining equation $S\circ T(u)$

$=S(T(u))$. Composition is particularly useful in the case $U=V=W$, that is, for *operators*. It then makes sense, for example, to form *powers* of an operator $T: U \to U$ by $T^1 = T$, $T^2 = T\circ T$, $T^3 = T\circ T\circ T$, etc.; one also puts $T^0 = \mathbf{1}$. Now one can form polynomials in an operator, for instance, $3T^2-2T+5$ (where the 5 means $5\cdot\mathbf{1}$).

We note a definition (paralleling an earlier definition for matrices): An operator T is *nilpotent* if some power T^k is 0. Thus, for any vector u we must have $T(T(\ldots(Tu)\ldots))=0$(with k "factors" T). Nilpotent operators may seem very special, but will play an important role later (for the Jordan form).

We note $T\circ\mathbf{1} = T = \mathbf{1}\circ T$ for any $T: U\to U$, as well as $T\circ 0 = 0 = 0\circ T$; thus, if we consider $\circ$ as a multiplication of operators, $\mathbf{1}$ functions as "unity" and 0, as "zero."

We translate into coordinate language; that is, we represent linear transformations by matrices: If T is represented by A, then (relative to the same bases) the linear transformation $r\cdot T$ is represented by $r\cdot A$. If T_1 and T_2 are two linear transformations from U to V, represented by A_1 and A_2 (with respect to the *same* bases, of course), then T_1+T_2 is represented by A_1+A_2. (PROOF. $(T_1+T_2)(u_j) = T_1u_j + T_2u_j = \Sigma_i a^1_{ij}v_i + \Sigma_i a^2_{ij}v_i = \Sigma_i(a^1_{ij}+a^2_{ij})$ v_i.) If $S: V\to W$ is a linear transformation, $\delta = \{w_1,\ldots,w_p\}$ a basis for W and S represented by the matrix B (with repsect to γ and δ), then the composition $S\circ T: U\to W$ is represented (with respect to β, δ) by the product matrix $B\cdot A: S\circ T(v_j) = S(Tu_j) = S(\Sigma_i a_{ij}v_i) = \Sigma_i a_{ij}Sv_i$ $= \Sigma_i a_{ij}\cdot(\Sigma_k b_{ki}w_k) = \cdots = \Sigma_k(\Sigma_i b_{ki}a_{ij})\cdot w_k$. We see that $\Sigma_i b_{ki}a_{ij}$ is the (k,j)-entry of the representing matrix, but that is precisely the (k,j)-entry of $B\cdot A$. In a sense, this last fact is the *reason* for defining matrix multiplication the way we did way back. "Composition corresponds to matrix multiplication."

Example

Let T:$\mathbf{F}^2\to\mathbf{F}^2$ be defined by $E^1\mapsto E^2$, $E^2\mapsto 0$. What is T^2? $T^2(E^1)$ $=T(T(E^1))=T(E^2)=0$ and $T^2(E^2)=T(T(E^2))=T(0)=0$ results in T^2 as the operator 0. (We could have written $E^1 \overset{T}{\mapsto} E^2 \overset{T}{\mapsto} 0$; $E^2 \overset{T}{\mapsto} 0 \overset{T}{\mapsto} 0$.) As a "generic" example, consider T_A: $\mathbf{F}^n\to\mathbf{F}^m$ and T_B: $\mathbf{F}^m\to\mathbf{F}^p$, via matrices A and B (see iv) in Section 1, for a definition). The composition $T_B\circ T_A$: $\mathbf{F}^n\to\mathbf{F}^p$ sends X to $B\cdot A\cdot X$, namely T_A sends X to $A\cdot X$; T_B sends any Y in $\mathbf{F}^m$ to $B\cdot Y$; in particular, it sends $A\cdot X$ to $B\cdot(A\cdot X) = B\cdot A\cdot X$. Thus $T_B\circ T_A = T_{B\cdot A}$.

To illustrate the concepts, we give the abstract version of the rank–inequality (Chapter 6, Section 3) and add a fact about nullity.

5.1 THEOREM. *Let T: $U \to V$ and S: $V \to W$ be linear transformations.*

a. $\rho_{S \circ T} \leqslant \min(\rho_S, \rho_T)$ and
b. $\nu_{S \circ T} \leqslant \nu_S + \nu_T$.

PROOF. a. $\operatorname{im}(S \circ T)$ is a subspace of $\operatorname{im} S$, by $S \circ T(u) = S(T(u))$ (every $S \circ T$-image vector is also S-image); therefore, $\rho_{S \circ T} \leqslant \rho_S$. On the other hand, $\operatorname{im}(S \circ T)$ is S-image of $\operatorname{im} T$ (again by $S \circ T(u) = S(T(u))$), and so is spanned by the S-images of a basis of $\operatorname{im} T$; therefore, $\rho_{S \circ T} \leqslant \rho_T$.

b. The definition $S \circ T(u) = S(T(u))$ shows that $\ker(S \circ T)$ consists precisely of those u that are sent into $\ker S$ by T (in particular, it contains $\ker T$). Writing T' for the restriction $T|\ker(S \circ T)$, we have thus $\operatorname{im} T' \subset \ker S$ (in fact $\operatorname{im} T' = \operatorname{im} T \cap \ker S$), and so $\rho_{T'} \leqslant \nu_S$. Next, it is clear that $\ker T' = \ker T$, and so $\nu_{T'} = \nu_T$. Finally, rank-nullity for T' yields the relation $\dim \ker(S \circ T) = \rho_{T'} + \nu_{T'}$. Putting all this together we get $\nu_{S \circ T} \leqslant \nu_S + \nu_T$, as promised. We also see that equality holds if and only if $\operatorname{im} T' = \ker S$, that is, if $\operatorname{im} T$ contains $\ker S$. ■

PROBLEMS

1. Given T: $U \to V$ and S: $V \to W$, show that $\operatorname{im} S \circ T$ is a subspace of $\operatorname{im} S$. Construct an example where the two are different, and one where the two are equal.

2. With the data of Problem 1 show that $\ker T$ is a subspace of $\ker S \circ T$. Again, construct examples with inequality and with equality.

3. For given U and V, find the dimension of the space $\mathcal{L}(U, V)$ of all linear transformations from U to V. (*Hint*. Represent the T's by matrices with respect to some chosen bases of U and V.)

4. Suppose T: $U \to U$ has $\ker T = \operatorname{im} T$. What can one say about T^2?

5. For the T: $P^3 \to P^3$, given by $Tp(x) = p(x+1)$, determine the operator $S = 3T^2 - T + 2$ (here 2 means, of course, $2 \cdot \mathbf{1}$). Find the matrices A and B for S and T with respect to the basis $\{1, x, x^2, x^3\}$, and check the relation $A = 3B^2 - B + 2I$.

6. With the data of Problem 1, show that the relation $S \circ T = 0$ is equivalent to the relation $\operatorname{im} T \subset \ker S$.

7. Suppose T is nilpotent, and S commutes with T (i.e., $S \circ T = T \circ S$). Show that then $S \circ T$ is also nilpotent.

6. INJECTIONS; SURJECTIONS; ISOMORPHISMS

We come to three important classes of linear transformations, characterized by special behavior of image and/or kernel. Let T: $U \to V$ be a linear transformation.

i. If $\operatorname{im} T = V$, that is, if every v in V comes up as $T(u)$ for suitable u, we call T **surjective** or **onto**. It is, of course, enough to have $\dim \operatorname{im} T$ $(=\rho_T) = \dim V$. In the geometric example of the map from the first plane to the second one is surjective, unless the projecting direction happens to be parallel to the first plane (in which case $\operatorname{im} T$ is only a line). As another example, let D: $P^{17} \to P^{17}$ be the operator "differentiation". Then $\operatorname{im} D \neq P^{17}$ since x^{17} (or any polynomial x^{17} + lower terms) cannot be gotten as derivative of a polynomial of degree $\leqslant 17$. Thus, D is not surjective.

In terms of Proposition 1.1, "surjective" means that the images $v_1, \ldots, v_n$ of the basis vectors $u_1, \ldots, u_n$ *span* V, $((v_1, \ldots, v_n)) = V$. (Recall that the v_i do not have to be independent.)

It follows from the rank–nullity law (Section 4) that for a surjective T the relation $\dim U \geqslant \dim V$ holds (namely $\dim U - \dim V = \nu_T = \dim \ker T \geqslant 0$).

ii. If $\ker T = 0$, that is, if the only u in U with T-image 0 is 0 itself, T is called **injective** or **one-to-one**. This means that different u's have different T-images: If $T(u) = T(u')$, then $T(u - u') = 0$; therefore, $u - u'$ must be 0, that is, $u = u'$. (In terms of the notion "inverse image," for any v, the linear variety $T^{-1}(v)$, if not empty (i.e., for v in $\operatorname{im} T$) consists of exactly *one* vector (is of dimension 0).) The effect of T in this case is so to speak to put U bodily, unchanged, into V as a subspace. As a consequence of the rank–nullity law, we get $\dim U \leqslant \dim V$.

Example

Any T_A: $\mathbf{F}^n \to \mathbf{F}^m$ for which the solution space of $A \cdot X = 0$ consists of 0 only, that is, $\nu_A = 0$.

iii. We can combine i and ii. If $\operatorname{im} T = V$ *and* $\ker T = 0$, we call T an **isomorphism**.

This means then that every v in V comes up as Tu, and moreover the u is uniquely determined. We might say that V is simply another copy of U *via* T. We can go "backwards" and assign to *each* v in V the *unique* u in U with $Tu = v$; we denote this u as $T^{-1}(v)$. Thus we have a *function* T^{-1} from V to U; by its very definition it satisfies $T(T^{-1}v) = v$ for any v in V, but we also have $T^{-1}(Tu) = u$ for any u in U, since u is *the* vector whose T-image is Tu. Furthermore, T^{-1}, as defined, is a *linear transformation* from V to U. Take v and v' in V; $u = T^{-1}v$ and $u' = T^{-1}v'$ are *the* vectors that have $Tu = v$ and $Tu' = v'$. Then $u + u'$ has $T(u + u') = Tu + Tu' = v + v'$, and so $T^{-1}(v + v') = u + u'$, which equals $T^{-1}v + T^{-1}v'$. The procedure is similar for $\cdot$. This is the special case, mentioned in Section 3.iii, where T^{-1} exists as a linear transformation.

We rewrite the two relations $T(T^{-1}v)=v$ and $T^{-1}(Tu)=u$ as $T\circ T^{-1}(v)=v$ (for all v in V) and $T^{-1}\circ T(u)=u$ (for all u in U), or simpler: $T\circ T^{-1}=\mathbf{1}_V$, $T^{-1}\circ T=\mathbf{1}_U$. This suggests a definition: A linear transformation T: $U\to V$ is called **invertible** if there exists a linear transformation S: $V\to U$ (in the opposite direction) such that $S\circ T=\mathbf{1}_U$ and $T\circ S=\mathbf{1}_V$. Such an S is called the inverse to T (we will see below that it is unique; we write T^{-1} for it). The argument above showed that an isomorphism T *has* an inverse. We show the converse: Suppose S is inverse to T. We have then $S(Tu)=u$ for all u; if Tu is 0, it follows that $u=S(0)=0$ is itself 0, that is, $\ker T=0$. Similarly, $v=T(S(v))$ shows that v is of the form $T(u)$, namely with $u=S(v)$; and so $\operatorname{im} T=V$. Therefore, T is an isomorphism. Thus, T is an isomorphism if and only if it is invertible (i.e., has an inverse).

What does all this mean in terms of coordinates? Clearly, a linear transformation T: $U\to V$ is invertible precisely if its matrix A (with respect to bases β and γ) is invertible. (Work out the details. For example, if S: $V\to U$ is inverse to T with matrix B (with respect to γ and β), then $S\circ T=\mathbf{1}_U$ has the consequence $B\cdot A=I$.)

As noted above, if T is an isomorphism, then V is "another copy" of U. More specifically, we have the following facts.

6.1. PROPOSITION. *Let T: $U\to V$ be an isomorphism with inverse S; let $u_1,\ldots,u_k$ be any vectors in U; put $v_i=Tu_i$. Then* (a) *the u_i are independent exactly if the v_i are independent*; (b) *the u_i span U exactly if the v_i span V. In particular, the u_i are a basis of U exactly if the v_i are one of V, and so* $\dim U=\dim V$.

Conversely, if T sends a basis of U to a basis of V, then it is an isomorphism.

PROOF. (a) Suppose the v_i are dependent; some relation $\Sigma a_i v_i=0$ holds with not all a_i vanishing. Simply applying S we get $\Sigma a_i S(v_i)=\Sigma a_i u_i=0$ with the same a_i, so that the u_i are also dependent. Using T we can argue in the other direction. (b) Suppose the u_i span U. Any v can be written as Tu by $\operatorname{im} T=V$, and u is a suitable $\Sigma b_i u_i$. Thus $v=Tu=T(\Sigma b_i u_i)=\Sigma b_i Tu_i=\Sigma b_i v_i$; this shows that the v_i span V. We can go the other way with S. Briefly, anything that is true in U is also true in V, and the other way around. Similar reasoning establishes the converse part. ■

If we assume that U and V have the same dimension, we can do a bit better; we need only *one* of the two conditions for isomorphism.

6.2 PROPOSITION. *Let T: $U\to V$ be a linear transformation. If* $\dim U=\dim V$ *and either* $\ker T=0$ *or* $\operatorname{im} T=V$, *then T is an isomorphism.*

PROOF. The rank–nullity law shows that if $\ker T=0$, that is, $\nu_T=0$, then

ρ_T, the dimension of $\operatorname{im} T$, equals $\dim V$ ($=\dim U$), but then $\operatorname{im} T$ must be all of V (Chapter 3, Section 2.4). If $\operatorname{im} T = V$, that is, $\rho_T = \dim V = \dim U$, then $\nu_T = 0$, that is, $\ker T = 0$. The matrix analog of this is 3.2.2, applied to the columns of a square matrix. ■

Finally, the analog of Theorem 6.3.3 follows.

6.3. THEOREM. *Let $T: U \to V$ be a linear transformation. If* $\dim U = \dim V$ *and there is a linear transformation $S: V \to U$ such that either $S \circ T = \mathbf{1}_U$ or $T \circ S = \mathbf{1}_V$, then T is an isomorphism, and S equals the T^{-1} defined above.*

PROOF. Suppose $S \circ T = \mathbf{1}_U$. We saw above that then $\ker T = 0$. Suppose $T \circ S = \mathbf{1}_V$. We also saw above that then $\operatorname{im} T = V$. Thus, we can apply Proposition 6.2 to show that T is an isomorphism. And further, if $S \circ T = \mathbf{1}_U$, then $S \circ T \circ T^{-1} = \mathbf{1}_U \circ T^{-1}$ or $S \circ \mathbf{1}_V = T^{-1}$ (using $T \circ T^{-1} = \mathbf{1}_V$), and so $S = T^{-1}$; the procedure is similar for $T \circ S = \mathbf{1}_V$ ("apply" T^{-1} on the left).

To add a piece of notation, an isomorphism of a vector space U with itself, or, what amounts to the same, an invertible *operator* $T: U \to U$, is called an **automorphism** of U.

PROBLEMS

1. Suppose $T: \mathbf{R}^4 \to \mathbf{R}^3$ has nullity 2. Can T be surjective?

2. Prove in detail that if $T: U \to V$ sends a basis of U to a basis of V, then T is an isomorphism.

3. Suppose $T: U \to V$ is injective, show that $\dim V \geqslant \dim U$. Show with an example that "greater than" (and not "equal to") can occur here.

4. Suppose that $T: U \to V$ is injective, show that there exists $S: V \to U$ such that $S \circ T$ is an automorphism of U (in fact, such that $S \circ T$ is the identity operator $\mathbf{1}_U$).

5. Restate Problem 4 in matrix language ("suppose the $m \times n$ matrix M has nullity 0 (and therefore rank n); then there exists…").

6. Let V and V' be two subspaces of U of the same dimension. Show that there exist automorphisms T of U with $T(V) = V'$.

7. Again V and V' are subspaces of U, but this time we have $\dim V' < \dim V$. Show that there still exist operators T on U with $T(V) = V'$, but no such T can be an automorphism of U.

8. Is the operator $T_M: \mathbf{C}^3 \to \mathbf{C}^3$ with

$$M = \begin{pmatrix} 1-i & i & 1+i \\ i & 1-i & -i \\ 2-i & 1+i & 1+2i \end{pmatrix}$$

invertible?

9. Show that T nilpotent implies T singular (i.e., $T^k=0$ *and* $T\circ S=\mathbf{1}$ together are impossible).

10. Suppose $T^2=0$, show that $\mathbf{1}-T$ is invertible, in fact, has $\mathbf{1}+T$ as inverse. Verify for $\begin{pmatrix} 10 & -4 \\ 25 & -10 \end{pmatrix}$.

11. Generalize Problem 10. Suppose $T^3=0$. Show that the inverse of $\mathbf{1}-T$ is given by $\mathbf{1}+T+T^2$. Guess the law for the inverse of $\mathbf{1}-T$, if T^k is zero for any $k>0$.

7. SOME EXAMPLES

1. This illustrates (2) of Section 1. The kernel of T consist of those polynomials that have $x_1,\ldots,x_k$ as roots; equivalently, they are divisible by $(x-x_1)\cdot(x-x_2)\cdots(x-x_k)$, or of the form $p(x)=(x-x_1)\cdots(x-x_k)\cdot q(x)$ with $q(x)$ of degree $\leqslant n-k$. Of course, for $k>n$ there is no such $p(x)$ except $p(x)=0$ (the 0 polynomial). Our result is $\nu_T=0$, if $k>n$; otherwise $\nu_T=n-k+1$ ($=$dim of the space P^{n-k} of the $q(x)$'s). Thus T is injective only for $k>n$. As for $\operatorname{im} T$, by rank–nullity we get $\rho_T=n+1$ if $k>n$; $\rho_T=k$ if $k\leqslant n$. Thus, T is surjective for $k\leqslant n+1$. (This means that if we have *at most* $n+1$ points $x_1,\ldots,x_k$, we can find a polynomial in P^n with prescribed values at these points—so-called "interpolation." Study the case $n=1$!) And for $k=n+1$ we get an isomorphism from P^n to F^{n+1}. ■

2. This illustrates (3) of Section 1. The existence and uniqueness theorem says, briefly, that the values $f(0), f'(0),\ldots,f^{(n-1)}(0)$ determine the solution f uniquely. (Any other value of x, in place of 0, would do.) Thus, for $k\leqslant n$ T is surjective, for $k\geqslant n$ it is injective, for $k=n$ it is an isomorphism (from the space of solutions to R^n). ■

3. This illustrates (4) of Section 1. $T_A:\mathsf{F}^n\to\mathsf{F}^m$, matrix A. Injective means $\nu_A=0$ or $\rho_A=n$. Surjective means $\rho_A=m$ (since $\operatorname{im} T$ is the column space of A). Isomorphism requires $m=n$ (so that T_A is an *auto*morphism of F^n), and $\nu_A=0$ or $\rho_A=n$; this is the same as "A is (square and) invertible." ■

4. This illustrates (2) in Section 2. The kernel consists of all $a\cdot\sin x$ (since $f(0)=b, f(\pi)=-b$); thus $\nu_T=1$. By rank–nullity we must also have $\rho_T=1$. In fact, $\operatorname{im} T=\{\text{set of all } [b,-b]\}$, as just noted, or im T is the subspace (line) of R^2 with equation $x_1+x_2=0$. ■

Now we look at a general result. It serves to illustrated the concepts, and we will use it later.

Let $T:U\to U$ be an operator. We consider its powers $\mathbf{1},T,T^2,T^3,\ldots$. Their kernels, ker $\mathbf{1}(=0)$, $\ker T$, $\ker T^2,\ldots$, form an *increasing* sequence: $\ker T^k\subset\ker T^{k+1}$, since $T^kv=0$ implies $T^{k+1}v=T(T^kv)=T0=0$. Similarly, the images $\operatorname{im}\mathbf{1}$ $(=U)$, $\operatorname{im}T$, $\operatorname{im}T^2,\ldots,$ form a decreasing sequence: $\operatorname{im}T^k \supset \operatorname{im}T^{k+1}$, since a vector of the form $T^{k+1}v$ is also of the form $T^k w$ (with $w = Tv$). Since the dimension of $\ker T^k$ can go up only a finite number of times (all spaces are subspaces of U), all the $\ker T^k$ must eventually (for large k) be equal to each other, and thus equal to a fixed subspace of U; we call this space the **stable kernel** (or also the "**nil space**") of T and write $\ker^s T$. Similarly, we have the **stable image** $\operatorname{im}^s T$ $(=\operatorname{im}T^k$ for large k). All the spaces mentioned here are, of course, T-invariant.

Actually the behavior of the sequence of ker's and of im's is quite simple: As soon as one ker equals the next, then the succeeding are all equal; the same is true for im.

7.1. PROPOSITION. *If* $\ker T^k = \ker T^{k+1}$, *then also* $\ker T^{k+1} = \ker T^{k+2}$; *the same relations hold for* im *instead of* ker.

PROOF. "$\ker T^k = \ker T^{k+1}$" means that if $T^{k+1}v = 0$, for some v, then already $T^k v = 0$. Suppose now u is such that $T^{k+2}u = 0$, that is, $T^{k+1}(Tu) = 0$. Then we already have, as just noted, $T^k(Tu) = 0$, that is, $T^{k+n} = 0$; in other words, $\ker T^{k+2} \subset \ker T^{k+1}$. Since, of course, $\ker T^{k+1} \subset \ker T^{k+2}$, the two must be equal. Similarly for im, we take any $T^{k+1}v$, and rewrite it as $T(T^k v)$. By assumption $T^k v$ can also be written as $T^{k+1}u$ with a suitable u. Then $T^{k+1}v = T^{k+2}u$, showing that $\operatorname{im}T^{k+1}$ is contained in $\operatorname{im}T^{k+2}$. It follows easily that the two spaces are equal. ■

Recall that in general $\ker T$ and $\operatorname{im}T$ can have nonzero intersection. This is not so for the stable kernel and image.

7.2. PROPOSITION. $\ker^s T$ *and* $\operatorname{im}^s T$ *have* U *as direct sum.*

PROOF. We are talking about $\ker T^k$ and $\operatorname{im}T^k$ for some large k. By rank–nullity their dimensions add up to $n = \dim U$. It is, therefore, enough to show that their intersection is 0. Thus, suppose v is of the form $T^k u$ and *also* satisfies $T^k v = 0$, so that $T^{2k}u = 0$. By stability we have $\ker T^{2k} = \ker T^k$, so that already $T^k u = 0$, that is, $v = 0$. ■

The behavior of T on the stable kernel and image is very simple (we regard T as operator on each of the two spaces).

7.3. PROPOSITION. $T|\ker^s T$ *is nilpotent*; $T|\operatorname{im}^s T$ *is invertible.*

PROOF. $\ker^s T$ is $\ker T^k$ for some (large) k; and then, by definition, T^k is 0 on this space. That $T|\operatorname{im}^s T$ is invertible comes from $\operatorname{im}^s T \cap \ker T = 0$ (by Proposition 7.2 and $\ker T \subset \ker^s T$); this says that the operator is injective (no vector of $\operatorname{im}^s T$ is sent to 0). Such an operator is invertible (by what earlier result?).

We make a final comment: In practice the "stable" k is usually very small, something like one, two, or three, unless one has a space of large dimension. If T is invertible, then $\ker^s T = 0$ and $\operatorname{im}^s T = U$. If T is nilpotent, then $\ker^s T = U$ and $\operatorname{im}^s T = 0$. ■

PROBLEMS

1. Determine the sequence of kernels and images of the iterates and the stable kernel and the stable image for $T_A: \mathbf{R}^4 \to \mathbf{R}^4$ with

$$A = \begin{pmatrix} 1 & 1 & 1 & 0 \\ -1 & -1 & -1 & 1 \\ 0 & 0 & 0 & 1 \\ 0 & 0 & 1 & 0 \end{pmatrix}.$$

8. THE TRANSPOSE OF A LINEAR TRANSFORMATION

This section is in the nature of an appendix. The results presented here are necessary to complete the theory, but we chall not make any use of them in the rest of the book.

Let U and V be two vector spaces over $\mathbf{F}$. Then there are the dual spaces U' and V', consisting of the linear functionals on U and V, respectively. Let T: $U \to V$ be a linear transformation. For any φ in V' (i.e., φ is a linear functional on V) we can then construct a linear functional $\varphi \circ T$ on U *by composition with* T. To be explicit, the value of the linear functional $\varphi \circ T$ at a vector u of U is, by definition, the scalar $\varphi(Tu)$; here Tu is a vector of V, and we are talking about the value of the linear functional φ at that vector. We can describe $\varphi \circ T$ by the diagram $U \xrightarrow{T} V \xrightarrow{\varphi} \mathbf{F}$. That $\varphi \circ T$ is linear is a special case of the linearity of the composition $S \circ T$ of any two (composable) linear transformations (see Section 5). The assignment $\varphi \to \varphi \circ T$ is a function from V' to U'. We call this function the **transpose** of T and denote it by T^t; so that $T^t(\varphi)$ is, *by definition*, $\varphi \circ T$. And we prove that this function T^t from V' to U' is in fact a *linear transformation*; this amounts to a quite trivial check of definitions; the most difficult point is to convince oneself that there is something to be proved: $T^t(\varphi + \psi) = (\varphi + \psi) \circ T = \varphi \circ T + \psi \circ T$ (see Section 5) $= T^t\varphi + T^t\psi$; the procedure is similar for scalar multiple. Note that T^t goes so to speak in the direction opposite to that of T from V' to U' (but not from V to U), and that T and T^t are coupled by the "transpose relation": for u in U and φ in V' we have

$$\varphi(Tu) = T^t\varphi(u) \tag{Tr}$$

(or, with more parentheses, $\varphi(T(u)) = (T^t\varphi)(u)$).

Example

Let $T=T_M$ be the linear transformation from F^n to F^m, defined by the $m\times n$ matrix M. Let A be a linear functional on F^m (which is a row vector in $(\mathsf{F}^m)'$); what is T^tA? It is the row vector AM in $(\mathsf{F}^n)'$. There are two ways to see this: If we think of A as the matrix of a linear transformation from F^m to F, then, by general rules, AM represents the composition with T_M (see Section 5). Or T^tA is determined by the transposition rule $T^tA(X) = A(TX)$ for all X in F^n." In matrix language (and writing B for T^tA) this means $B\cdot X=A\cdot M\cdot X$ for all X, which implies, of course, $B=AM$.

We come to the connection with matrix transposition (in fact, what follows is the "real reason" for introducing the notion of transpose of a matrix).

8.1. PROPOSITION. *Let T: $U\to V$ be linear transformation with transpose $T^t: U'\to V'$. Let β and γ be bases for U and V, and let β' and γ' be the corresponding dual bases for U' and V'. Then the matrix for T^t with respect to β' and γ' is transpose of the matrix for T with respect to β and γ.*

PROOF. We set $\beta=\{u_j\}$, $\gamma=\{v_i\}$, $\beta'=\{\varphi_r\}$, and $\gamma'=\{\psi_s\}$. The matrix M for T is defined by $Tu_j=\Sigma_i m_{ij}v_i$. The matrix N for T^t is defined by $T^t\psi_s=\Sigma_r n_{rs}\varphi_r$. We write down the transpose relation $T^t\psi_s(u_j)=\psi_s(Tu_j)$, which by substituting yields $\Sigma_r n_{rs}\varphi_r(u_j)=\Sigma_i m_{ij}\psi_s(v_i)$. Now we use the relations $\varphi_r(u_j)=\delta_{rj}$ and $\psi_s(v_i)=\delta_{si}$, expressing the duality of the bases. Each side of our equation reduces to a single term, and we find $n_{js}=m_{sj}$, which just says $N=M^t$. ■

Finally, we come to the behavior of T and T^t with respect to the notions kernel and image. The statement of the results is very brief. It involves the notion "annihilator" (see Chapter 4, Section 7; Chapter 5, Section 1).

8.2. THEOREM. (a) $\ker(T^t)=(\operatorname{im} T)^\perp$; (b) $\operatorname{im}(T^t)=(\ker T)^\perp$.

Note that the vector spaces in (a) are subspaces of V', those in (b) subspaces of U'.

PROOF. For (a), φ is in $\ker T^t$, if $T^t\varphi=0$. This *means* $T^t\varphi(u)=0$ for all u in U. By the transpose relation this becomes $\varphi(Tu)=0$, but this says precisely that is in $(\operatorname{im} T)^\perp$, namely that it annihilates the whole subspace $\operatorname{im} T$ of V.

For (b) we use the equivalent (by Theorem 7.3 of Chapter 4 or Chapter 5, Section 1) formulation $(\operatorname{im} T^t)^\perp=\ker T$. Thus we have to show that u in U has $Tu=0$ precisely if $T^t\varphi(u)=0$ for *all* φ in V'. By the transpose relation this means that $\varphi(Tu)=0$ for all φ in V'. But then the vector Tu in V must be 0. ■

Using Theorems 7.3 and 7.3′ of Chapter 4 or their abstract analogs in Chapter 5, Section 1, and Theorem 1.1 of Chapter 5 ($\dim U = \dim U'$), we get from (a) and (b) the dimension relations $\dim \operatorname{im} T = \dim \ker T^t = m$ ($= \dim V$); $\dim \ker T + \dim \operatorname{im} T^t = n$ ($= \dim U$). And then rank–nullity (Theorems 7.2 of Chapter 4 and 4.1 of Chapter 8) shows $\dim \operatorname{im} T = \dim \operatorname{im} T^t$; in other words, Theorem 8.2 implies that the ranks of T and T^t are equal, $\rho_T = \rho_{T^t}$. (This is the abstract form of the matrix law 7.1 (Chapter 5): column rank equals row rank.)

In the case of an operator ($U = V$), part (a) of Theorem 8.2 is often called the **Fredholm "alternative"**: If $\ker T = 0$, then $\operatorname{im} T = U$, and in general u is in $\operatorname{im} T$ precisely if it is annihilated by all the φ in $\ker T^t$. (Note that $\dim \ker T = \dim \ker T^t$ for an operator.)

9. AFFINE TRANSFORMATIONS

An operator T on a vector space U always holds the origin fixed; we know $T(0) = 0$. But obviously, thinking, for instance, about the plane there are sensible transformations that *shift* the origin, like moving the plane in itself in the x- or y-direction or rotating it around some point *than* the origin. How does one handle such transformations?

First, one introduces the notion of translation of U by a given vector u_0; this is the function t_{u_0} from U to itself that sends any u to $u + u_0$; thus, $t_{u_0}(u) = u + u_0$. Intuitively, we move U parallel by the "amount" u_0. This is not a linear transformation as we defined that term; it is neither additive nor homogeneous, and it does not preserve 0 (where does 0 go under t_{u_0}?). Nevertheless, it is quite close to being linear; it does not (in general) send subspaces to subspaces, but it does send linear varieties to linear varieties: A linear variety, of the type of a subspace V through a point v, goes to a linear variety, of the type V again, but through $u_0 + v$.

We now go a step further: Let T be an operator on U, and as before let u_0 be a vector in U. We define a function τ from U to U by $\tau(u) = Tu + u_0$, for any u in U; we call τ the **affine operator** determined by T and u_0—one can write τ as $t_{u_0} \circ T$. To spell this out, we first apply T to U, then we shift everything by u_0. Again, τ is no more linear; however, given a subspace V, it sends any flat of type V through v to the flat of type $T(V)$ through $v + u_0$.

There are simple laws (similar to additivity and homogeneity for linear transformations) that characterize affine operators: A function τ from U to U is affine (i.e., of the form $t_{u_c} \circ T$ for some T and u_0), if it satisfies $\tau(ru + (1-r)v) = r \cdot \tau u + (1-r) \cdot \tau v$, for any scalar r and any u and v in U. To derive this, put $\tau(0) = u_0$ and show that the map $u \mapsto \tau u - u_0$ is linear. We omit the details.

How does one describe such a τ in coordinates relative to some basis $\{u_1,\ldots,u_n\}$? Let A be the matrix representing T and let Y_0 be the column vector for u_0. Writing, as usual, X for the column vector representing u, and Y for the column vector representing $\tau(u)$, we clearly have

$$Y = AX + Y_0. \tag{Af}$$

There is a scheme to convert this into just matrix multiplication, without (in appearance at least) the addition of the extra term Y_0. First, we replace any vector X by the $(n+1)$-vector $[X,1]=[x_1,\ldots,x_n,1]$, denoting the latter by $\tilde{X}$, say. Next we replace A by the $(n+1)\times(n+1)$ matrix $\tilde{A}=\begin{pmatrix} A & Y_0 \\ 0 & 1 \end{pmatrix}$, a block matrix, where 0 is $1\times n$. And now our equation (Af) can be rewritten simply as

$$\tilde{Y} = \tilde{A}\tilde{X}. \tag{$\tilde{\text{A}}$f}$$

Check the multiplication; check, in particular, that the last coordinate of $\tilde{Y}$ automatically comes out as 1.

A word of warning about the "replacement" $X \to \tilde{X}$ above: If X goes to $\tilde{X}$, then rX does *not* go to $r\tilde{X}$ (unless $r=1$), and if X_1, X_2 go to $\tilde{X}_1, \tilde{X}_2$, then X_1+X_2 does *not* go to $\tilde{X}_1+\tilde{X}_2$; the relation is *not* linear (the last coordinate does not behave right). The relation is in fact an affine transformation from R^n to R^{n+1}, a special case of the general idea of an **affine transformation** from one vector space U to another one V. What one means by that is, of course, the natural generalization of the idea of affine operator, namely a function τ from U to V of the form $t_{v_0}\circ T$ for some linear transformation $T: U \to V$ and some vector v_0 of V (so that $\tau(u) = Tu + v_0$ for any u in U). Such affine maps are characterized by the same law replacing additivity and homogeneity mentioned above. The matrix representation is essentially the same as above, except that we also need a basis for V to introduce coordinates in V. If $\dim V = m$, then A is $m\times n$, and $\tilde{A}$ is $(m+1)\times(n+1)$ now.

10. APPENDIX: QUOTIENT SPACE

Let U be a vector space with a subspace V. We have seen (Section 4) that there are linear transformations of U into suitable spaces, whose kernel is precisely V; and we know (Section 3.v) that then for any vector in the image space the inverse image is a flat parallel to V. This suggests the following construction.

With U and V as before, we *define* a vector space, denoted by U/V and called **quotient space** of U by V: The vectors of U/V are the linear varieties parallel to V in U (thus a "vector" of U/V is in real life a set of the form $u+V$ in U); the "sum" of two vectors of U/V is the well-determined flat in which the sums of vectors in the two flats lie (the sum of any vector in $u+V$ and any vector in $u'+V$ is always in the flat $(u+u')+V$). Similarly, the "product" of a flat by a scalar r is *the* flat containing the products by r of the vectors of the given one (the products by r of vectors in $u+V$ all lie in $ru+V$). It is a straightforward matter to verify that the vector space axioms hold (e.g., the 0 of U/V is the flat V itself; the negative of $u+V$ is, of course, $-u+V$). To abbreviate we write $\bar{u}$ for the flat (=vector of U/V) $u+V$.

Furthermore, U/V comes equipped with a "canonical" linear map, say $R: U \to U/V$ (**reduction modulo** V), defined by $u \mapsto \bar{u} = u+V$ (every vector of U is sent to the flat parallel to V in which it lies); it is clear from the definition of U/V that this is indeed a linear map. And it is also clear from the definition that (a) R is surjective, $\operatorname{im} R = U/V$, and that (b) the nullspace of R is precisely V, $\ker R = V$.

U/V and R are an example of the situation described in the first paragraph. They are in fact a "universal" example in the following sense: Let $T: U \to W$ be any linear transformation whose kernel *contains* V. We can then define a linear transformation $T': U/V \to W$ as follows: Since $T(V)=0$, all vectors in a flat $u+V$ go to one and the same vector under T (we have $T(u+v) = Tu + Tv = Tu + 0 = Tu$ if v is in V); we let that vector be $T'(u+V)$. (This amounts to $T'\bar{u} = Tu$.) It is easy to see that T' is linear. T' is called "induced" by Y.

The definition is so set up that we have the relation $T' \circ R = T$; the two steps "from u to $Ru = \bar{u}$ to $T'\bar{u}$" give the same as the one step "from u to Tu." We have "factored T through R." This is often symbolized with the diagram

$$\begin{array}{ccc} U & \xrightarrow{T} & W \\ & {}_R\searrow \quad \nearrow_{T'} & \\ & U/V & \end{array}.$$

The image space of T', $\operatorname{im} T'$, equals that of T (by $T'\bar{u} = Tu$).

Suppose the kernel of T is *equal* to V. Then T' is *injective* (since the only flat $u+V$ whose vectors go to 0 under T is V itself). Combining the last two sentences, we see that if T is surjective and has kernel V, then T' is an isomorphism of U/V with W.

11. APPENDIX: EXACTNESS

We introduce some concepts that are widely used in present-day mathematics. Let $S: U \to V$ and $T: V \to W$ be two linear transformations.

We describe this by the diagram $U \xrightarrow{S} V \xrightarrow{T} W$. We say that the diagram is **of order two** if $T \circ S = 0$ (or, equivalently, $\operatorname{im} S \subset \ker T$), and that it is **exact** if $\operatorname{im} S = \ker T$. If we represent S and T by matrices A and B (relative to bases in U, V, W), then "order 2" translates into "$BA = 0$," and "exact" translates into "$BA = 0$ and $\rho_A + \rho_B = \dim V$" (by rank–nullity the second condition says $\dim \operatorname{im} S = \dim \ker T$). We can think of this as follows: If $BA = 0$, then the columns of A are solutions of the system $BY = 0$ (here Y is a column vector representing vectors in V) and exactness means that the columns of A actually *span* the solution space of the system $BY = 0$. The terminology is also applied to longer (possible infinite) sequences $\cdots \to U_i \xrightarrow{T_i} U_{i+1} \xrightarrow{T_{i+1}} U_{i+2} \to \cdots$ of spaces and maps; such a diagram is of order 2, respectively, exact if every section of length three is so. (We note one implication of exactness in such a diagram: If the map T_i happens to be the map 0, then T_{i+1} is injective (its kernel is $\operatorname{im} T_i = \operatorname{im} 0 = 0$) and T_{i-1} is surjective (its image is $\ker T_i = \ker 0 = U_i$).)

To any sequence of spaces and maps one can form the *dual* sequence: Each space has a dual space (i.e., space of linear functionals), and each map has a *transpose* (Section 8), going between the dual spaces, but in the opposite direction.

We prove one result (which in a sense is the abstract version of "row rank = column rank").

11.1. THEOREM. *Let the diagram $U \xrightarrow{S} V \xrightarrow{T} W$ be of order 2. Then the dual diagram $W' \xrightarrow{T'} V' \xrightarrow{S'} U'$ is also of order 2. The same holds with "exact" in place of "of order 2."*

PROOF. Order 2 is immediate: Since $S^t \circ T^t$ is the transpose of $T \circ S$, it is 0 if $T \circ S$ is 0. (In matrix language, if $BA = 0$, then $A^t B^t = (BA)^t = 0$.) For exactness, since row rank equals column rank (abstractly, $\rho_T = \rho_{T^t}$, by Section 8) and the dimension of a vector space equals that of the dual space, we have $\rho_{A^t} + \rho_{B^t} = \rho_A + \rho_B = \dim V = \dim V'$. ∎

We take a moment to look at another version of this: In addition to the obvious $S^t \circ T^t = 0$ we have to show that an element φ of V' is of the form $T^t\psi$ for some ψ in W', if it satisfies $S^t\varphi = 0$. Now $S^t\varphi = 0$ means that φ vanishes on the subspace $\operatorname{im} S$ of V (by $S^t\varphi(u) = \varphi(Su)$). But $\operatorname{im} S = \ker T$ by exactness; thus φ takes the same value at any two vectors of V with the same T-image. This means that there is a well-defined and obviously linear functional, say $\bar{\psi}$, on the space $\operatorname{im} T$ that satisfies the relation $\bar{\psi}(Tv) = \varphi(v)$ for all v in V. Now $\bar{\psi}$ can be extended to a linear functional ψ on W (by Proposition 1.1 of Chapter 4, extend a basis of $\operatorname{im} T$ to one of W). The resulting relation $\psi(Tv) = \varphi(v)$ for all v in V means precisely $\varphi = T^t\psi$. ∎

9
EIGENVECTORS, EIGENVALUES

This chapter introduces the important notion of **eigenvector** and **eigenvalue** of an operator; this means a vector that is sent to a scalar multiple of itself by the operator—the scalar factor being the eigenvalue. These concepts have a wide range of theoretical and practical applications. Many problems, particularly those having to do with oscillations of any sort, lead to them (the eigenvalues correspond to the basic frequencies that appear in a complex oscillation). The eigenvalues are found as roots of the **characteristic polynomial** (of the operator or matrix). The topic can be viewed as the process (or the attempt) to make a given matrix *diagonal* (by a suitable change of basis).

1. BASIC DEFINITIONS

Let U be a vector space over the field F, and let $T: U \to U$ be an operator. We recall that a subspace W of U is *T-invariant* if $T(W) \subset W$. The case of *one*-dimensional invariant subspace is particularly important: For a vector u to span such a subspace, the image Tu must be in $((u))$, that is, it must be of the form $r \cdot u$ with some scalar r. Traditionally, the scalar factor here is usually denoted by λ instead of r. We are ready for the main definition.

1.1. DEFINITION. *A nonzero vector u is called an eigenvector of the operator T if $Tu = \lambda u$ with some scalar λ. λ is called the eigenvalue of u, with respect to T.*

(In place of *eigen*vector and *eigen*value the following terms are also used: *proper, characteristic, latent*—"eigen" is German for proper.) Of course, λ has to belong to the field F entering into the definition of U. (*Warning*. The concept of eigenvalue—but not that of eigenvector—will be broadened somewhat later on.)

Analogously, we have the notion of eigenvalue and eigenvector of a (square) matrix A by taking as T the operator $T_A\colon \mathsf{F}^n \to \mathsf{F}^n$. What this amounts to then is a number λ and a *nonzero* vector X in F^n satisfying $A \cdot X = \lambda X$; usually one takes $\mathsf{F} = \mathsf{C}$ here.

Examples

1. $A = \begin{pmatrix} 1 & 1 & 1 \\ 1 & 1 & 1 \\ 1 & 1 & 1 \end{pmatrix}$. $T = T_A\colon \mathsf{R}^3 \to \mathsf{R}^3$. One computes for $X_0 = [1,1,1]$ that $T_A(X_0) = A \cdot X_0 = 3X_0$; so $[1,1,1]$ is eigenvector of T_A with eigenvalue 3.
2. Let $T\colon U \to U$ with $\ker T \neq 0$, and let u be any *nonzero* vector in $\ker T$; then $Tu = 0 = 0 \cdot u$ and so u is eigenvector of T with eigenvalue 0.

We emphasize that *by definition* an eigenvector of T is *nonzero*. (The reason is that then the eigen*value* is determined by $Tu = \lambda u$; whereas, for 0 we have $T(0) = 0 = \lambda \cdot 0$ for any λ whatsoever.)

If u satisfies $Tu = \lambda u$, so does any multiple $r \cdot u$ (with the same λ!); if u' also satisfies $Tu' = \lambda u'$ (again with the *same* λ), then so does $u + u'$: $T(u + u') = Tu + Tu' = \lambda u + \lambda u' = \lambda(u + u')$. Clearly, we have the following.

1.2. DEFINITION AND THEOREM. *Given $T\colon U \to U$ and a scalar λ, the set of all eigenvectors to λ, together with 0, form a subspace of U, called the eigenspace of λ (with respect to T), denoted by U_λ. The dimension r_λ or U_λ is called the geometric multiplicity of λ (with respect to T).*

Note that there might not be *any* eigenvector to λ, and in that case U_λ is simply the 0 subspace.

Examples

1. This example is trivial, but nevertheless important. Let $T = \mathbf{1}_U$, then $Tu = u = 1 \cdot u$ for *every* u in U, and so $U_1 = U$. The situation is similar for $T = 0$: $U_0 = U$. The geometric multiplicity is n in both cases.
2. In example 1 in the text the three components of $A \cdot X$ are clearly equal to each other for *any* X; therefore, $TX = \lambda X$ requires either $\lambda = 0$, that is, X in $\ker T$, or that X be a multiple of $[1,1,1]$, in which case $\lambda = 3$, as we saw. By standard computation we find $\ker T = (([-1,1,0], [-1,0,1]))$. We obtain $U_0 = \ker T$, $U_3 = (([1,1,1]))$, and for any other λ: $U_\lambda = 0$. The geometric multiplicities are $r_0 = 2$ for $\lambda = 0$, $r_3 = 1$ for $\lambda = 3$, and $r_\lambda = 0$ for all other λ.

PROBLEMS

1. Let T be an operator on U. Show that im and ker of T^2 are invariant subspaces.

2. On P^n consider the operator D (differentiation). Show that D does not have any eigenvectors with any *nonzero* eigenvalue λ. Furthermore, there is, up to a scalar factor, only one eigenvector of D to eigenvalue 0. Which polynomials satisfy $Dp(x) = 0 \cdot p(x)$ $(=0)$?

3. Let T be a rotation, in the usual sense, of ordinary space. It is known from kinematics or otherwise that such a T rotates space around a certain axis through a certain angle. Are there any eigenvectors? To what eigenvalues?

4. Show: If λ is eigenvalue for T, then λ^2 is eigenvalue for T^2. How did you find the associated eigenvector? What about higher powers of T?

5. Show: If u_1 and u_2 are eigenvectors of T to two *different* eigenvalues λ_1 and λ_2, then $u_1 + u_2$ is *not* eigenvector of T.

6. Show: If all nonzero vectors of U are eigenvectors for T, then T is a scalar operator $(= r \cdot \mathbf{1})$.

7. Show that 2 is an eigenvalue of the matrix $M = \begin{pmatrix} 2 & 0 & 3 \\ 0 & 2 & 0 \\ 0 & 0 & 1 \end{pmatrix}$ by verifying that the system of equations $MX = 2X$ (with $X = [x_1, x_2, x_3]$) has a nontrivial solution. (Write out the three equations, and transport all terms to the left-hand side.)

2. THE CHARACTERISTIC POLYNOMIAL

The main idea for *finding* eigenvalues and eigenvectors consists in rewriting $Tu = \lambda u$ as $(T - \lambda\mathbf{1})(u) = 0$; this shows that the eigenvectors to λ are the nonzero vectors (if any) in the kernel of the operator $T - \lambda\mathbf{1}$. (Recall that one often writes $T - \lambda$ for $T - \lambda\mathbf{1}$.) Thus, if a scalar λ wants to have an eigenvector associated with it, $\ker(T - \lambda\mathbf{1})$ must not be just $\underset{\sim}{0}$, that is, $T - \lambda$ must be *not injective*, it must be *singular*. This turns out to be a polynomial condition on λ. To establish this, we translate everything into matrix language: Let $u_1, \ldots, u_n$ be a basis for U, then T is represented by a matrix A (see Chapter 8, Section 2, where $Tu_j = \Sigma a_{ij} u_i$). $T - \lambda\mathbf{1}$ is then represented by $A - \lambda I$ (and an eigenvector u is represented by a column X satisfying $A \cdot X = \lambda X$). From the determinant criterion (Chapter 7, Section 3.b) we find the following proposition.

2.1. PROPOSITION. *A scalar* λ *in* F *is eigenvalue of* T *exactly if* $\det(A - \lambda I) = 0$.

Note that is does not matter what basis for U we use. If we change to another basis $\{u_i'\}$, with transition matrix P, then by Chapter 8, Section 2, the new matrix for $T - \lambda\mathbf{1}$, $A' - \lambda I$, will equal $P^{-1} \cdot (A - \lambda I) \cdot P$, that is, be similar to $A - \lambda I$. But similar matrices have equal determinants.

This leads to the following: Let x be a variable; given a matrix A, form $\det(A - x\mathbf{1})$; from formula (D) in Chapter 7, Section 2, one sees that this is a *polynomial* in x, called the **characteristic polynomial** of A and written $\chi_A(x)$. At the same time we define the **characteristic polynomial** $\chi_T(x)$ of the operator T as that of any matrix representing T (again the choice of basis does not matter: If $A' = P^{-1} \cdot A \cdot P$, then $A' - xI = P^{-1} \cdot (A - xI) \cdot P$—note that $P^{-1} \cdot xI \cdot P = xI$!—and so they have the same determinant). We now restate Proposition 2.1.

2.1′ PROPOSITION. *A scalar λ in* F *is eigenvalue of T if and only if it is a root of the characteristic polynomial $\chi_T(x)$, that is, satisfies $\chi_T(\lambda) = 0$.*

Example

Consider the matrix A of 1) in Section 1 (we saw already that the eigenvalues are 0 and 3).

$$\chi_A(x) = \det(A - xI) = \det\begin{pmatrix} 1-x & 1 & 1 \\ 1 & 1-x & 1 \\ 1 & 1 & 1-x \end{pmatrix} = \cdots = 3x^2 - x^3.$$

Clearly, 0 and 3 are roots of this, and there are no other roots—we can factor χ_A as $x \cdot x \cdot (3 - x)$.

We make a comment on the role of F in Propositions 2.1 and 2.1′: If F = C, then the eigenvalues of T are exactly all the roots of $\chi_T(x)$. But if F = R, we have to be a little careful: The polynomial $\chi_T(x)$, although its coefficients are real, might have strictly complex roots ($\lambda = a + ib$ with $b \neq 0$); we *continue* to call such roots eigenvalues of T, although there cannot be any u with $Tu = \lambda u$ for such λ (because for U over R the operation $\lambda \cdot u$ with complex λ is not defined). However, the corresponding *matrix* equation $A \cdot X = \lambda X$ still makes sense, but any solution ($\neq 0$) X must involve complex numbers ($X = X' + iX''$ with X' and X'' real and $X'' \neq 0$), since A is real, but λ is *not*.

We note some coments on $\chi_A(x)$: If A is $n \times n$ ($\dim U = n$), then $\chi_A(x)$ is of degree n, and the coefficient of x^n is $(-1)^n$: $\chi_A(x) = a_0 + a_1 x + a_2 x^2 + \cdots + (-1)^n x^n$. One way to see this is to simply put $A = 0$. Then $A - xI$ reduces to $-xI$, and $\det(-xI) = (-1)^n x^n$ as is easily seen. Another way is to use the explicit formula for $\det(A - xI)$. One notes that the diagonal entries of $A - xI$ are $a_{11} - x, a_{22} - x, \ldots, a_{nn} - x$; no other entry has x in it. Therefore, no power of x beyond n can appear, and x^n can only come from the principal term $(a_{11} - x) \cdot (a_{22} - x) \cdots (a_{nn} - x)$ of det; expanding, one gets $(-1)^n x^n +$ lower powers of x.

The following proposition is a consequence from what one knows about polynomials.

2.2. PROPOSITION. *There are at most n different eigenvalues of A (or T), with A $n\times n$ (or $\dim U=n$).*

The "at most" comes from the fact that polynomials can have multiple roots (can factor off not only $x-\lambda$, but some power $(x-\lambda)^s$). Suppose we factor $\chi_A(x)$ *completely* so that

$$\chi_A(x)=(-1)^n\cdot(x-\lambda_1)^{s_1}\cdot(x-\lambda_2)^{s_2}\cdots(x-\lambda_t)^{s_t}.$$

Here $\lambda_1,\ldots,\lambda_t$ are the *different* eigenvalues of χ_A; the exponents $s_1,\ldots,s_t$ are called the **algebraic multiplicities** of the λ_i. This factorization (equivalent to finding all the roots of $\chi_A(x)$) is, of course, a difficult algebraic problem. (For quadratic, cubic, and fourth-order polynomials there are *formulae* for the roots. In general, one has to be lucky to find any *rational* roots; beyond that there are only numerical methods which usually give the roots with any desired degree of approximation, but never exactly. We leave this whole area to the algebraists and the computers.)

There is an important special case: For a triangular, and in particular for a diagonal, matrix the characteristic polynomial reduces to the principal term $(a_{11}-x)\cdot(a_{22}-x)\cdots(a_{nn}-x)$; it is thus completely factored, and the diagonal entries a_{ii} are the eigenvalues.

(We note and apologize for a notational inconsistency: Above we wrote $\lambda_1,\ldots,\lambda_t$ for the *different* eigenvalue, each having a certain multiplicity s_i. Occasionally, we shall write $\lambda_i,\ldots,\lambda_n$ for *all* the eigenvalues, admitting the possibility that several λ_i are equal to each other.)

The coefficient a_0 in $\chi_A(x)$ has a simple meaning: It is just $\det A$ itself (put $x=0$!). The coefficient of x^{n-1} is also simple: From the discussion of the principal term above one easily sees that, up to the sign $(-1)^{n-1}$, the coefficient is $a_{11}+a_{22}+\cdots+a_{nn}$, the *sum* of the diagonal entries of A (this is called the **trace** of A and denoted by $\operatorname{tr} A$). The other coefficients have similar, but slightly more complicated interpretations.

We recapitulate: To find the eigenvalues and eigenspaces of an operator, represented by a matrix A: (1) Compute the polynomial $\chi_A(x)=\det(A-x\mathbf{1})$. (2) Find the roots of $\chi_A(x)$, say $\lambda_1,\ldots,\lambda_t$, with (algebraic) multiplicities $s_1,\ldots,s_t$. (3) For each λ_i *separately* find the solution space U_{λ_i} of $A\cdot X=\lambda_i X$, that is, of $(A-\lambda_i I)\cdot X=0$; "finding the space" means, as usual, finding a basis for it. This includes finding the dimensions r_i of the U_{λ_i}, the geometric multiplicities. (Note that if the scalars are $\mathbf{R}$, then complex eigenvalues cannot be used since their solution column vectors will not be real.)

PROBLEMS

1. Find characteristic polynomial, eigenvalues, and eigenvectors (bases for the different eigenspaces) for:

$$\begin{pmatrix} 3 & 2 & 4 \\ 2 & 0 & 2 \\ 3 & 2 & 3 \end{pmatrix}, \begin{pmatrix} 1 & 2 & 3 \\ 0 & 2 & 3 \\ 0 & 0 & 3 \end{pmatrix}, \begin{pmatrix} 0 & 1 \\ 1 & 0 \end{pmatrix}, \begin{pmatrix} 0 & -1 \\ 1 & 0 \end{pmatrix}, \begin{pmatrix} 3 & 2 & -6 \\ 2 & 2 & -5 \\ 2 & 1 & -4 \end{pmatrix},$$

$$\begin{pmatrix} -2 & 0 & -6 & -2 \\ 0 & 2 & 0 & 0 \\ 1 & 0 & 3 & 0 \\ 0 & 0 & 0 & 2 \end{pmatrix}, \begin{pmatrix} \sqrt{3}/2 & -1/2 \\ 1/2 & \sqrt{3}/2 \end{pmatrix}, \begin{pmatrix} \cos\alpha & -\sin\alpha \\ \sin\alpha & \cos\alpha \end{pmatrix} \text{ for an arbitrary } \alpha.$$

(Eigenvalues and vectors may be complex!)

2. Construct a matrix whose characteristic polynomial is $(x-2)^2 \cdot (x^2-1)$.

3. Let T be the operator on $\mathbf{R}^3$ defined by the matrix

$$\begin{pmatrix} 0 & -1 & 0 \\ 1 & 0 & 0 \\ 0 & 0 & 1 \end{pmatrix}.$$

Find eigenvalues and eigenvectors (be precise about reality questions).

4. Determine the characteristic polynomial of the "general" 2×2 matrix $\begin{pmatrix} a & b \\ c & d \end{pmatrix}$.

5. Let $T: P^3 \to P^3$ be defined by $Tp(x) = p(x+1)$ (the effect of T is substituting $x+1$ for x). Find χ_T, the eigenvalues for T and the corresponding eigenspaces (first find the matrix for T with respect to, say, the basis $\{1, x, x^2, x^3\}$ of P^3). Note that since x is the variable for the polynomials in P^3, it is better to use a different variable, say t, for χ_T. Interpret your result directly: For what λ and $p(x)$ can $p(x+1)$ equal $\lambda \cdot p(x)$?

6. Let T be an operator. Show that $\det T$ equals the product of the eigenvalues of T. (Instead of "operator T" we could say "matrix M" here.) (*Hint*. Consider the characteristic polynomial in completely factored form.)

7. (Continuation of Problem 6.) Show that the trace $\operatorname{tr} T$ (or $\operatorname{tr} M$) equals the sum of the eigenvalues of T (or M).

8. Let A and B be two matrices of shapes $m \times n$ and $n \times m$ (with $m = n$ allowed). Write out $\operatorname{tr} AB$ and $\operatorname{tr} BA$ in terms of the entries a_{ij} and b_{ji}, and show that the two are equal to each other. Use this to show that two similar (square) matrices have equal traces.

9. Let A be a (square) matrix. Show that A and A^t have the same eigenvalues (in fact, the same characteristic polynomial).

3. DIAGONALIZATION

We start with two examples.

Examples

1. In the matrix $\begin{pmatrix} 1 & 1 & 1 \\ 1 & 1 & 1 \\ 1 & 1 & 1 \end{pmatrix}$ above, we found $\chi(x)=x^2\cdot(3-x)$. Thus, we have two roots, $\lambda_1=0$, twice, $s_1=2$; $\lambda_2=3$, once, $s_2=1$. The eigenspace to λ_1, consists of the solutions of $(A-0\cdot I)X=0$, that is, of $A\cdot X=0$. Clearly, this reduces to $x_1+x_2+x_3=0$; x_2 and x_3 are free and a basis for U_0 is $X_1=[-1,1,0]$, $X_2=[-1,0,1]$.

For the eigenspace to λ_2, we need the solutions of $(A-3I)\cdot X$ or

$$\begin{pmatrix} -2 & 1 & 1 \\ 1 & -2 & 1 \\ 1 & 1 & -2 \end{pmatrix}\cdot\begin{pmatrix} x_1 \\ x_2 \\ x_3 \end{pmatrix}=0.$$

The row-echelon form of the matrix is, for example,

$$\begin{pmatrix} 1 & 1 & -2 \\ 0 & 1 & -1 \\ 0 & 0 & 0 \end{pmatrix};$$

the equations are $x_1+x_2-2x_3=0$ and $x_2-x_3=0$; x_3 are free and a basis for U_3 is $X_3=[1,1,1]$.

2.

$$B=\begin{pmatrix} 3 & 2 & -6 \\ 2 & 2 & -5 \\ 2 & 1 & -4 \end{pmatrix}\cdot\chi(x)=\det\begin{pmatrix} 3-x & 2 & -6 \\ 2 & 2-x & -5 \\ 2 & 1 & -4-x \end{pmatrix}=\cdots$$

(use scheme for 3×3 determinants)$=-(x-1)^2(x+1)$. Thus $\lambda_1=1$ and $s_1=2$; $\lambda_2=-1$ and $s_2=1$. For U_1: $B-I=\begin{pmatrix} 2 & 2 & -6 \\ 2 & 1 & -5 \\ 2 & 1 & -5 \end{pmatrix}$, we have two equations $x_1+x_2-3x_3=0$ and $x_2-x_3=0$; here x_3 is free and a basis for U_1 is $Z_1=[2,1,1]$. For U_{-1}: $B+I=\begin{pmatrix} 4 & 2 & -6 \\ 2 & 3 & -5 \\ 2 & 1 & -3 \end{pmatrix}$; the row-echelon form of the matrix is $\begin{pmatrix} 2 & 1 & -3 \\ 0 & 1 & -1 \\ 0 & 0 & 0 \end{pmatrix}$; we have equations $2x_1+x_2-3x_3=0$ and $x_2-x_3=0$; and a basis for U_{-1} is $Z_2=[1,1,1]$.

Now note that in the first example we got three vectors altogether: $X_1, X_2,$ and X_3; they form a basis for $\mathbf{R}^3$. In the second we got only two vectors, Z_1 and Z_2. We consider the matter in more detail: In the first example, the double eigenvalue $\lambda_1 = 0$ brought with it two independent eigenvectors X_1 and X_2. In the second example, the double eigenvalue 1 brought with it only *one* independent eigenvector Z_1.

We see that the eigenvectors of an operator sometimes span the space and sometimes do not. This is related to the question whether an operator can be represented (with respect to a suitable basis) by a *diagonal* matrix, or, in matrix terms, whether a (square) matrix A can be **diagonalized**; that is, whether one can find a matrix such that $P^{-1} \cdot A \cdot P$ is diagonal. (Operators or matrices that can be diagonalized are also called *semisimple*.) One likes operators to be "diagonalizable," since diagonal matrices are about as simple as matrices can be. We state the main result for operators and its companion for matrices.

3.1. PROPOSITION. (a) *The matrix A of $T: U \to U$ with respect to a basis $\beta = \{u_1, \ldots, u_n\}$ is diagonal if and only if all the u_i are eigenvectors of T*; (b) *T can be diagonalized* (i.e., *its matrix with respect to some basis is diagonal*) *if and only if the eigenvectors of T span U.*

3.1′. PROPOSITION. *The $(n \times n)$ matrix A can be diagonalized* (i.e., *is similar to a diagonal matrix*) *if and only if the eigenvectors of A span $\mathbf{F}^n$. If $X_1, \ldots, X_n$ are n independent eigenvectors, then the matrix $P = (X_1, \ldots, X_n)$ diagonalizes A; that is, $P^{-1} \cdot A \cdot P$ is then diagonal* (*with the eigenvalues λ_i of A on the diagonal*) *and can be written* $\operatorname{diag}(\lambda_1, \ldots, \lambda_n)$.

PROOF. For Proposition 3.1 (a) one just looks at the definition of the matrix A representing T with respect to β, namely $Tu_j = \Sigma a_{ij} u_i$. For u_i to be an eigenvector, the coefficients a_{ij} must be 0 except for the jth one, which equals the eigenvalue. For (b) all we need is that a basis spans and that from a spanning set a basis can be extracted. For Proposition 3.1′ we apply Proposition 3.1 to the operator T_A on $\mathbf{F}^n$, and use the fact that P is precisely the transition matrix from the standard basis to the basis $\{X_1, \ldots, X_n\}$. ■

We interpolate another interpretation of Proposition 3.1′, of a somewhat formal, but more direct, nature. The relation that we are after, $P^{-1}AP = \operatorname{diag}(\lambda_1, \ldots, \lambda_n)$, can be rewritten as $AP = P \cdot \operatorname{diag}(\lambda_1, \ldots, \lambda_n)$; we are trying to find an *invertible* P that satisfies the last equation. Now (Exercise!), if P $(= (X_1, \ldots, X_n))$ has column vectors X_i, then $P \cdot \operatorname{diag}(\lambda_1, \ldots, \lambda_n)$ is the matrix $(\lambda_1 X_1, \lambda_2 X_2, \ldots, \lambda_n X_n)$ with column vectors $\lambda_i X_i$. On the other hand, the column vectors of AP are simply the vectors $AX_1, AX_2, \ldots, AX_n$. Thus our matrix equation amounts to the n relations $AX_1 = \lambda_1 X_1, \ldots, AX_n = \lambda_n X_n$; it says precisely that the columns of P are eigenvectors of A, with eigenvalues λ_i. Invertibility of P is equivalent to independence of the X_i. ■

We look at an example of a matrix that can be diagonalized: the matrix A of Example 1, in Section 1.1. We found two independent eigenvectors $X_1=[-1,1,0]$ and $X_2=[-1,0,1]$ to $\lambda=0$, and one vector $X_3=[1,1,1]$ to $\lambda=3$; thus there are "enough" eigenvectors. X_1, X_2, and X_3 form a basis for $\mathbf{R}^3$. The matrix $P=(X_1,X_2,X_3)$ should then diagonalize A; that is, $P^{-1}\cdot A\cdot P$ should be diagonal ($=\operatorname{diag}(0,0,3)$, in fact). Verify this observation, say in the form $A\cdot P=P\cdot\operatorname{diag}(0,0,3)$. On the other hand, the matrix B of Example 2 above cannot be diagonalized. There are only two independent eigenvectors.

Another, simpler, example of a matrix that cannot be diagonalized is $C=\begin{pmatrix}0&0\\1&0\end{pmatrix}$. If it could be diagonalized, we would get

$$P^{-1}\cdot C\cdot P=\operatorname{diag}(\lambda_1,\lambda_2)=0,$$

since both eigenvalues of C are 0. But that is, of course, impossible; if a matrix is similar to 0, it *is* 0.

Remark. We will see in Chapter 10, Section 6, that in a certain sense "most" matrices are diagonalizable. To have something to go on, we also state (part of) the main result of Chapter 13: Every real *symmetric* matrix is diagonalizable.

PROBLEMS

1. Diagonalize the matrices of Section 2, Problem 1, if possible.

2. Let T be the operator $D^2+D+\mathbf{1}$ on P^n (i.e., $Tp(x)=p''(x)+p'(x)+p(x)$ for every vector of P^n). Can T be diagonalized?

3. Diagonalize

$$\begin{pmatrix}2&0&1\\-1&0&-1\\1&2&2\end{pmatrix},$$

if possible.

4. Diagonalize

$$\begin{pmatrix}-4&4&4&-1\\-3&4&3&-1\\-5&4&5&-1\\-5&4&5&0\end{pmatrix},$$

if possible.

4. FURTHER FACTS

We recapitulate: To diagonalize a matrix A, find all eigenvalues $\lambda_1,\ldots,\lambda_t$; for each λ_i find a basis for the eigenspace (solutions of $(A-\lambda_i\cdot I)\cdot X=0$). *If* these bases *together* span F^n, they will give the P of Proposition 3.1′. If they do not, then A *cannot* be diagonalized. To complete the argument, we need one more fact.

4.1. THEOREM. *The eigenspaces* U_{λ_i}, $i=1,\ldots,t$, *of* T *are independent, that is, a relation* $v_1+\cdots+v_t=0$ *with each* v_i *in* U_{λ_i} *can hold only if all the* v_i *are* 0. (*It follows then that bases for the* U_{λ_i} *together give a basis for the span of the* U_{λ_i}.)

4.2. COROLLARY. *If* T *is semisimple, then* U *is direct sum of the eigenspaces* U_{λ_i}, *and conversely.*

PROOF. Applying T to the relation, we get $\lambda_1 v_1+\lambda_2 v_3+\cdots+\lambda_t v_t=0$. Subtracting $\lambda_t\cdot(v_1+\cdots+v_t)=0$, we get

$$(\lambda_1-\lambda_t)\cdot v_1+(\lambda_2-\lambda_t)\cdot v_2+\cdots+(\lambda_{t-1}-\lambda_t)\cdot v_{t-1}=0;$$

the factors $\lambda_i-\lambda_t$ here are all *nonzero*. Thus we got a *shorter* relation; we can iterate the procedure; when we get down to a single vector, we find $v_1=0$; the procedure is similar for the other v_i. ■

Next, we establish a theorem to show that one never gets "too many" independent vectors.

4.3. THEOREM. *For each eigenvalue* λ_i *of* $T:U\to U$ *the geometric multiplicity* $r_i=\dim U_{\lambda_i}$ *is at most equal to the algebraic multiplicity* s_i *(see Section 2). (Recall* $\sum s_i=n$, *from the degree of* χ_A.)

PROOF. Let $u_1,\ldots,u_{r_1}$ be a basis for U_{λ_1}; extend to a basis $u_1,\ldots,u_n$ of U. Then for the first r_1 columns of the matrix A all elements not on the diagonal are 0, and the diagonal entries all equal λ_1 (see the proof of Proposition 3.1). Writing out $A-x\cdot I$ and computing $\det(A-xI)$ by expansion along the first column (and iterating), one gets $\det(A-xI)=(\lambda_1-x)^{r_1}\cdot q(x)$, with some polynomial $q(x)$. This shows that $\chi_A(x)$ is divisible *at least* by $(x-\lambda_1)^{r_1}$; that is, $s_1\geqslant r_1$. The procedure is similar for the other λ_i. ■

If $r_i=s_i$ for some i, one calls the eigenvalue λ_i semisimple or diagonalizable.

Example

As a case with $r_i<s_i$ take $A=\begin{pmatrix}1&1\\0&1\end{pmatrix}$. We have $\chi_A=(x-1)^2$, that is, 1 is double eigenvalue; $s_1=2$. But the eigenspace, the kernel of $A-I=\begin{pmatrix}0&1\\0&0\end{pmatrix}$,

is given by the equation $x_2=0$; a basis is $[1,0]$, and we see $r_1=1$.

We record a particular case of "diagonalizability."

4.4. PROPOSITION. *If the eigenvalues of T (or A) are pairwise different, then T is diagonalizable (A is similar to a diagonal matrix) (but see Note 2 below).*

This follows at once from Theorem 4.1 and Proposition 3.1: There are now enough independent eigenvectors to form a basis for U (or F^n).

Note 1. In algebra one shows that the condition of Proposition 4.4 can be checked without finding the eigenvalues. It is equivalent to the condition that the greatest common divisor of χ_T and its derivative χ_T' be of degree 0 (a constant) or, equivalently, that χ_T and χ_T' be relatively prime.

Note 2. Take a word of caution to make Proposition 4.4 more precise: Everything is all right if our field is C. But if it is R, then complex (nonreal) eigenvalues give trouble, as noted above; for such an eigenvalue there cannot be an eigenvector. Then, of course, T is not diagonalizable. (However, a representing matrix can be diagonalized, using a *complex* P.) It could even happen that χ_T has n different (complex) eigenvalues, but there is not a single eigenvector.

Example

Define an operator by $E^1 \mapsto E^2$ and $E^2 \mapsto -E^1$ on R^2 (the "90° rotation"). Obviously, there is no eigenvector—no vector stays on its own line. $\chi_T(x)=x^2+1$ has roots $\pm i$.

For a real matrix A this trouble, if present, is less visible: If λ is a nonreal eigenvalue, then $A\cdot X=\lambda X$ still has (nonzero) solutions X. But they are necessarily *complex* in C^n; $X=X'+iX''$ with X' and X'' in R^n and $X''\neq 0$. This can be reinterpreted: With $\lambda=\alpha+i\beta$ the equation $A\cdot X=\lambda\cdot X$ multiplied out, becomes $A\cdot X'+iA\cdot X''=\alpha X'-\beta X''+i(\beta X'+\alpha X'')$; on separating real and imaginary part, we get

$$A\cdot X'=\alpha X'-\beta X'' \qquad \text{and} \qquad A\cdot X''=\beta X'+\alpha X''.$$

Note that everything is real now. But this means that the subspace $((X',X''))$ of R^n is *invariant* under T_A. We claim that $\dim((X',X''))=2$; that is, X' and X'' are independent (in R^n): Suppose, say, $X'=r\cdot X''$ with real r, then $X'+iX''=(r+i)\cdot X''$ and $A\cdot X=\lambda\cdot X$ would become $(r+i)\cdot A\cdot X''=(r+i)\lambda\cdot X''$. On cancelling $r+i$ ($\neq 0!$) we get $A\cdot X''=\lambda X''$, but then the *real* vector X'' would be eigenvector to λ, which is impossible. In the usual way we can translate this back to an abstract operator T on a *real* vector space U, and see that in that situation a nonreal eigenvalue gives rise *not* to an eigenvector, but to a *two*-dimensional T-invariant subspace. ■

The next theorem is a fact that appears frequently in applications, "simultaneous diagonalization of a commuting family of diagonalizable operators or matrices."

4.5. THEOREM. *Let $S_1, S_2, \ldots$ be semisimple operators on U that commute pairwise ($S_1 \circ S_2 = S_2 \circ S_1$,* etc.). *Then U is spanned by simultaneous eigenvectors of the S_i* (i.e., *by vectors that are at the same time eigenvectors of S_1 and of S_2,* etc.).

4.6. COROLLARY. *The sum of two commuting semisimple operators is also semisimple.*

4.5′. THEOREM. (*matrix version of Theorem* 4.5). *Let $A_1, A_2, \ldots$ be diagonalizable matrices that commute pairwise. Then there is a simultaneous diagonalization; that is, there is a matrix P such that all $P^{-1}A_1P$, $P^{-1}A_2P, \ldots$ are diagonal.*

4.7. COROLLARY. *$P^{-1}(A_1 + A_2)P$ is diagonal.*

PROOF. For two operators (the general case follows by easy iteration). From $S_1 u = \lambda u$, we get $S_1(S_2 u) = S_2(S_1 u) = S_2(\lambda u) = \lambda S_2 u$, that is, $S_2 u$ is again λ-eigenvector of S_1 (or 0); S_2 maps each eigenspace U_{λ_i} of S_1 to itself. It follows (see Problem 6 below) for each eigenvector of S_2 that its components in the various U_{λ_i} (see Corollary 4.2) are again eigenvectors of S_2. This shows that there are enough simultaneous eigenvectors of S_1 and S_2 to span U.

For the matrix version assume A_1 diagonal to begin with; show that the relation $A_1A_2 = A_2A_1$ implies that A_2 consists of blocks along the diagonal corresponding to the eigenvalues of A_1—size of block equal to multiplicity of eigenvalue. Now diagonalize each such block individually (why is this possible?). ∎

We end this chapter with an application: One knows that the differential equation $dx/dt = a \cdot x$ is solved by $x = e^{at}$ (or more generally by $x = c \cdot e^{at}$ for any c). We generalize to systems of linear differential equations in several dependent variables $x_1, \ldots, x_n$. Each x_i is now a function of t; naturally we combine them into a *vector function* $X(t) = [x_1(t), \ldots, x_n(t)]$. With the notation $df/dt = \dot{f}$ we form the derivative vector $\dot{X} = [\dot{x}_1, \ldots, \dot{x}_n]$. (For instance, if $X = [t, 1 + t^2, t^3, \sin t]$, then $\dot{X} = [1, 2t, 3t^2, \cos t]$.) Let now A be an $n \times n$ matrix (of scalars); we form the *linear system* of differential equations $\dot{X} = AX$. Written out, this reads $\dot{x}_i = \Sigma_j a_{ij} x_j$. Our aim is to solve the system, to find functions x_i of t that satisfy these n relations.

We try to simplify the system by changing coordinates, with the help of a transition matrix P (scalar entries); see Chapter 8, Section 2. Writing $Y = [y_1, \ldots, y_n]$ for the new coordinates (X' would be confusing), we get from $X(t)$ the new function $Y(t)$ with $X(t) = P \cdot Y(t)$, and we have $\dot{X} = P\dot{Y}$

since P is constant. The differential system becomes $P\dot{Y}=APY$ or $\dot{Y}=P^{-1}APY$; that is, under a change of coordinates the matrix of the system changes to a similar one.

Suppose now that we are lucky and that A can be diagonalized. With a suitable P the system then becomes $\dot{Y}=DY$, where $D=\operatorname{diag}(\lambda_1,\ldots,\lambda_n)$ is diagonal (with the eigenvalues of A on the diagonal); written out this reads $\dot{y}_i=\lambda_i y_i$. The dependent variables are now cleanly separated. For each y_i we have an equation of the simple type $\dot{x}=ax$; the solutions are simply $y_i=c_i e^{\lambda_i t}$, for $i=1,\ldots,n$. Of course, once we know Y, we get X as PY; thus the x_i will be certain linear combinations of the $e^{\lambda_i t}$; the coefficients involve the arbitrary constants c_i.

What if A cannot be diagonalized? The situation is then more complicated but can still be handled. The answer involves the *Jordan form*; we describe this in Chapter 10, Section 9.

Example

$$\dot{x}_1=-8x_1+6x_2 \quad \text{and} \quad \dot{x}_2=-12x_1+9x_2.$$

$$A=\begin{pmatrix} -8 & 6 \\ -12 & 9 \end{pmatrix}.$$

$$\chi_A=x^2-x.$$

$$\lambda_1=0 \quad \text{and} \quad \lambda_2=1.$$

$$X_1=[3,4] \quad \text{and} \quad X_2=[2,3].$$

$$P=\begin{pmatrix} 3 & 2 \\ 4 & 3 \end{pmatrix}.$$

We have the new system $\dot{y}_1=0\cdot y_1=0$ and $\dot{y}_2=1\cdot y_2=y_2$. The solutions are $y_1=c_1$ and $y_2=c_2e^t$, respectively. Finally,

$$x_1=3c_1+2c_2e^t \quad \text{and} \quad x_2=4c_1+3c_2e^t,$$

from $X=PY$.

If an eigenvalue λ is complex of the form $\alpha+i\beta$, then $e^{\lambda t}$ becomes $e^{\alpha t}\cdot e^{i\beta t}=e^{\alpha t}\cdot(\cos\beta t+i\sin\beta t)$. If A is real, $\bar{\lambda}=\alpha-i\beta$ is also eigenvalue; one gets then linear combinations of $e^{\alpha t}\sin\beta t$ and $e^{\alpha t}\cos\beta t$ as solutions.

PROBLEMS

1. Diagonalize, if possible, $\begin{pmatrix} 3 & 2 & 2 \\ 1 & 4 & 1 \\ -2 & -4 & -1 \end{pmatrix}$ and $\begin{pmatrix} 2 & -i & 0 \\ i & 2 & 0 \\ 0 & 0 & 2 \end{pmatrix}$.

2. For $A=\begin{pmatrix}3 & 5 & 5\\ -2 & -4 & -3\\ 0 & 1 & 0\end{pmatrix}$, verify that there is one real and two complex (conjugate to each other) eigenvalues; find the corresponding eigenvectors and invariant plane of $T_A: \mathsf{R}^3 \to \mathsf{R}^3$ (the two conjugate eigenvalues determine the same plane!). Find the matrix for T_A with respect to a basis consisting of the eigenvector and a basis for the invariant plane.

3. Let T be an operator on U. Suppose a vector u can be written as $u_1+u_2+\cdots+u_r$ with the u_i denoting eigenvectors of T to *different* eigenvalues $\lambda_1,\lambda_2,\ldots,\lambda_r$. Show that the u_i can be obtained as linear combinations of the vectors u, Tu, T^2u, $T^3u,\ldots,T^{r-1}u$. (*Hint*. Study the case $r=2$. Try an induction argument for the general case.)

4. For the matrices in Problem 1 find the gcd of χ_A and χ_A'.

5. Let $A=\operatorname{diag}(\lambda_1,\lambda_2,\ldots,\lambda_n)$ with no two λ's equal to each other. Show that if a matrix B commutes with A, then B is also diagonal.

6. Suppose $U=V\oplus W$ is direct sum of two subspaces and the operator $T: U\to U$ has V and W as invariant subspaces. Show: If u is eigenvector of T to λ, then the two components v and w of u in V and W are eigenvectors of T to λ (provided they are not 0). (There is an obvious converse.)

7. With the data of Problem 6 show that if T is diagonalizable, so are the operators $T|V$ and $T|W$. (Show that there are "enough" eigenvectors.)

8. Interpret Problem 7 in matrix language. (Use block matrices.)

9. Let $A=\begin{pmatrix}7 & 1 & -8\\ 6 & 2 & -8\\ 6 & 1 & 7\end{pmatrix}$ and $B=\begin{pmatrix}1 & 1 & -2\\ 10 & 2 & -12\\ 2 & 1 & -3\end{pmatrix}$.

a. Show that A and B are semisimple.

b. Show $AB=BA$.

c. Find a simultaneous diagonalization of A and B (i.e., find a basis for F^3, consisting of vectors that are eigenvectors of A and of B, and thus define a matrix P such that $P^{-1}AP$ and $P^{-1}BP$ are *both* diagonal).

10. Let T be an operator on U, and suppose V is an invariant subspace. Show that the characteristic polynomial χ_T of T on U factors as $\psi(x)\cdot\varphi(x)$, where $\psi(x)$ is the characteristic polynomial of T on V. (*Hint*. Use a suitable basis (see Chapter 8, Section 3, Problem 8), and apply Problem 8 of Chapter 7, Section 3.)

11. Let $T: U\to U$ be diagonalizable, and suppose there is a T-invariant subspace V.

12. With T and V as in Problem 11, prove that there exists a complementary T-invariant subspace W (so that $U=V\oplus W$ and $T(W)\subset W$). (*Hint*. Use eigenvector bases). We note that this is often taken as *definition* of semisimplicity.

10

MINIMUM POLYNOMIAL; JORDAN FORM

This chapter treats two questions left over from the last chapter.

I. How to decide whether a matrix can be diagonalized (i.e., is semisimple) without finding the eigenvalues. This is done with the help of the *minimum polynomial* of a matrix (or operator), the polynomial of lowest degree annihilated by the matrix (or operator).

II. What to do if a matrix cannot be diagonalized. The point is that one can still do something: One is lead to a "normal form," involving the eigenvalues, which is more complicated than the diagonal form, the *Jordan form*. This relies heavily on the analysis of *nilpotent* operators (operators some power of which is 0).

Some special topics, triangular form and projections, attach themselves naturally here.

1. THE CAYLEY–HAMILTON THEOREM; MINIMUM POLYNOMIAL

Let A be an $n \times n$ matrix (over $\mathsf{F}=\mathsf{R}$ or C). All $n \times n$ matrices form a vector space of dimension n^2 (Chapter 6, Section 1); thus the n^2+1 matrices $I, A, A^2, \ldots, A^{n^2}$ must be dependent: There must be constants $c_0, \ldots, c_{n^2}$, not all 0, such that $c_0 I + c_1 A + \cdots + c_{n^2} A^{n^2} = 0$. This means that A *satisfies* (is "matrix root" of) a (nontrivial) polynomial $c_0 + c_1 x + \cdots + c_{n^2} x^{n^2}$, of degree at most n^2 (note that c_0 becomes $c_0 I$ under the substitution $x \to A$). For example, the matrix $A=\begin{pmatrix} 0 & -1 \\ 1 & 0 \end{pmatrix}$ has $A^2=\begin{pmatrix} -1 & 0 \\ 0 & -1 \end{pmatrix}=-1\cdot I$, so that $I+A^2=0$; A is a root of x^2+1. Writing down such a polynomial for *any* matrix A turns out to be much easier than one might expect.

1.1. THEOREM. *(The Cayley–Hamilton Theorem). Every matrix A is root of its own characteristic polynomial $\chi_A(x)$*

$$\chi_A(A)=0. \tag{CH}$$

For the example above, $\det\begin{pmatrix} -x & -1 \\ 1 & -x \end{pmatrix}=x^2+1$, and $A^2+I=0$.

PROOF. We work with matrices whose *entries* are polynomials in x; $A-x\cdot I$ is such a matrix. Let B be its adjunct (Chapter 7, Section 3.e), not the adjunct of A, but of $A-xI$; the entries of B are also polynomials in x. We have $(A-xI)\cdot B=\det(A-xI)\cdot I=\chi_A(x)\cdot I$ by the basic property of the adjunct. We rewrite B in the form $B_0+xB_2+x^2B_2+\cdots+x^kB_k$ with matrices $B_0, B_1, \ldots$, whose entries are scalars. Putting

$$\chi_A(x)=a_0+a_1x+\cdots+a_nx^n,$$

we write the relation $(A-xI)\cdot B=\chi_A(x)I$ in the form

$$(A-xI)\cdot(B_0+xB_1+\cdots+x^kB_k)=a_0I+a_1xI+,\cdots+a_nx^nI. \tag{*}$$

Now we note that in multiplying out the left side to get the right side, we only use the simple laws of Chapter 6, Section 1; in particular, we use the "associative" law, $A\cdot(x^rB_r)=x^rA\cdot B_r$ and $I\cdot x^rB_r=x^rI\cdot B_r=x^rB_r$. It should be clear then that the equation (*) will remain true if for x we substitute any quantity that obeys these laws. This happens to hold in particular for $x=A$ (since $A\cdot(A^rB_r)=A^{r+1}B_r=A^r\cdot AB_r$ and $I\cdot A^rB_r=A^rB_r$). Thus, we may put $x=A$ into (*). But then the left side is 0 and the right side is precisely $\chi_A(A)$. ■

A consequence of (CH) is that we never have to go to degree n^2 to find a polynomial with root A; n is certainly enough. Can one go lower? Sometimes yes; for example, for $A=I$, the polynomial $x-1$ has I as root, whereas $\chi_I(x)=(1-x)^n$. One defines the **minimum polynomial** $m_A(x)$ of A as the (monic) polynomial of lowest possible (positive) degree that has A as root–monic means that the coefficient of the highest power of x is 1. This is well defined: If one had two such polynomials, then the difference, if not identically 0, would be a polynomial of still lower degree, with A still as root.

It follows from the usual long division process that any polynomial $p(x)$ that has A as root is *divisible* by $m_A(x)$: We can write

$$p(x)=q(x)\cdot m_A(x)+r(x)$$

with (partial) quotient $q(x)$ and remainder $r(x)$, where $r(x)$, if not 0, has lower degree than $m_A(x)$. Since $m_A(A)=0$, we get $0=p(A)=r(A)$, but $r(A)$ can be 0 only if $r(x)$ is the polynomial 0, since $m_A(x)$ is *minimal*. In particular, $m_A(x)$ must divide $\chi_A(x)$ by (CH). How does one find $m_A(x)$? (One way would be to find all factors of χ_A and test whether they have A as root.) Somewhat surprising is the fact one can find it by using only long divisions (but *many*, the process is long) and that one does not need find the eigenvalues of A first. We show this in Proposition 1.2 below.

Let $b(x)$ be the (monic) greatest common divisor (gcd) of the entries of the adjunct B of $A-xI$. Recall that the gcd of two polynomials $f(x)$ and $g(x)$ is found by the "Euclidean algorithm": Divide f by g, which gets $f(x)=q(x)\cdot g(x)+r_1(x)$ with degree of $r_1(x)$ lower than that of $g(x)$. Then divide $g(x)$ by $r_1(x)$ and get $g(x)=q_1(x)\cdot r_1(x)+r_2(x)$; then $r_1(x)=q_2(x)$ $\times r_2(x)+r_3(x)$, and so on. The first $r_i(x)$ that divides $r_{i-1}(x)$ is gcd of f and g. For the gcd of more than two polynomials $f_1, f_2, f_3, \ldots$, one iterates: First, the gcd g_1 of f_1 and f_2, then the gcd g_2 of g_1 and f_3, etc.

1.2. PROPOSITION. *$b(x)$ divides $\chi_A(x)$, and $m_A(x)=(-1)^n\cdot\chi_A(x)/b(x)$.*

PROOF. We can write $B=b(x)C$, where the gcd of the entries of C is 1 (no common factor of positive degree). From $(A-xI)\cdot B=\chi_A(x)I$, we get $b(x)(A-xI)\cdot C=\chi_A(x)I$. This shows that $b(x)$ divides $\chi_A(x)$ (the left side shows that all entries have $b(x)$ as factor). Putting $\chi_A(x)/b(x)=t(x)$ we get $(A-xI)\cdot C=t(x)\cdot I$. We can substitute A for x, by reasoning as in the proof of (CH). Therefore, $t(A)=0$, and so $m_A(x)$ must divide $t(x)$.

On the other hand, since $m_A(A)=0$, there is a factorization $m_A(xI)$ $=Q(x)\cdot(xI-A)$(analogous to $p(x)=q(x)\cdot(x-a)$) if $p(a)=0$, with suitable matrix $Q(x)$. Multiplying by C and using $(A-xI)\cdot C=t(x)I$, we get $m_A(x)C=-t(x)\cdot Q(x)$. Since $m_A(x)$ is the gcd of the entries of the matrix on the left, $t(x)$ must divide it. Thus, $m_A(x)$ and $t(x)$ divide each other and must then be equal up to a constant factor (which is $(-1)^n$, from $\chi_A(x)$ $=(-1)^n\cdot x^n+\cdots$). ■

As noted, this way of finding $m_A(x)$ is straightforward and involves only long division, but it is quite laborious. (There is no short way.) For completeness we give a slightly different scheme to find $m_A(x)$: For matrices with polynomial entries we consider the following row (and column) operations:

1. Interchange of two rows (or columns);
2. Multiplication of a row (or column) by a nonzero scalar (*not* by any

nonzero polynomial, since we want to be able to reverse the operation; $1/(p(x))$ is not a polynomial, unless the degree of $p(x)$ is 0); and

3. Adding a multiple of a row (column) to another row (column), where the factor can be any polynomial.

Applying these to $A-xI$ repeatedly, one can change $A-xI$ (quite radically) to the form $\operatorname{diag}(f_1(x),\ldots,f_n(x))$, such that each $f_i(x)$ divides the next term $f_{i+1}(x)$ (and where usually the first few $f_i(x)$ are simply equal to 1) (so-called "elementary divisor form"). Then

$$\chi_A(x) = \pm f_1(x)\cdot f_2(x)\cdots f_n(x)$$

and $m_A(x)=f_n(x)$. The proof of the last two relations is an exercise in determinants: One shows that although the matrix changes under the operations, the characteristic polynomial and the gcd of the adjunct do not change; and for a diagonal matrix as described, the gcd of the adjunct is $\Pi_1^{n-1} f_i(x)$. Here is the idea in getting to the elementary divisor form: One uses Operation 3 and division, starting from a nonzero entry of lowest degree in the matrix, to reduce the other entries in that row and column to even lower degree (or 0); one does this until one gets an entry that divides *all others*. Then one moves this entry to the (1, 1)-place, and produces 0 elsewhere in first row and column; and one operates similarly on the remaining $(n-1)\times(n-1)$ matrix.

Example

$$A=\begin{pmatrix} 1 & 1 \\ 1 & 1 \end{pmatrix}.$$

$$A-xI=\begin{pmatrix} 1-x & 1 \\ 1 & 1-x \end{pmatrix}\to\begin{pmatrix} 0 & 1 \\ 2x-x^2 & 1-x \end{pmatrix}\to\begin{pmatrix} 1 & 0 \\ 1-x & 2x-x^2 \end{pmatrix}$$

$$\to\begin{pmatrix} 1 & 0 \\ 0 & x(x-2) \end{pmatrix}.$$

Our first step is to multiply the second column by $1-x$ and to subtract if from the first.) Thus, $m_A(x)=x\cdot(x-2)$. This divides $\chi_A(x)$; in fact, it is equal to it (this happens quite often).

From the above we note that it is clear that $m_A(x)$ will have real coefficients if A is a real matrix. Could we have predicted that?

PROBLEMS

1. Verify (CH) for the matrices of Chapter 9, Section 3.

2. For $M=\begin{pmatrix}1&0&0\\1&1&0\\0&0&1\end{pmatrix}$ find the minimum polynomial m_M by finding the gcd for the adjunct of $M-xI$, as described above.

3. For the matrix $A=\begin{pmatrix}1&0&-1\\2&1&0\\1&-1&1\end{pmatrix}$, compute the adjunct of B of $A-xI$ and verify the relation $(A-xI)\cdot B=\chi_A(x)\cdot I$.

4. (Generalization of Chapter 6, Section 3, Problem 4.) Suppose $\det A\neq 0$. Show, using (CH), that A is invertible and that A^{-1} is given by a certain polynomial in A.

5. For $A=\begin{pmatrix}7&12&-12\\-2&-3&4\\2&4&-3\end{pmatrix}$, find χ_A and m_A by the second method described in the text.

6. Let $p(x)=x^n+a_{n-1}x^{n-1}+a_{n-2}x^{n-2}+\cdots+a_1x+a_0$ be a polynomial. Let $u_1,u_2,\ldots,u_n$ be a basis for a vector space U. Define a T: $U\to U$ by $Tu_1=u_2$, $Tu_2=u_3,\ldots,Tu_{n-1}=u_n$, and $Tu_n=-a_{n-1}u_n-a_{n-2}u_{n-1}-\cdots-a_1u_2-a_0u_1$.
a. Show that $p(T)=0$. (*Hint*. Substitute $u_2=Tu_1$, etc., into the equation for Tu_n.)
b. Write out the matrix A for T (this is called the *companion matrix* of $p(x)$).
c. Show that T cannot be root of a polynomial of degree less than n; conclude that (up to sign) $p(x)$ is the characteristic *and* the minimum polynomial of T and A.

2. SEMISIMPLICITY

Here is now our answer to Question I above.

2.1. THEOREM. *The matrix A is diagonalizable precisely if its minimum polynomial m_A has no repeated roots.*

Remark. We mentioned earlier that a polynomial $p(x)$ has no repeated roots exactly if the gcd of $p(x)$ and its derivative $p'(x)$ is 1 (i.e., of degree 0). Thus the problem of deciding whether A is diagonalizable is reduced completely to (a number of) long divisions; first to find m_A and then to find the gcd of m_A and m'_A. This does *not* involve finding the roots of $\chi_A(x)$. We recall, that in terms of eigenvalues, semisimplicity amounts to $r_i=s_i$ for all i (the geometric multiplicities equal the algebraic multiplicities); whereas, in general only $r_i\leqslant s_i$.

PROOF OF THEOREM 2.1. First, we note that similar matrices have not only the same characteristic polynomial, but also the same minimum polynomial: If $A' = P^{-1} \cdot A \cdot P$, then for any polynomial $p(x)$ we have $p(A') = P^{-1} \cdot p(A) \cdot P$; thus $p(A) = 0$ is equivalent to $p(A') = 0$.

a. This is the easy part. Suppose A is similar to a diagonal matrix D ($\operatorname{diag}(\lambda_1, \ldots, \lambda_2, \ldots, \lambda_t, \ldots)$); there are t λ_i's, each repeated s_i times. We claim that $m_D(x) = (x - \lambda_1)(x - \lambda_2) \cdots (x - \lambda_t)$; that is, each $x - \lambda_i$ only to first power (instead of s_ith power, as in $\chi_A(x)$). Consider

$$(D - \lambda_1 I) \cdot (D - \lambda_2 I) \cdots (D - \lambda_t I).$$

The factor $D - \lambda_i I$ (still diagonal) has 0's where D has λ_i; clearly the product is 0. And just as clearly, if any one factor is left out, then the product is not 0. This shows that the product of the $x - \lambda_i$ is the minimum polynomial, and it has no repeated roots.

b. We first note that if A has eigenvalue λ (and eigenvector X), then for any polynomial $p(x)$ the matrix $p(A)$ has $p(\lambda)$ as eigenvalue, with the same eigenvector X. Namely, if $AX = \lambda X$, then

$$A^2 X = A \cdot A \cdot X = A \cdot \lambda X = \lambda A \cdot X = \lambda \cdot \lambda X = \lambda^2 X;$$

similarly, $A^3 X = \lambda^3 X$, and so on. By adding terms we get $p(A) \cdot X = p(\lambda) X$.

In particular, if $p(x)$ has A as root (so that $p(A) = 0$), then $p(\lambda) = 0$ for any eigenvalue λ of A; and so $p(x)$ is divisible by $x - \lambda$. Thus, if $\lambda_1, \ldots, \lambda_t$ are the *different* eigenvalues of A, $m_A(x)$ must at any rate be divisible by $(x - \lambda_1) \cdots (x - \lambda_t)$. In general, each $(x - \lambda_i)$ might appear to a higher power. Our hypothesis now implies: $m_A(x) = (x - \lambda_1) \cdots (x - \lambda_t)$; therefore, $(A - \lambda_1 I) \cdots (A - \lambda_t I) = 0$. We now apply (repeatedly) the inequality of Chapter 9, Section 5.1(b): The nullity of a product $A \cdot B$ of matrices (or composition $S \circ T$ of linear transformations) is at most equal to the sum of the nullities of A and B (or S and T).

Now the nullity of $A - \lambda_i I$ is the geometric multiplicity r_i. The nullity of 0 being n, we see from $0 = (A - \lambda_1 I) \cdots (A - \lambda_t I)$ that $n \leqslant r_1 + r_2 + \cdots + r_t$. We also know that $r_i \leqslant s_i$ and $\Sigma s_i = n$ (Theorem 4.2); it follows that each r_i must equal s_i and so $\Sigma r_i = n$. From Theorem 4.1 we see that the eigenspaces of the λ_i span the whole space ($\mathbf{C}^n$ at present) (and in fact have $\mathbf{C}^n$ as direct sum). Proposition 3.1 now tells us that A is diagonalizable.

PROBLEMS

1. Determine whether $x^3 - 4x + 2$ and $x^4 - 3x + x^2 + 3x - 2$ have repeated roots, without finding the roots. (Fractional coefficients will appear.)

2. Is the matrix A of the example in Section 1 diagonalizable?

3. Show, using Theorem 2.1, that if the eigenvalues of A are pairwise different (i.e., all algebraic multiplicities equal 1), then A is diagonalizable. Show with an example that the converse is not true.

4. Suppose A is diagonalizable. Are the powers A^k diagonalizable? Conversely, suppose some A^k is diagonalizable. Does it follow that A is diagonalizable?

5. Let A be an idempotent matrix ($A^2=A$). Can A be diagonalized? If yes, what are the possibilities for the diagonal form?

3. THE JORDAN FORM

We come to Question II above and introduce the **Jordan form** or "Jordan normal form" of a matrix. If a matrix (or operator) is diagonalizable, the Jordan form is simply the diagonal matrix to which the matris is similar (this is well defined up to the *position* of the various eigenvalues on the diagonal). The Jordan form in general will tell us "how far one can go in diagonalizing a matrix if it cannot be diagonalized." (From now on $\mathsf{F}=\mathsf{C}$, so there is no "reality" trouble.) We first describe what a matrix in Jordan form is, then interpret this in operator language, and prove that every matrix is similar to one in Jordan form. A **basic Jordan block** is a matrix of type

$$\begin{pmatrix} \lambda & & & 0 \\ 1 & \ddots & & \\ & \ddots & \ddots & \\ 0 & & 1 & \lambda \end{pmatrix},$$

that is, a scalar λ, "the" eigenvalue, is repeated on the diagonal, with ones just below the diagonal and zeros everywhere else. Writing H (or H_n) for the matrix

$$\begin{pmatrix} 0 & & & \\ 1 & \ddots & & \\ & \ddots & \ddots & \\ & & 1 & 0 \end{pmatrix},$$

with $n-1$ ones just below the diagonal and zeros elsewhere, a basic Jordan block can be written as $\lambda I_n + H_n$. A **big Jordan block** is a matrix $\operatorname{diag}(K_1,\ldots,K_r)$, $r \geqslant 1$, where each K_i is a basic Jordan block and all of the K_i have the *same* eigenvalue λ. Finally, a matrix in **Jordan form** is one of type $\operatorname{diag}(M_1,\ldots,M_t)$, $t \geqslant 1$, where each M_i is a big Jordan block, to eigenvalue λ_i, the λ_i pairwise different. (Such a matrix is lower triangular in particular.)

Example

$$\text{basic: } \begin{pmatrix} 2 & & \\ 1 & 2 & \\ & 1 & 2 \end{pmatrix};$$

$$\text{big: } \left(\begin{array}{cc|ccc} 2 & & & & \\ 1 & 2 & & & \\ \hline & & 2 & & \\ & & 1 & 2 & \\ & & & 1 & 2 \end{array}\right);$$

$$\text{Jordan: } \left(\begin{array}{cc|ccc|c|cc} 2 & & & & & & & \\ 1 & 2 & & & & & & \\ \hline & & 2 & & & & & \\ & & 1 & 2 & & & & \\ & & & 1 & 2 & & & \\ \hline & & & & & -3 & & \\ \hline & & & & & & -3 & \\ & & & & & & 1 & -3 \end{array}\right).$$

(All blanks are 0.) Note that a basic block could be 1×1: (λ) and that $\begin{pmatrix} \lambda & & 0 \\ & \ddots & \\ 0 & & \lambda \end{pmatrix}$ is a big block, made up of 1×1 blocks. Now we state the main result.

3.1. THEOREM. *Any (square) matrix A is similar $(P^{-1}AP)$ to a Jordan matrix, and this matrix is unique except for the arrangement of the blocks along the diagonal.*

The elements on the diagonal are, of course, the eigenvalues of A, and the size of each big block equals the algebraic multiplicity of the eigenvalue.

We describe this now in operator language. Recall that an operator S:

$V \to V$ is *nilpotent* if some power S^k, $k \geqslant 1$, is 0; it is called nilpotent of *height* k, if S^{k-1} is not yet 0 . Note that a nilpotent operator has only 0 as an eigenvalue: $Sv = \lambda v$ implies $0 = S^k v = \lambda^k v$; thus $\lambda = 0$; and the characteristic polynomial $\chi_S(x)$ is $\pm x^n$ with $n = \dim V$, since it has only 0 as (repeated) root.

If $R: V \to V$ is any operator, the V is called R-**cyclic** (or just *cyclic*) if there is some ("cyclic") vector u_0, such that the vectors $u_0, Ru_0, R^2u_0, \ldots$ span V; one also calls R cyclic. Here is an important special case: With R: $V \to V$, suppose v is a vector such that for some $k \geqslant 1$, we have $R^k v = 0$ (but $R^{k-1}v \neq 0$); then the subspace $W = ((v, Rv, R^2v, \ldots, R^{k-1}v))$ of V is R-invariant, $R|W$ is nilpotent on W, and W is cyclic with v as cyclic vector. That W is R-invariant should be clear (each of the spanning vectors $R^i v$ has R-image $R^{i+1}v$ or 0, for $i = k-1$). $R|W$ is nilpotent since applying it at most k times to any of the spanning vectors we get 0. That W is cyclic is evident.

More is true: The vectors $v, Rv, \ldots, R^{k-1}v$ are automatically *independent,* a basis for W. The argument is an important trick: Suppose

$$a_1 v + a_2 Rv + \cdots + a_k R^{k-1} v = 0.$$

Apply R^{k-1} to this equation; the result will be $a_1 R^{k-1} v = 0$ (since $R^{k-1}(Rv) = R^k v = 0$, etc.) But $R^{k-1}v \neq 0$; thus $a_1 = 0$. Now apply R^{k-2} to the original relation, and get $a_2 R^{k-1} v = 0$ and so $a_2 = 0$, and so on.

Finally, the matrix of the nilpotent operator $R|W$ on the vector space W with respect to the basis $v, Rv, \ldots, R^{k-1}v$ is precisely the H_k above:

$$\begin{pmatrix} 0 & & & \\ 1 & \ddots & & \\ & \ddots & \ddots & \\ & & 1 & 0 \end{pmatrix}.$$

The expansion of $R(R^i v)$, with respect to this basis, is simply

$$R^{i+1}v \; (= 0 \cdot v + 0 \cdot Rv + \cdots + 1 \cdot R^{i+1}v + 0 \cdot R^{i+2}v + \cdots),$$

or just 0 in case $i = k-1$.

The Jordan form of a matrix has the following interpretation in operator language for an operator T on a vector space U: Each basic Jordan block corresponds to a T-invariant subspace W, such that T (or better: $T|W$), as operator on W, is of the form $\lambda\mathbf{1} + N$ with N cyclic nilpotent (represented by $\lambda I + H$ with respect to a basis of the type described). The arrangements of the blocks along the diagonal corresponds to the direct sum of the corresponding subspaces, and to prove that any matrix can be brought into Jordan form amounts to showing that for any $T: U \to U$ the vector space U is direct sum of such subspaces.

We shall do this in two parts: First, we show that U can be written as direct sum $U^{\lambda_1} \oplus \cdots \oplus U^{\lambda_t}$, where the U^{λ_i} are certain T-invariant subspaces, the "associated" spaces, one for each eigenvalue λ_i (each U^{λ_i} contains the eigenspace U_{λ_i}, but is usually larger), such that T on U^{λ_i} is of the form $\lambda_i \mathbf{1}$ + nilpotent; second, each U^{λ_i} will be split into a direct sum of (T-invariant) subspaces W, such that T on each W is $\lambda_i \mathbf{1}$ + *cyclic* nilpotent (this second part is a general theorem about nilpotent operators).

PROBLEMS

1. Let $\dim U = n$. Show: If the operator T on U is nilpotent, then $T^n = 0$. (Thus for nilpotency one never has to go beyond n with the exponent; if $T^{n+k} = 0, k > 0$, then already $T^n = 0$. Often some lower power of T is already 0.)

2. Suppose T is cyclic and nilpotent on U and $\dim U = n$. Show that T^k is *not* 0 for $k < n$.

4. PRIMARY DECOMPOSITION

We start with the first part: Let λ be a scalar; a nonzero vector u in U is called *T-associated to* λ, if $(T - \lambda\mathbf{1})^k u = 0$ for some $k \geq 1$; *all* these together with 0 form the *associated* (strictly, *T-associated to* λ) *space* U^λ, a subspace of U. (The terms *generalized eigenvector* and *generalized eigenspace*, to λ as eigenvalue, are also used for these objects.) U^λ contains the λ-eigenvectors; they have $(T - \lambda\mathbf{1})^k u = 0$ for $k = 1$ already. We never have to take k greater than n; namely, if $(T - \lambda\mathbf{1})^k u = 0$ for some $k > n$, then already $(T - \lambda\mathbf{1})^n u = 0$, where $n = \dim U$; otherwise we would get too many independent vectors $u, (T - \lambda\mathbf{1})u, \ldots,$ see Section 3. U^λ is T-invariant: $(T - \lambda\mathbf{1})^k u = 0$ implies $(T - \lambda\mathbf{1})^k (Tu) = T(T - \lambda\mathbf{1})^k u = 0$. (Justify the first "=".) Thus $(T - \lambda\mathbf{1})|U^\lambda$ is a nilpotent operator on U^λ; or, in other words T, as operator on U^λ, is of the form $\lambda\mathbf{1} + N_\lambda$, where N_λ is nilpotent. Being nilpotent, $T - \lambda\mathbf{1}$ has only 0 as eigenvalue, and so T has *only* λ as eigenvalue on U^λ. In fact, U^λ is the largest subspace of U on which $T - \lambda\mathbf{1}$ is nilpotent.

We now prove the "primary decomposition."

4.1. THEOREM. *U is direct sum $U^{\lambda_1} \oplus \cdots \oplus U^{\lambda_t}$ of the associated spaces to the different eigenvalues $\lambda_1, \ldots, \lambda_t$ of T; and* $\dim U^{\lambda_i}$ *equals the algebraic multiplicity s_i; of λ_i.*

PROOF. We apply the concepts "stable kernel" and "stable image" developed in Chapter 8, Section 7, as well as the facts established there, but instead of T we use the operator $T-\lambda_1\mathbf{1}$, where λ_1 is an eigenvalue of T. The space U^{λ_1} is nothing but the stable kernel of $T-\lambda_1\mathbf{1}$. Therefore, U splits as $U^{\lambda_1}\oplus W$, where W is the stable image of $T-\lambda_1\mathbf{1}$. W is invariant under $T-\lambda_1\mathbf{1}$, and therefore (why?) also under T; also $\dim W<\dim U$, since $\dim U^{\lambda_1}\geqslant \dim U_{\lambda_1}>0$. Thus we can apply induction (since Theorem 4.1 is certainly true for a space of dim 0 or 1): W can be written as $W^{\mu_1}\oplus\cdots\oplus W^{\mu_s}$, where μ_i are the eigenvalues of $T|W$, and the W^{μ_i} are the associated spaces; thus $U=U^{\lambda_1}\oplus W^{\mu_1}\oplus\cdots\oplus W^{\mu_s}$. We claim that the μ_j are just the other λ_i ($\neq\lambda_1$), and the W^{μ_j} the U^{λ_i}. That the μ_j are eigenvalues of T is trivial. They are different from λ_1, since U^{λ_1} by construction contains *all* eigenvectors to λ_1. There are no eigenvalues beyond $\lambda_1,\mu_1,\ldots,\mu_s$: For an eigenvector in a T-invariant direct sum (of two or more subspaces) each of its components, in the various subspaces, must be an eigenvector, and they all have the same eigenvalue; so that in the present situation an eigenvector must lie in only one subspace. (See Problem 6, Chapter 9, Section 4.) Thus we now have $U=U^{\lambda_1}\oplus\cdots\oplus U^{\lambda_t}$. Finally, we discuss the statement of Theorem 4.1 about the algebraic multiplicities: We could use, from the next section, the fact that the matrix for $T|U^{\lambda_i}$ is a big Jordan block with λ_i on the diagonal. Or more directly, choosing a basis in *each* U^{λ_i}, we get a matrix A_i for $T|U^{\lambda_i}$; the matrix A for T is just $\operatorname{diag}(A_1,\ldots,A_t)$. A little bit of work with determinants shows that $\chi_A(x)$ is the *product* of the $\chi_{A_i}(x)$. Furthermore, $\chi_{A_i}(x)$ is just $\pm(x-\lambda_i)^{n_i}$ with $n_i=\dim U^{\lambda_i}$ (since $T|U^{\lambda_i}$ has only λ_i as eigenvalue). Thus $\chi_A(x)=\pm\prod(x-\lambda_i)^{n_i}$, which shows that the n_i are our old s_i.

We recapitulate: Each eigenspace U_{λ_i} is subspace of the associated space U^{λ_i}; the dimensions are r_i (geometric multiplicity) and s_i (algebraic multiplicity). The sum $U_{\lambda_1}+\cdots+U_{\lambda_t}$ *is direct* but might be not all of U (namely, if at least one r_i is smaller than s_i), whereas *always* $U=U^{\lambda_1}\oplus\cdots\oplus U^{\lambda_s}$ (every u in U splits uniquely as $u_1+\cdots+u_t$, where u_i is annihilated by *some power* of $T-\lambda_i\mathbf{1}$). To find U^{λ_i}, one computes the increasing sequence of kernels of $T-\lambda_i,(T-\lambda_i)^2,(T-\lambda_i)^3,\ldots$,until it becomes constant.

PROBLEMS

1. For the matrix $A=\begin{pmatrix}1&1&2\\0&0&-1\\0&0&1\end{pmatrix}$ find the associated spaces U^{λ_i} in F^3.

5. NILPOTENT OPERATORS

We come to the second part for the Jordan form. By construction, $T-\lambda_i\mathbf{1}$, which we denote by N_i, is nilpotent on U^{λ_i}. Thus we are reduced to the study of a nilpotent operator N on a vector space V. We prove the following proposition.

5.1. PROPOSITION. *V is direct sum (not uniquely) of N-invariant subspaces $V^{(1)}, V^{(2)}, \ldots$ such that N restricted to any $V^{(i)}$ is cyclic nilpotent. Equivalently, a nilpotent matrix is similar to a matrix in Jordan form, that is, one consisting of matrices of type H_k along the diagonal.*

PROOF. The *height* of a vector v (relative to N) is the r with $N^{r-1}v \neq 0$, but $N^r v = 0$. We sharpen Proposition 5.1 by adding the following: Let v_i be a cyclic vector for the space $V^{(i)}$, then the v_i of largest height (of height equal to the height h of N) can be taken to be any set of vectors whose N^{h-1}-images form a basis for the space $\operatorname{im} N^{h-1}$. The proposition is trivially true for $h=1$.

We proceed by induction on $\dim V$; clearly, Proposition 5.1 holds if $\dim V = 1$ (in fact N is then 0). Choose $v_1, \ldots, v_t$ as indicated in the addition to Proposition 5.1, namely, so that their N^{h-1}-images form a basis for $\operatorname{im} N^{h-1}$. From rank–nullity we know that $\ker N^{h-1}$ and $((v_1, \ldots, v_t))$ have V as direct sum (see Chapter 8, Section 4). By hypothesis Proposition 1 holds for the space $W = \ker N^{h-1}$ and the operator $N|W$ (we have $\dim W < \dim V$, since $N^{h-1} \neq 0$); thus W is direct sum of cyclic subspaces $W^{(i)}$. Clearly $N|W$ is of height $h-1$. The vectors $Nv_1, \ldots, Nv_t$ are in W (by $N^h = 0$); their N^{h-2}-images are independent (they are just the N^{h-1}-images of the v_i). By the sharper form of Proposition 1 we may assume that $Nv_1, \ldots, Nv_t$ are cyclic vectors for some of the $W^{(i)}$, say for $W^{(1)}, \ldots, W^{(t)}$. We define cyclic subspaces $V^{(i)}$ of V as follows: For $1 \leqslant i \leqslant t$ we take $V^{(i)}$ as the span of v_i and $W^{(i)}$; for the remaining i we put $V^{(i)}$ equal to $W^{(i)}$. It should be clear that V is direct sum of the $V^{(i)}$ and that the cyclic vectors of precisely the first t of them are of height h. ■

We noted that this splitting of V into cyclic subspaces is not unique (in the proof this appears when we chose the vectors v_i—they are not unique). However, some uniqueness is left: The *number*, say d_r, of cyclic subspaces of any given dimension r is unique, that is, the same for any two splittings. In fact, the d_r can be computed from the nullities of N and its powers: The nullity of N equals the total number of cyclic subspaces in the splitting $V = V^{(1)} \oplus V^{(2)} \oplus \cdots$, since each $V^{(i)}$, being cyclic, contributes exactly one vector to the kernel (the last one in the sequence $v_i, Nv_i, N^2v_i, \ldots$). Next we consider the nullity of N^2: Here we get a contribution of 2 from each $V^{(i)}$ whose dimension is at least 2, but only one from the one-dimensional ones. For ν_{N^3} we get 3 from each $V^{(i)}$ of dimension at least three, 2 from those of dimension two, 1 from those of dimension one.

$$\nu_N = d_1 + d_2 + d_3 + \cdots$$

$$\nu_{N^2} = d_1 + 2d_2 + 2d_3 + 2d_4 + \cdots$$

$$\nu_{N^3} = d_1 + 2d_2 + 3d_3 + 3d_4 + \cdots$$

$$\vdots$$

These equations determine the d_i; see Problem 4.

PROBLEMS

1. Suppose that T is cyclic nilpotent on U; let $\dim U = n$. Show that for every k with $0 \leqslant k \leqslant n$, there is a T-invariant subspace of U.

2. With the data of Problem 1, show that none of these spaces for $k \neq 0, n$ has a T-invariant complementary subspace (it is impossible to have $U = V \oplus V'$ with $T(V) \subset V$ *and* $T(V') \subset V'$).

3. Suppose the nilpotent T on U has $\nu_T = 4$, $\nu_{T^2} = 8$, $\nu_{T^3} = 10$, $\nu_{T^4} = 11$, and $\nu_{T^5} = 11$. Find the dimensions of the cyclic subspace of U (as in Proposition 5.1) and the dimension of U.

4. Solve the equations at the end of the section for the d_r.

6. JORDAN FORM CONCLUDED

It should be clear that Theorem 4.1 and Proposition 5.1 (applied to $T - \lambda_i \mathbf{1}$ on U^{λ_i}) establish the main Theorem 3.1, the existence of the Jordan form.

We could, of course, restate this for operators: For any operator T: $U \to U$ there is a basis for U so that the representing matrix is in Jordan form (unique up to order of the big J-blocks). For a more abstract approach to this see Section 9 (6).

We still have to prove the uniqueness statement. The discussion at the end of the last section showed that for each eigenvalue the *number* of basic Jordan blocks of given size (and with that eigenvalue on the diagonal) is independent of how one brings the matrix into Jordan form (e.g., if the matrix is diagonalizable, then basic J-blocks of size 2 or more will never appear, no matter how hard one tries). But that is precisely the uniqueness statement.

All this is over $\mathbf{C}$. For reality questions see again Section 9.b.

It also follows that one can determine the Jordan form of a matrix by computing the nullities of the powers of $A-\lambda_i I$ for the various eigenvalues λ_i; they determine, as just indicated, the *number* of blocks $\lambda_i I+H$ of a given size.

It is somewhat more tricky, because of the nonuniqueness, to actually find the subspaces of U corresponding to these blocks: For any eigenvalue λ one has to find, sucessively, the kernels of $A-\lambda I, (A-\lambda I)^2, (A-\lambda I)^3, \ldots,$ adding basis vectors at each step, until there is no more increase. The vectors in the last step generate each a cyclic subspace of largest dimension. One then finds vectors of the last-but-one step that are independent of the $(A-\lambda I)$-images of the last group (some readjustment is necessary here); they will generate cyclic spaces of the next lower dimension, etc. For a matrix of any size this can be quite a chore.

Example

Find the Jordan form of $A=\begin{pmatrix} 1 & -1 & -1 \\ -3 & -4 & -3 \\ 4 & 7 & 6 \end{pmatrix}$. $\chi_A(x)$, after some fiddling, $=-x^3+3x^2-4$, equal to $-(x+1)(x-2)^2$ (*Hint*. -1 is a root; divide by $x+1$); the eigenvalues are $\lambda_1=-1$ and $\lambda_2=2$; and the algebraic multiplicities are $s_1=1$ and $s_2=2$. So there are two possibilities for the Jordan form: $\begin{pmatrix} -1 & & \\ & 2 & \\ & & 2 \end{pmatrix}$ or $\begin{pmatrix} -1 & & \\ & 2 & \\ & 1 & 2 \end{pmatrix}$. To decide which, we determine the nullity (or the rank) of $A-2I=\begin{pmatrix} -1 & -1 & -1 \\ -3 & -6 & -3 \\ 4 & 7 & 4 \end{pmatrix}$; the row-echelon scheme yields $\rightarrow \begin{pmatrix} 1 & 1 & 1 \\ 0 & 3 & 0 \\ 0 & 3 & 0 \end{pmatrix} \rightarrow \begin{pmatrix} 1 & 1 & 1 \\ 0 & 3 & 0 \\ 0 & 0 & 0 \end{pmatrix}$; $\rho=2$, and $\nu=1$. $\nu=1$ says: The total number of basic Jordan blocks to λ_2 is 1. Therefore, the second case above is the correct one. (The first, diagonal, matrix has *two* blocks for λ_2, each of size 1×1!; equivalently, $A-2I$ would have nullity 2.) Now we note the actual vectors: For λ_1 that is just the kernel of $A+I=\begin{pmatrix} 2 & -1 & -1 \\ -3 & -3 & -3 \\ 4 & 7 & 7 \end{pmatrix}$; the solution, which can be obtained by row-echelon form, etc., is X_1

$=[0,1,-1]$. For $\lambda_2, (A-2I)\cdot X=0$ has one solution $X_2=[-1,0,+1]$. But now we must consider the squared operator: $(A-2I)^2\cdot X=0$ has *another* solution in addition to X_2;—$(A-2I)^2=\begin{pmatrix} 0 & 0 & 0 \\ 9 & 18 & 9 \\ -9 & -18 & -9 \end{pmatrix}$; the solutions are $X_2=[-1,0,+1]$ *and* $X_3=[-2,1,0]$. Thus X_3 is cyclic vector of $A-2I$ for a two-dimensional subspace, and $(A-2I)\cdot X_3$ must be (a multiple of) X_2. In fact, $(A-2I)\cdot X_3=-X_2$. We put $-X_2=X_2'$. The matrix of the operator T_A on R^3 with respect to the basis $\{X_1, X_3, X_2'\}$ (in this order!) is, or should be, $\begin{pmatrix} -1 & & \\ & 2 & \\ & 1 & 2 \end{pmatrix}$; this says that $A\cdot X_1=-X_1$, $A\cdot X_3=2X_3+X_2'$, and $A\cdot X_2'=2X_2'$, and checks.

We add a comment which, at first sight, seems to say that all this work on the Jordan form is really quite unnecessary: "Most" matrices are diagonalizable; "not diagonalizable" is quite exceptional. As evidence for this we note that any matrix can be "made" diagonalizable by an arbitrarily small change of the entries; one can even make all the eigenvalues distinct. One can see this plainly in the Jordan form; just change the diagonal terms. However, not diagonalizable matrices come up often enough, even in applications, so that for solid understanding of the matter one simply needs the Jordan form.

PROBLEMS

1. Bring the following matrices into Jordan form; in each case determine the various cyclic subspaces and their generating vectors:

a.

$$\begin{pmatrix} 2 & 0 & 1 \\ 0 & 1 & 2 \\ 0 & 0 & 2 \end{pmatrix} \begin{pmatrix} 2 & 10 & 5 \\ -2 & -4 & -4 \\ 3 & 5 & 6 \end{pmatrix} \begin{pmatrix} 0 & 0 & -1 \\ 1 & 0 & -3 \\ 0 & 1 & -3 \end{pmatrix}.$$

b.

$$\begin{pmatrix} 2 & -1 & 1 & -1 \\ -1 & 2 & -2 & 1 \\ 0 & 1 & 1 & 1 \\ 0 & -1 & 1 & 0 \end{pmatrix}.$$

2. Let T be cyclic nilpotent on U; $\dim U=n$. What is the minimum polynomial of T?

3. Let A be given in Jordan form. What is the minimum polynomial of A? Precisely what information does one need from the Jordan form for this question?

4. P^3 is the vector space of polynomials, in t of degree $\leqslant 3$. $T: P^3 \to P^3$ is defined by $Tp(t) = Tp(t+1)$.
a. Find the matrix for T with respect to the basis $\{1, t, t^2, t^3\}$.
b. Find the eigenvalues and eigenspaces of A and T.
c. What must the Jordan form for T look like, and what is the minimum polynomial?

5. Suppose the operator T satisfies the relation $T^2 = T$ (idempotent). What are the possibilities for the minimum polynomial and the Jordan form of T? ($\dim U = n$.)

7. TRIANGULAR FORM

The "triangular form" is much simpler (in theory) than the Jordan form, but is still quite useful.

7.1. THEOREM. *Every (square) matrix A is similar (over $\mathbf{C}$) to an upper (or lower) triangular matrix (which is not unique, but the diagonal elements are, of course, the eigenvalues of A).*

This is contained in the Jordan form which is lower triangular; however, we give a direct proof. (We need $\mathbf{C}$ since eigenvalues could be complex. If A and all eigenvalues are real, one can stay with $\mathbf{R}$.) Let λ_1 be an eigenvalue and X_1, a λ_1-eigenvector. Extend to a basis $\{X_1, X_2, \ldots, X_n\}$ of $\mathbf{C}$ (the other X_i do not have to be eigenvectors of A!); put $P = (X_1, \ldots, X_n)$. Then the matrix $A' = P^{-1}AP$ has $[\lambda_1, 0, \ldots, 0]$ as first column, since X_1 is eigenvector with eigenvalue λ_1(i.e., $T_A(X_1) = \lambda_1 X_1$); thus we are getting closer to a triangular matrix.

We use induction (for 1×1 matrices the theorem is all right!): Let A'_{11} be the $(n-1)\times(n-1)$ matrix obtained from A' by removing the first row and column. By induction assumption we can find an invertible $(n-1)\times(n-1)$ Q' such that $(Q')^{-1} \cdot A'_{11} \cdot Q'$ is (upper) triangular. Put

$$Q = \operatorname{diag}(1, Q'), = \left(\begin{array}{c|c} 1 & 0 \cdots 0 \\ \hline 0 & \\ \vdots & Q' \\ 0 & \end{array}\right).$$

One verifies that $A'' = Q^{-1} \cdot A' \cdot Q$ is triangular; its first column is still

$[\lambda_1, 0, \ldots, 0]$, and the matrix A_{11}'' is precisely $(Q')^{-1} \cdot A_{11}' \cdot Q'$.

$$\left(\begin{array}{c|c} 1 & 0 \cdots 0 \\ \hline 0 & \\ \vdots & (Q')^{-1} \\ 0 & \end{array}\right) \cdot \left(\begin{array}{c|c} \lambda_1 & * \cdots * \\ \hline 0 & \\ \vdots & A_{11}' \\ 0 & \end{array}\right) \cdot \left(\begin{array}{c|c} 1 & 0 \cdots 0 \\ \hline 0 & \\ \vdots & Q' \\ 0 & \end{array}\right) = \left(\begin{array}{c|c} \lambda_1 & * \cdots * \\ \hline 0 & \\ \vdots & (Q')^{-1} A_{11}' Q' \\ 0 & \end{array}\right)$$

$$= \begin{pmatrix} \lambda_1 & & & \\ 0 & \lambda_2 & & * \\ \vdots & 0 & \ddots & \\ 0 & & & \lambda_n \end{pmatrix}.$$

What is behind this is the idea of **partitioned** or **block** matrices: Let

$$A = \left(\begin{array}{c|c} A' & A'' \\ \hline A''' & A^{iv} \end{array}\right) \quad \text{and} \quad B = \left(\begin{array}{c|c} B' & B'' \\ \hline B''' & B^{iv} \end{array}\right)$$

be partitioned, then

$$A \cdot B = \left(\begin{array}{c|c} A'B' + A''B''' & A'B'' + A''B^{iv} \\ \hline A'''B' + A^{iv}B''' & A'''B'' + A^{iv}B^{iv} \end{array}\right);$$

here A and B do not have to be square, but, of course, their shapes and those of the submatrices have to be such that all the products indicated make sense. Note that we wrote $\lambda_1, \ldots, \lambda_n$ for the eigenvalues, with possible repetitions. ■

Bringing a matrix into triangular form, although theoretically simple, is a bit of mess, because of all the choices and the induction.

The existence of the triangular form is used for the description of the eigenvalues of the powers of a matrix, including their multiplicities.

7.2. PROPOSITION. *Let $\lambda_1, \ldots, \lambda_2, \ldots, \lambda_t, \ldots$ be the eigenvalues of A (with λ_i repeated s_i times); then for any integer $r \geqslant 1$, the eigenvalues of A^r are $\lambda_1^r, \ldots, \lambda_2^r, \ldots, \lambda_t^r, \ldots$.*

Note that it can happen that $\lambda_i^r = \lambda_j^r$, even if $\lambda_i \neq \lambda_j$; for example, in the 2×2 case, if A has eigenvalues 1 and -1, then A^2 has eigenvalues 1 and 1.

PROOF. Transform A into (upper) triangular form A' $(=P^{-1}\cdot A\cdot P)$, then A^r transforms to A'' (since

$$(P^{-1}\cdot A\cdot P)^r = P^{-1}\cdot A\cdot P\cdot P^{-1}\cdot A\cdot P\cdots P^{-1}\cdot A\cdot P = P^{-1}A^rP);$$

but the diagonal elements of the (triangular) A'' are exactly the rth powers of those of A'.

Addendum. If A is invertible, then Proposition 7.2 remains true for ($r=0$ and) $r<0$; in particular, the eigenvalues of A^{-1} are just the inverses of those of A. (Recall that A is not invertible, that is, singular, precisely if it has 0 as eigenvalue.

We bring a geometric interpretation of the triangular form.

7.3. PROPOSITION. *Let $T: U \to U$ be an operator. The matrix A of T with respect to a given basis $\beta = \{u_1, \ldots, u_n\}$ is triangular exactly if for each $j, 1 \leqslant j \leqslant n$, the space $V_j = ((u_1, \ldots, u_j))$ is T-invariant.*

What this says is that in the expansion $Tu_j = \Sigma_1^n a_{ij}u_i$ for each j only the vectors $u_1, \ldots, u_j$ occur; the sum goes in effect only from 1 to j. (In particular, u_1 is an eigenvector.) The V_j form an increasing sequence $0 \subset V_1 \subset V_2 \subset \cdots \subset V_n = U$ with $\dim V_j = j$. Such a system of spaces is sometimes called a **flag** in U—suggested by $0 \subset V_1 \subset V_2$, a plane through a line. ■

A flag $\{V_j\}_1^n$ is T-**invariant** if for each j we have $T(V_j) \subset V_j$. The theorem on triangular form of matrices can now be restated in the following form.

7.4. THEOREM. *If $T: U \to U$ is an operator, then there exist T-invariant flags.*

(This is over $\mathbf{C}$; if U and all eigenvalues are real, it works over $\mathbf{R}$.) From such a flag one gets a basis as for Proposition 7.3 by taking u_1 nonzero in V_1, extending to a basis $\{u_1, u_2\}$ for V_2, extending this to a basis $\{u_1, u_2, u_3\}$ of V_3, etc.

We comment on how this looks in practice: Making A triangular involves finding the eigenvalues (in the triangular form they are the diagonal elements). If A can be made *diagonal* (i.e., A semisimple), one might as well diagonalize—it is no more work. In the opposite case, it is still a good idea to first find all eigenvectors (bases for the U_{λ_i}).

PROBLEMS

1. Let $p(x)$ be the characteristic polynomial of the matrix $A = \begin{pmatrix} 2 & -5 & 5 \\ 2 & -3 & 8 \\ 3 & -8 & 7 \end{pmatrix}$. Find

a polynomial with integral coefficients, whose roots are the squares of the roots of $p(x)$.

2. Show that the product of two lower triangular matrices is again lower triangular.

3. Show that every 2×2 matrix $\begin{pmatrix} a & b \\ c & d \end{pmatrix}$ with $a \neq 0$ can be written as product of a lower and an upper triangular matrix. Justify the restriction $a \neq 0$, and give an example of a matrix with $a = 0$ that cannot be so factored.

8. PROJECTIONS; THE EQUATION $E^2 = E$

Let U be a vector space, with two subspaces V and W that have U as *direct sum*; we write $U = V \oplus W$. The operator **projection**, say E, of U *onto* V *along* W is defined as follows: Take u in U; write it (uniquely) as $v + w$ with v in V and w in W; put $Eu = v$ "the part of u in V." In the case where W has dimension one, this is precisely the picture of Chapter 8, Section 1.i. We note that W is equal to the kernel of E; namely, $Eu = 0$ means that the splitting of u is $0 + w$. In other words, u must lie in W. Similarly, V is the image space: By definition, Eu is a vector in V; and *any* v in V can come up as Eu, simply taking $u = v + w$ with any w in W, for example, 0!; in particular, we see $Ev = v$ for any v in V (one says that v is fixed by E or is a fixed vector for E). Thus E determines V and W, and conversely; the only requirement is that V and W have U as direct sum (i.e., $V \cap W = 0$ and $V + W = U$). We also see that E has 0 and 1 as eigenvalues with eigenspaces W and V; since $U = V \oplus W$, E can be diagonalized and has diagonal form $\mathrm{diag}(1, \ldots, 1, 0, \ldots, 0)$.

Next we note that such a projection E has the property $E^2 (= E \circ E) = E$; it equals its own square (such an operator is called **idempotent**). From $u = v + w$ we have by definition $Eu = v$. To get $E^2 u = E(Eu)$, we have to split Eu according to $V \oplus W$ and take the V-part, but that splitting is, of course $Eu = v + 0$ (v is in V and 0 is in W). Consequently, $E(Eu) = v = Eu$. ■

Conversely, if an operator E satisfies $E^2 = E$, it is a projection (with $W = \ker E$ and $V = \mathrm{im}\, E$); $E^2 - E = 0$ tells us that the minimum polynomial is $x^2 - x$, that is, $x(x-1)$ or a factor of it. There are no repeated roots, so E is diagonalizable (Theorem 2.1); the eigenvalues are 0 and 1 (roots of $x(x-1)$); and so U is direct sum of the two eigenspaces U_0 and U_1. But U_0 by definition is $\ker E$. And U_1 consists of the u with $Eu = u$; any such u is clearly in $\mathrm{im}\, E$; and any vector in $\mathrm{im}\, E$ satisfies $Eu = u$ because of $E^2 = E$ (if $u = Eu'$, then $Eu = E^2 u' = Eu' = u$).

Here is another way to establish this: Any u can be written as $v + w$, putting $v = Eu$ and $w = (u - Eu)$. But then $Ev = E^2 u = Eu = v$ and Ew

$= Eu - E^2u = Eu - Eu = 0$; that is, v is in U_1 and w is in U_0; and so we see $U = U_1 + U_0$. Furthermore, take u in $U_1 \cap U_0$ then $u = Eu = 0$, We see that $U = U_1 \oplus U_0 = \operatorname{im} E \oplus \ker E$.

Remark. The rank ρ_E of a projection, (which equals $\dim V$), can be anything between 0 and $n = \dim U$; for $\rho_E = 0$ we have $E = 0$ and for $\rho_E = n$, $E = \mathbf{1}_U$. The in-between cases are, naturally, the interesting ones. Frequently one is only given $\operatorname{im} E$ or $\ker E$ and has to *choose* the other; note that there are many projections with the same $\operatorname{im} E$ or $\ker E$.

Example

Project R^2 onto $(([1,2]))$. We take $((E^2))$ as a complement and determine A from the equations $A \cdot E^2 = 0$ and $A \cdot [1,2] = [1,2]$. They give $a_{12} = a_{22} = 0$, $a_{11} \cdot 1 = 1$, and $a_{21} \cdot 1 = 2$. $A = \begin{pmatrix} 1 & 0 \\ 2 & 0 \end{pmatrix}$.

Final Remark. The operator $\mathbf{1} - E$ is the projection of U onto W along V.

PROBLEMS

1. In R^4 construct a projection E with $\operatorname{im} E = (([2,-1,3,3],[-1,1,-1,-1])) = V$, say. (You have to take a (basis for *a*) subspace W, complementary to V, preferably some standard vectors E^i.)

2. The matrix $A = \begin{pmatrix} -1 & -2 & -2 \\ 1 & 2 & 1 \\ 0 & 0 & 1 \end{pmatrix}$ defines a projection in F^3. Verify that this is so. Find (bases for) $\ker T_A$ and $\operatorname{im} T_A$; verify the equations $AX = X$ for X in $\operatorname{im} T_A$.

3. Suppose that E and F are projections in U such that $\ker E \subset \ker F$ *and* $\operatorname{im} F \subset \operatorname{im} E$. Prove $F \circ E = E \circ F = F$. (Check separately the vectors in $\ker E$ and $\operatorname{im} E$, as well as those in $\ker F$ and $\operatorname{im} F$.)

4. Let E be a projection U. Let β be a basis consisting of a basis for $\operatorname{im} E$ together with one for $\ker E$. What is the matrix of E relative to β?

9. COMMENTS ON THE JORDAN FORM

a. At the end of Chapter 8, Section 4, we considered linear systems $\dot{X} = AX$ of differential equations. We now take up the case left open there, namely that of a nondiagonalizable A. We know that we may replace A by any similar matrix. Thus, by Theorem 3.1 we may take A in Jordan form. Clearly, the system then splits into little systems, one for each basic J-block. If the block is 1×1, we are back to $\dot{y} = \lambda \cdot y$, with λ an eigenvalue of A, and so $y=e^{\lambda t}$. Suppose we have a 2×2 block $\begin{pmatrix} \lambda & \\ 1 & \lambda \end{pmatrix}$. Then the two equations are $\dot{y}_1=\lambda y_1$ and $\dot{y}_2=y_1+\lambda y_2$. This is solved, by "inspection," by taking $y_1=e^{\lambda t}$ and $y_2=t\cdot e^{\lambda t}$: $\dot{y}_1=\lambda\cdot y_1, \dot{y}_2=dt/dt\cdot e^{\lambda t}+t\cdot de^{\lambda t}/dt=e^{\lambda t}+t\cdot\lambda e^{\lambda t}=y_1+\lambda y_2$; thus, $[\dot{y}_1,\dot{y}_2]=\begin{pmatrix} \lambda & \\ 1 & \lambda \end{pmatrix}[y_1,y_2]$ A second, simpler, solution is $y_1=0$ and $y_2=e^{\lambda t}$.

The situation is similar for an $r\times r$ block λI_r+H_r. Here is a solution: $[y_1,y_2,\ldots,y_r]=[e^{\lambda t},t\cdot e^{\lambda t},t^2/2\cdot e^{\lambda t},\ldots,(t^{r-1}/(r-1)!)e^{\lambda t}]$. In addition to this most elaborate solution there are $r-1$ others:

$$\left[0,e^{\lambda t},te^{\lambda t},\frac{t^2}{2}e^{\lambda t},\ldots,\frac{t^{r-2}}{(r-2)!}e^{\lambda t}\right],\left[0,0,e^{\lambda t},\ldots,\frac{t^{r-3}}{(r-3)!}e^{\lambda t}\right],\ldots,[0,\ldots,0,e^{\lambda t}].$$

If λ happens to be 0, then $e^{\lambda t}$ is just 1.

Altogether we have found n different solution vectors $Y(t)$ (r for each $r\times r$ block); one can show (we omit this) that any solution is linear combination, with constant coefficients, of these. The important point is that the exponential function $e^{\lambda t}$, with the eigenvalues of A as factor of t, and the related functions $t^k\cdot e^{\lambda t}$ appear. (For complex λ the $e^{\lambda t}$ brings in sin and cos, that is, oscillations.)

b. An abstract view of the J-form. The J-form of A has a *diagonal* part, say S', and an *off-diagonal* part N'; $A'=S'+N'$. Going back to $A=P\cdot A'\cdot P^{-1}$, we get $A=S+N$, where now $S(=P\cdot S'\cdot P^{-1})$ is no longer diagonal, but still diagonalizable, and N, just as N', is nilpotent. In addition, and that is crucial, S and N commute not only with each other (and therefore with A), but they have the more complicated property of commuting with every matrix that commutes with A. One can see that in several ways: Back in the J-form, $A'=S'+N'$, one can work out by direct computation that any matrix that commutes with A' has to split into blocks, arranged along the diagonal, that correspond in size to the big J-blocks of A' (i.e., to the *different* eigenvalues). Then the commuting becomes clear, because each big block of S' is just λI. Another approach consists in showing that in fact S and N are given by certain polynomials in A. In fact, more is true: Each *projection* E_{λ_i} of U onto the associated space U^{λ_i}, along the direct sum of the other U^{λ_j}, is a polynomial in A (and $S=\lambda_1E_{\lambda_1}+\lambda_2E_{\lambda_2}+\cdots$): This reduces to showing in Proposition 7.3 of Chapter 8, that the projection of U onto either $\ker^s T$ or $\operatorname{im}^s T$, along the other space, is a polynomial in T.

We prove the latter: The operator T^k, for some large k, is 0 on $\ker^s T$ and invertible on $\operatorname{im}^s T$; that is not too far from being a projection. Writing, temporarily, R for $T^k|\operatorname{im}^s T$, we claim that there is a polynomial $p(R)$ giving R^{-1} (compare Problem 3 of Chapter 10, Section 1): First, there is formula (CH) of Section 1: $a_0 I + a_1 R + \cdots + a_r R^r = 0$ ($r = \dim \operatorname{im}^s T$); here $a_0 = \det R \neq 0$. We can divide by a_0 and rewrite as $I - R\cdot(b_1 I + \cdots + b_r R^{r-1}) = 0$; the polynomial $p(R)$ is now apparent. If we form $T^k \cdot p(T^k)$, we get an operator on U that is 0 on $\ker^s T$ and the *identity* on $\operatorname{im}^s T$. That is just the projection we wanted. Now we state the abstract version of the Jordan form. *Any operator $T: U \to U$ can be written as $S + N$, with S semisimple, N nilpotent, and S and N commuting with each other* (*or with* T); *this splitting is unique*.

We still have to show the uniqueness; suppose also $T = S_1 + N_1$ similarly. S_1 and N_1 commute with each other, and so with T, and so with S and N (since those are polynomials in T). It follows that $N - N_1$ is nilpotent (its powers are sums of terms $N^r \cdot N_1^{k-r}$; if N and N_1 did not commute, we would get only $N \cdot N_1 \cdot N \cdot N_1 \cdots N$, etc.!). Furthermore, $S_1 - S$ is semisimple; namely, the sum of two semisimple *commuting* operators is again semisimple (Corollary 4.6 of Chapter 9). Now $S + N = S_1 + N_1$ gives $S_1 - S = N - N_1$. However, if an operator is both semisimple and nilpotent, it is 0 (all eigenvalues are 0!), and so $S_1 = S$ and $N = N_1$.

c. We look at one more application of the J-form: Let us call a (square) matrix A *power-stable*, if the powers $A, A^2, A^3, \ldots$ are bounded; that is, if all the entries of all the A^k are, in absolute value, less than some fixed constant. It is not easy, in general, to compute *all* powers A^k. But one can decide the question of power-stability from the J-form. First, we note that because of $P^{-1}A^2P = P^{-1}\cdot A \cdot P \cdot P^{-1} \cdot A \cdot P = (P^{-1}AP)^2$, etc., A is power-stable if and only if $P^{-1}\cdot A \cdot P$ is. Thus we may take A in J-form. And here the result is simple.

9.1. PROPOSITION. *A is power-stable exactly if*

a. *every eigenvalue λ_i is $\leqslant 1$ in absolute value, and*

b. *any eigenvalue λ_i with $|\lambda_i| = 1$ is semisimple, that is, the big J-block to λ_i is diagonal* ($= \lambda_i I$).

PROOF. If any $|\lambda_i|$ is > 1, then the powers A^k (A in J-form) have $\lambda_i, \lambda_i^2, \lambda_i^3, \ldots$ as entries, and this sequence goes to ∞; this is (a). For (b) we consider only the 2×2 case (the higher ones are similar): The powers of $\begin{pmatrix} \lambda & 0 \\ 1 & \lambda \end{pmatrix}$ are

$$\begin{pmatrix} \lambda^2 & \\ 2\lambda & \lambda^2 \end{pmatrix}, \begin{pmatrix} \lambda^3 & 0 \\ 3\lambda^2 & \lambda^3 \end{pmatrix}, \begin{pmatrix} \lambda^4 & 0 \\ 4\lambda^3 & \lambda^4 \end{pmatrix}, \ldots;$$

clearly, if $|\lambda| \geqslant 1$, the $(2,1)$-entry goes to ∞; and if $|\lambda|<1$, then it stays bounded, in fact, goes to 0 (and the same holds for the other entries). The "applied" interest of this is that the powers of A correspond to repeated applications of some physical process. Nonstability of A means that the process gets wilder (and in practice blows up somewhere along the line).

d. Finally, we look at the Jordan form for real matrices: Let A be a real (square) matrix, with Jordan decomposition $S+N$ (semisimple part + nilpotent part). We claim that S and N are automatically real (in spite of the fact that the eigenvalues may well be complex). We can prove this by taking the complex-conjugate and getting $A=\bar{S}+\bar{N}$. Clearly, $\bar{S}$ is semisimple (diagonalizable by $\bar{P}$, if P diagonalizes S); $\bar{N}$ is nilpotent, of course; and finally, $\bar{S}$ and $\bar{N}$ commute (since S and N do). It follows now from the uniqueness of the J-decomposition, proved above, that $\bar{S}=S$ and $\bar{N}=N$; that is, S and N are real.

PROBLEMS

1. Solve, that is find all solutions of, the system $\dot{X}=AX$, where $X=X(t)=[x_1(t),x_2(t),x_3(t)]$ and A is the first matrix in Problem 1 of Section 6.

2. Prove: If an operator T is semisimple and also nilpotent, then it is 0.

11

QUADRATIC FORMS

In this chapter linear algebra reaches beyond itself, from linear, first order, expressions to **quadratic forms**, second order terms. One way these arise is as the second order term in the Taylor expansion of a function of several variables. For instance, the potential energy of a (physical or other) system can in this sense often be taken as a quadratic function, particularly if one is interested in the neighborhood of a critical point (one where the first derivatives of the function vanish: maximum, minimum, saddle point). We concentrate on the geometry and discuss the generalizations of the conics (ellipse and hyperbola) to higher dimensions, the **quadrics**. They are classified through the **indices** of **positivity, negativity,** and **degeneracy**. The main result is Sylvester's theorem, which says essentially that these indices of a quadratic form are well defined. Consideration of the tangent plane of a quadric gives rise to the notion of **conjugacy** with respect to the quadric, a useful relation between points and hyperplanes.

1. BILINEAR FUNCTIONS

It may seem peculiar that in *linear* algebra (we have always stressed that that means *first* order, where exponents are 1) one studies quadratic expressions, such as $3x_1^2 - 4x_1x_2 + x_2^2$. A partial justification is that "quadratic" is closely related to "bilinear," which means, roughly, a formula or expression that is linear *separately* in two different sets of variables, for instance (related to the quadratic expression above) $3x_1y_1 - 2x_1y_2 - 2x_2y_1 + x_2y_2$; this is linear in the x's (if the y's are held constant) and linear in the y's (if the x's are held constant). If we put $y_1 = x_1$ and $y_2 = x_2$, we get back the **quadratic** expression above.

We generalize, first to the case where the number of x's is different from the number of y's, and then to the abstract situation (the field F, as always, is R or C).

1.1. DEFINITION. *A bilinear function (also bilinear form), say φ, defined on $\mathsf{F}^m \times \mathsf{F}^n$ is an expression of the form $\sum_{i,j} a_{ij} \cdot x_i \cdot y_j$; each term is product of a factor a_{ij}, an x_i, and a y_j; the a_{ij} are given scalars in F, for $i=1,\ldots,m$ and $j=1,\ldots,n$,—the x_i are the components of a (variable) vector X in F^m; the y_j are those of a vector Y in F^n. φ is a function in the sense that for any specific (numerical) X and Y we obtain a (scalar) value $\varphi(X,Y) = \sum a_{ij} x_i y_j$.*

Example

$m=3$ and $n=2$;

$$\varphi(X,Y) = 2x_1y_1 + x_1y_2 - x_2y_1 - 3x_2y_2 - x_3y_1 + 4x_3y_2.$$

$$\varphi([1,2,-1],[-1,3]) = 2\cdot 1\cdot(-1) + 1\cdot 3 - 2\cdot(-1) - 3\cdot 2\cdot 3$$
$$-(-1)\cdot(-1) + 4\cdot(-1)\cdot 3 = \cdots = -28.$$

We note that the coefficients a_{ij} form a matrix $A = (a_{ij})$, $m \times n$, called the matrix of φ relative to the standard bases of F^m and F^n; the sum $\sum_{i,j} a_{ij} x_i y_i$ can be written as X^tAY in matrix notation (X^t is the *row* $(x_1,\ldots,x_m)$); clearly, φ and A determine each other. For the example we have
$A = \begin{pmatrix} 2 & 1 \\ -1 & -3 \\ -1 & 4 \end{pmatrix}$. ■

Now we consider the abstract version of the definition.

1.1′. DEFINITION. *Let U and V be vector spaces over F of dimensions m and n. A bilinear function, say φ, on U and V (or better, on $U \times V$) is a function that assigns to each ordered pair (u,v) (with u in U, v in V) a scalar value $\varphi(u,v)$, such that the linearity relations*

BL_1: $\varphi(u_1+u_2,v) = \varphi(u_1,v) + \varphi(u_2,v)$; $\varphi(u,v_1+v_2) = \varphi(u,v_1) + \varphi(u,v_2)$.

BL_2: $\varphi(ru,v) = r\cdot\varphi(u,v) = \varphi(u,rv)$.

hold for vectors u, u_1, and u_2 in U, v, v_1, and v_2 in V, and scalar r.

BL_1 and BL_2 can be expressed by saying that for each fixed v in V, $\varphi(u,v)$ is a linear function of u; similarly, for each fixed u in U, $\varphi(u,v)$ is linear in v; φ is, thus, "bilinear." We comment on the "ordered pair": U and V could have a nonzero intersection, could even be identical; if u and v are both in U *and* in V, we can consider the pair (u,v) and also the pair (v,u)—they count as being different.

Remark. The set $U \times V$ of ordered pairs (u,v) (i.e., the "cartesian product" of U and V) that appears here is *not* a vector space. This is a flaw that can be corrected by the introduction of the *tensor product* $U \otimes V$; we shall not enter into a discussion of that concept. ■

As before, the abstract situation more or less reduces to the concrete one by the use of *bases*.

1.2. PROPOSITION. (*Construction principle*). *Let* $\beta = \{u_1, \ldots, u_m\}$ and $\gamma = \{v_1, \ldots, v_n\}$ *be bases for U and V. Then a bilinear function* φ *is completely determined by the* $m \cdot n$ *values* $\varphi(u_i, v_j)$; *furthermore, given a* $m \times n$ *matrix* $A = (a_{ij})$, *there is a (unique) bilinear function* φ *with* $\varphi(u_i, v_j) = a_{ij}$.

The proof will show, moreover, how abstract bilinear functions reduce to concrete ones.

Each u in U has a coordinate vector $X = [x_1, \ldots, x_m]$ with respect to $\beta (u = \Sigma x_i u_i)$; similarly, $v \underset{\gamma}{\leftrightarrow} Y$ for v in V. Then $\varphi(u, v) = \varphi(\Sigma x_i u_i, \Sigma y_j v_j) = \Sigma_{i,j} x_i y_j \varphi(u_i, v_j)$ by repeated use of linearity (BL_1 and BL_2); this shows that $\varphi(u, v)$ can be computed once one knows the $\varphi(u_i, v_j)$. Putting $\varphi(u_i, u_j) = a_{ij}$ and $A = (a_{ij})$, the formula can be written as $\varphi(u, v) = X^t A Y$.

Conversely, given the a_{ij}, we *define* $\varphi(u, v)$ as $\Sigma_{i,j} x_i \cdot y_j \cdot a_{ij}$; clearly, this is linear in X and in Y (e.g., $X^t \cdot A \cdot (Y_1 + Y_2) = X^t \cdot A \cdot Y_1 + X^t \cdot A \cdot Y_2$), and so $\varphi(u, v)$ is linear in u and v. $\varphi(u_i, v_j)$ is precisely a_{ij}; $(E^i_{(m)})^t \cdot A \cdot E^j_{(n)}$ reduces to a_{ij}. ■

Extending earlier notation, we write $\varphi \underset{\beta,\gamma}{\leftrightarrow} A$; once bases are chosen, we can *represent* bilinear functions by matrices.

How does the matrix change if we change bases? This is answered with the help of the notion of transition matrix and the formula (T) of Chapter 6, Section 4: If β' and γ' are "new" bases for U and V, we have transition matrices $P = T^{\beta}_{\beta'}$ in U and $Q = T^{\gamma}_{\gamma'}$ in V; coordinates of vectors change by $X = P \cdot X'$ in U and $Y = Q \cdot Y'$ in V. And now $\varphi(u, v) = X^t \cdot A \cdot Y$ becomes $(P \cdot X')^t \cdot A \cdot Q Y' = (X')^t \cdot P^t \cdot A \cdot Q \cdot Y'$, and we read off

(CB) $$A' = P^t \cdot A \cdot Q \cdot = \left(T^{\beta}_{\beta'}\right)^t \cdot A \cdot T^{\gamma}_{\gamma'}.$$

This is *different* from formula (CL) for change of the matrix of a linear transformation (Chapter 8, Section 2), not so much because P and Q have reversed their position, but mainly because we have P^t (transpose) here, where the other has Q^{-1} (inverse).

We note the special, but important, case $U = V$, so that A is now square; we then take $\beta = \gamma$, as usual (and write $\varphi \underset{\beta}{\leftrightarrow} A$), as well as $\beta' = \gamma'$, so that $P = Q$. The formula becomes

(C) $$A' = P^t \cdot A \cdot P.$$

Note again the difference from the case of operators, Chapter 8, Section 2; there we have $P^{-1}\cdot A\cdot P$. A and A' are then said to be **congruent** (via P; here P has to be invertible, so that one also has $A=(P^{-1})^t\cdot A'\cdot P^{-1}$).

PROBLEMS

1. For $U=V=\mathsf{R}^2$ and $A=\begin{pmatrix}2 & -1\\ 1 & 3\end{pmatrix}$, we write out the bilinear form X^tAY. Find the matrix representing the form with respect to the new basis $X_1=[1,2]$ and $X_2=[1,3]$ for both U and V.

2. Write out the matrix for the bilinear form $2x_1y_1-x_1y_2-3x_1y_3-x_1y_4+4x_2y_1-3x_2y_2+x_2y_3-2x_2y_4$ on the space $\mathsf{F}^3\times\mathsf{F}^4$ (note the dimensions!).

3. $U=V=P^3$. Define φ by putting $\varphi(p(t),q(t))=\int_0^1 t\cdot p(t)\cdot q(t)\,dt$ for any two polynomials p and q. Find the matrix of φ with respect to the basis $\{1,t,t^2,t^3\}$ and also with respect to $\{1,t-1,(t-1)^2,(t-1)^3\}$.

4. With the data of Problem 3, take $p=t^2-1$ and find q such that $\varphi(t^2-1,q)=0$. (There are many choices for q.)

2. QUADRATIC FORMS

Bilinear functions with $U\neq V$ are not terribly important for us. We shall concentrate on the case $U=V$; we speak then of a bilinear function defined *on* U. There are two special types of such functions that come up frequently: **symmetric** ones, satisfying $\varphi(u,v)=\varphi(v,u)$ for any u, v in U, and **skew symmetric** (or just **skew**) ones, satisfying $\varphi(u,v)=-\varphi(v,u)$. The matrices representing such functions with respect to a basis are then *symmetric* ($A=A^t$) or *skew* ($A=-A^t$), and conversely. For instance, suppose $X^tAY=Y^tAX$ for all X and Y; we note that $Y^t\cdot A\cdot X$ is 1×1, therefore, equal to its transpose $(Y^t\cdot A\cdot X)^t$ which in turn equals $X^t\cdot A^t\cdot Y$ (we used $(A\cdot B)^t=B^t\cdot A^t$ and $(Y^t)^t=Y$). Thus $X^t\cdot A\cdot Y=X^t\cdot A^t\cdot Y$ or $X^t\cdot(A-A^t)\cdot Y=0$ for all X and Y; clearly (uniqueness in Proposition 1.1!) this implies $A-A^t=0$ or $A=A^t$.

Abstracting a little, to any bilinear function φ on U we can assign another one, called its **transpose** φ^t, by the formula $\varphi^t(u,v)=\varphi(v,u)$; that is, we switch the two variables around. Then, clearly, if φ is represented by A, φ^t is represented by A^t (since $\varphi^t(u_i,u_j)=\varphi(u_j,u_i)$; φ is symmetric (or skew) if $\varphi=\varphi^t$ (or $\varphi=-\varphi^t$).

We come to the concept that we are really after: "quadratic function on U." A **quadratic function** (or form) q (on U) is a function from U to F of the form $q(u)=\varphi(u,u)$ for a suitable *bilinear* function φ on U (and all u in U).

Concretely, on F^n, this says that a quadratic function is obtained from some X^tAY by taking $X=Y$, so that we have $q(X)=X^tAX=\sum a_{ij}x_ix_j$ (a sum-of-squares x_i^2 and "cross products" x_ix_j, $i\neq j$, with scalar factors a_{ij}).

By definition q is determined by φ; conversely, *if* φ (or A) is also symmetric, φ (or A) is determined by q. We can compute $\varphi(u,v)$ (or similarly X^tAY), if we know the values $\varphi(w,w)$ (or Z^tAZ) that φ takes if the two variables are equal to each other. This is done by the so-called "**polarization** formula": We expand $\varphi(u+v,u+v)$ by linearity, getting $\varphi(u,u)+\varphi(u,v)+\varphi(v,u)+\varphi(v,v)$; the two middle terms are equal, by assumption, and so

$$\varphi(u,v)=\tfrac{1}{2}(\varphi(u+v,u+v)-\varphi(u,u)-\varphi(v,v))$$

$$\text{(P)} \qquad =\tfrac{1}{2}(q(u+v)-q(u)-q(v)).$$

In the matrix version this says that from a quadratic expression (involving squares x_i^2 and cross products x_ix_j) we can read off uniquely a *symmetric* matrix A so that the expression equals X^tAX. One should understand that a term like $5x_2x_3$ has to be thought of $\frac{5}{2}x_2x_3+\frac{5}{2}x_3x_2$, giving rise to entries $\frac{5}{2}$ and $\frac{5}{2}$ at the (2,3)- and (3,2)-position. Or, in terms of polarization, given a quadratic expression q, one can find a *unique* bilinear expression X^tAY, with $A=A^t$, that reduces to the given one upon setting $Y=X$.

Example

$2x_1^2-3x_1x_2+x_2^2+4x_1x_3+6x_2x_3-x_3^2$ polarizes to

$$2x_1y_1-\tfrac{3}{2}x_1y_2-\tfrac{3}{2}x_2y_1+x_2y_2+2x_1y_3+2x_3y_1+3x_2y_3+3x_3y_2-x_3y_3;$$

the matrix is

$$A=\begin{pmatrix} 2 & -\frac{3}{2} & 2 \\ -\frac{3}{2} & 1 & 3 \\ 2 & 3 & -1 \end{pmatrix}.$$

Remark. Formula (P) can be used to give a definition of "quadratic function" which is in some ways more direct: A function q from U to F is "quadratic" if (a) the function of two variables defined by $\varphi(u,v)=\frac{1}{2}(q(u+v)-q(u)-q(v))$ is bilinear symmetric, and (b) q satisfies $q(ru)=r^2\cdot q(u)$ (in fact this is needed only for $r=2$). ■

From now on quadratic forms X^tAX are automatically assumed to have a symmetric matrix, $A^t=A$. (And, incidentally, we shall say very little about skew matrices or bilinear forms.)

We mention a case where quadratic forms appear "naturally": In calculus, in the Taylor expansion of a function of any number of variables the second order terms constitute a quadratic form (with $x_i - a_i$ taking the role of our present x_i, where $[a_1, \ldots, a_n]$ is the point around which the function is expanded).

PROBLEMS

1. Polarize the quadratic function $x_1^2 - 4x_1x_2 - 2x_2^2 + 5x_1x_3 + 3x_3^2 - 6x_1x_4 + 7x_2x_3 - 4x_2x_4$ on $\mathbf{R}^4$, and write out the representing (symmetric!) matrix.
2. Write out the matrix for the quadratic form $2x_1^2 + x_1x_2 - x_2^2$.
3. Work out the Remark in the text. (Is $q(u) = \varphi(u,u)$?)
4. Write out the matrix of the skew form $\psi(X, Y) = \det(X, Y)$ on $\mathbf{R}^2$, with respect to the standard basis and also with respect to the basis $\{[1,2],[1,3]\}$. Check the relation $A' = P^tAP$ (here P equals the transition matrix).
5. Show: If A is symmetric (or skew), then P^tAP is again symmetric (or skew) for any P.
6. Let ψ be a *skew* form on U. What can one say about the values of the "quadratic" form $q(u) = \psi(u,u)$? Can one reconstruct the (bilinear) form ψ from the "quadratic" form?

3. COMPLETING SQUARES

The simplest quadratic form (on $\mathbf{F}^n$) that one can think of is surely $x_1^2 + x_2^2 + \cdots + x_n^2$; only slightly less simple are the forms $x_1^2 + x_2^2 + \cdots + x_r^2 - x_{r+1}^2 - \cdots - x_n^2$ (any number of minus signs), with no "cross-terms" x_ix_j, $i \neq j$; among the latter is the famous form $x^2 + y^2 + z^2 - t^2$ of relativity theory. Our first aim is to show that these forms are in fact typical, in the sense that any quadratic form can be reduced to one of these by a change of variables (with possibly some of the x_i^2 missing). We call this "reduction to sum-of-squares."

3.1. THEOREM. *Let $q(X) = X^tAX$ be a quadratic form (over $\mathbf{R}$ or $\mathbf{C}$). There exists (not uniquely) a change of coordinates $X = PX'$, with invertible transition matrix P that reduces the form to a "sum of squares" $\Sigma \epsilon_i x_i'^2$ with $\epsilon_i = +1$ or -1 or 0 (for $\mathbf{F} = \mathbf{C}$: $\epsilon_i = 1$ or 0). Equivalently, to a symmetric matrix A there exists an invertible P such that P^tAP is a diagonal matrix, with diagonal entries $+1$ or -1 or 0 (1 or 0, if $\mathbf{F} = \mathbf{C}$). ($\epsilon_i = 0$ means that x_i^2 is missing from the sum.)*

Remark. We will see later that the number of 0's is always the same, no matter how one performs the reduction; similarly, if $\mathbf{F} = \mathbf{R}$, the number of $+1$'s and -1's is always the same (Sylvester's theorem, Section 6).

PROOF. The equivalence mentioned in the theorem comes, of course, from formula (CB) (Section 1). We come to the proof proper: We "get rid" of the cross-terms $x_i x_j$ with $i \neq j$ by the method of "completing the square"; this is very old, goes back to the Babylonians, and is, therefore, sometimes called the Babylonian method. First, we consider an example: The form $x^2+2xy+3y^2$ can be rewritten as $x^2+2xy+y^2+2y^2=(x+y)^2+2y^2$; we have "completed" the part x^2+2xy to a "complete square." We introduce $x'=x+y$ and $y'=\sqrt{2}\,y$ as new coordinates; our quadratic form appears then as $x'^2+y'^2$. (To get the coordinate change mentioned in the theorem, namely x and y in terms of x' and y', we just solve the two equations $x'=x+y$ and $y'=\sqrt{2}\,y$; the result is $x=x'-1/\sqrt{2}\cdot y'$ and $y=1/\sqrt{2}\cdot y'$.)

Now, in general, let $q=\Sigma a_{ij}x_i x_j$ be the form. Suppose $a_{11}\neq 0$. Collect all terms with x_1 as factor; they are

$$a_{11}x_1^2+2a_{12}x_1x_2+2a_{13}x_1x_3+\cdots+2a_{1n}x_1x_n.$$

($2a_{12}x_1x_2$ comes from $a_{12}x_1x_2+a_{21}x_2x_1$, since $a_{12}=a_{21}$.) Complete the square, that is, rewrite this as

$$a_{11}\cdot(x_1+(a_{12}/a_{11})x_2+\cdots+(a_{1n}/a_{11})x_n)^2$$

plus terms not involving x_1; the square is so designed that after multiplying-out the terms involving x_1 are precisely equal to those occurring in q, just described. Putting

$$x_1'=x_1+(a_{12}/a_{11})x_2+\cdots+(a_{1n}/a_{11})x_n$$

we get for q the new expression $a_{11}(x_1')^2+$ a quadratic form in $x_2,\ldots,x_n$.

We now work on the second part, and transform it into $a_2(x_2')^2$ plus a quadratic form in $x_3,\ldots,x_n$; here a_2 is a scalar factor, and x_2' is of the form x_2 plus linear terms in $x_3,\ldots,x_n$. Eventually we get to something of the form $a_1(x_1')^2+a_2(x_2')^2+\cdots+a_n(x_n')^2$. (Some of the a_i might be 0.) We have one more step (and here R and C part ways a bit). For $\mathsf{F}=\mathsf{C}$ we write each nonzero a_i as $(\sqrt{a_i}\,)^2$ and take $x_i''=\sqrt{a_i}\,x_i'$ as new variable, getting q to the form "sum of some $(x_i'')^2$" (if $a_i=0$, there is no $(x_i'')^2$). For $\mathsf{F}=\mathsf{R}$ we write a_i, if not 0, as $\pm\sqrt{|a_i|}^{\,2}$ and get q as "sum of some $\pm(x_i'')^2$". There is still a complication: What if, at the start, $a_{11}=0$? We simply start with some other x_i, whose factor a_{ii} is not 0. If all diagonal a_{ii} are 0, we must make a preliminary change: Find some a_{ij} that is not 0, substitute $x_i=x_i'+x_j'$ and $x_j=x_i'-x_j'$ (and leave all other x's unchanged). Now there will be a nonzero diagonal term, in fact two of them, because $x_i x_j$ becomes $(x_i')^2-(x_j')^2$. ■

Example

$$\begin{pmatrix} 1 & -2 & 2 \\ -2 & 1 & 2 \\ 2 & 2 & 1 \end{pmatrix}.$$

Here $a_{11}=1$ (good!). The terms with x_1 are $x_1^2-4x_1x_2+4x_1x_3$. Rewrite this as $(x_1-2x_2+2x_3)^2-4x_2^2-4x_3^2+8x_2x_3$; thus, $x_1'=x_1-2x_2+2x_3$. Adding all the other terms of q (the ones not involving x_1), we get $(x_1')^2-3x_2^2-3x_3^2+12x_2x_3$ for q. Now, the terms of the second part involving x_2 are $-3x_2^2+12x_2x_3$. Rewrite this as $-3(x_2-2x_3)^2+12x_3^2$. Put x_2' equal to x_2-2x_3 and get q in the form $(x_1')^2-3(x_2')^2+9x_3^2$. For completeness, also write $x_3'=x_3$. Thus the coordinate change $x_1'=x_1-2x_2+2x_3$, $x_2'=x_2-2x_3$, *and* $x_3'=x_3$ changes q to $(x_1')^2-3(x_2')^2+9(x_3')^2$. And, finally, putting $x_1''=x_1'$, $x_2''=\sqrt{3}\,x_2'$, and $x_3''=3x_3$, we get $(x_1'')^2-(x_2'')^2+(x_3'')^2$ for q. If we work with C, we could even put $x_2''=\sqrt{3}\,ix_2'$ and get $q=(x_1'')^2+(x_2'')^2+(x_3'')^2$. (Actually, one often suppresses these last steps.) Note that the equations for the x_i' in terms of the x_j should be written (in accordance with earlier notation) as $X'=P^{-1}\cdot X$ and can be solved to $X=P\cdot X'$. In the example then, $P^{-1}=\begin{pmatrix} 1 & -2 & 2 \\ 0 & 1 & -2 \\ 0 & 0 & 1 \end{pmatrix}$, and P (easily found for the x_j) $=\begin{pmatrix} 1 & 2 & 2 \\ 0 & 1 & 2 \\ 0 & 0 & 1 \end{pmatrix}$. One should now check $P^t\cdot A\cdot P=\operatorname{diag}(1,-3,9)$. Strictly speaking, the transition matrix of Theorem 3.1 is the matrix, say P', expressing the x_i in terms of the x_j'', and we should verify $P'^tAP'=\operatorname{diag}(1,-1,1)$ [or $=\operatorname{diag}(1,1,1)=I$, in the C-case].

Remark. For the abstract case, a quadratic form q on a vector space U (with associated symmetric bilinear from φ), we get from Theorem 3.1, together with Proposition 1.2 and the discussion there the following: There exists (although not uniquely) a basis $\beta=\{u_1,\ldots,u_n\}$ of U such that $\varphi(u_i,u_j)=0$ for $i\neq j$, and $\varphi(u_i,u_i)=+1$ or -1 or 0 [$=1$ or 0 in case $\mathsf{F}=\mathsf{C}$]. We shall come back to this later on (Section 7).

PROBLEMS

1. Reduce the following to sum-of-squares form
(a, b, and c are over R):

a. $$q(x,y,z)=x^2+5y^2+z^2-4xy+2xz-2yz.$$

b. $$x^2+2y^2+4z^2-4t^2-2xy+2xz-6yz+2yt-8zt.$$

c. $$x^2+y^2+3z^2+3t^2+2xz+2yz-2xt-2yt-2zt.$$

d. $$x^2-2ixy-2y^2 \qquad \text{(over } \mathsf{C}\text{)}$$

2. Show that any symmetric matrix can be written in the form M^tDM, with M invertible and D diagonal. Also show that the converse holds (quite trivial): any matrix of the form M^tDM with D diagonal is symmetric. (How is M related to the P of Theorem 3.1?)

3. Find the matrix P' for the example in the text.

4. TYPES OF FORMS

In this section we take $\mathsf{F}=\mathsf{R}$; the a_{ij} and x_i take real values only. Consequently, x_i^2 is necessarily nonnegative (whereas in the complex case it could be negative or in fact take any complex value); more precisely, x_i^2 is positive, unless x_i is 0. We extend this a bit: All values taken by the form $x_1^2+x_2^2+\cdots+x_n^2$, for any values of the x_i, are positive, except that one gets 0 if *all* the x_i are 0; whereas, any other sum-of-squares form $\Sigma_1^n \epsilon_i x_i^2$ (with $\epsilon_i=+1$ or -1 or 0, but not all of them $+1$) will come out 0 or negative for some nonzero X (if ϵ_i is 0 or -1, take $x_i=1$ and the other $x_j=0$). Now, since the values that a quadratic form takes do not depend on the coordinate system in which we write the form, we can characterize the forms that, via Theorem 3.1, reduce to $x_1^2+x_2^2+\cdots+x_n^2$ (all $\epsilon_i=+1$) by a *property*: They are those forms for which the value X^tAX (or $q(u)$ in the abstract case) is *positive* for every choice of X (or u)—except, of course, that for the 0-vector the value is 0. Quadratic forms of this type are called **positive definite** (often abbreviated to **p.d.**). In addition to these there are several other types of forms, defined in terms of the signs of the values taken by them: A form is **negative definite**, if *all* the values are negative—except, of course, 0 for the 0-vector. It is **positive semidefinite**, if it takes no negative value (but the form is allowed to be 0 for some nonzero vectors); similarly there is **negative semidefinite**. Finally, there is **indefinite**: The form takes positive values for some vectors and negative values for some other vectors (it "changes sign"). (These definitions also apply to symmetric matrices. Such a matrix A is called positive definite, if the quadratic form X^tAX is positive definite, etc.)

From the discussion above it should be clear what all this amounts to when one reduces a form (via Theorem 3.1) to sum-of-squares: Positive definite means all ϵ_i are $+1$; positive semidefinite means the ϵ_i are $+1$ or 0, with no -1 occurring; similarly, for negative definite, all ϵ_i are -1 and for semidefinite, the ϵ_i are -1 or 0; finally, indefinite means at least one ϵ_i equal to $+1$ and one equal to -1. (In the last case, if, say, $\epsilon_1=1$ and $\epsilon_2=-1$, then the form takes a positive value, $+1$, for $X=E^1$ and a negative value, -1, for $X=E^2$.)

We comment briefly on the role of $\epsilon_i=0$. For this purpose we consider the determinant. Formula 1.(C) shows that the determinant of the matrix of a quadratic form changes with the coordinate system: If we change coordinates by $X=PX'$, the det of the matrix changes by the factor $(\det P)^2$. Thus, strictly speaking, a quadratic form q does not have a determinant; there is no way to assign to q a value $\det q$ unambiguously. However, since $(\det P)^2$ is positive, the *sign* of $\det A'$ is always the same; it makes sense to say that q has a positive (or negative) determinant. Similarly, if $\det A=0$ in one coordinate system, then one gets 0 in any coordinate system; and so the statement "$\det q=0$" makes sense. A form q with $\det q=0$ is called **degenerate**. For the sum-of-squares form, degeneracy clearly means that one or more ϵ_i are 0 (one or more x_i^2 absent). ∎

Writing out q, reduced to a sum-of-squares, as

$$x_1^2+\cdots+x_r^2-x_{r+1}^2-\cdots-x_{r+s}^2$$

(with $r+s\leqslant n$), we have the following table:

Positive definite	Positive semidefinite	Indefinite
$r=n$	$s=0$	$r>0, s>0$

Negative semidefinite	Negative definite	Degenerate
$r=0$	$s=n$	$r+s<n$

Digression. An application of these ideas to functions of several variables (we mentioned this at the end of Section 2). A problem that occurs often is that of finding and studying the *extreme* values (maxima, minima, saddle points) of such functions $f=f(x_1,\ldots,x_n)$. They are *found* as the points where all n first partial derivatives $\partial f/\partial x_i$ vanish. Suppose we have such a point; for simplicity let it be the origin 0. The Taylor expansion is

$$f(x_1,\ldots,x_n)=f(0)+\tfrac{1}{2}\Sigma\,\partial^2 f/\partial x_i\partial x_j(0)\cdot x_i x_j+\cdots$$

(the linear terms are absent by hypothesis). We define a quadratic form $q(X)=\Sigma a_{ij}x_i x_j$, with $a_{ij}=\partial^2 f/\partial x_i\partial x_j(0)$; the function f is then approximately (up to higher terms) given by the simple expression $f(0)+1/2q(X)$; this is a good approximation for small x_i-values. The nature of f near 0 is (more or less) determined by the type of the form q. If q is positive definite, 0 is a minimum point of f; similarly negative definite corresponds to a local maximum. Indefinite (nondegenerate) q's correspond to saddle points of various kinds. For a degenerate q one has to take higher order terms into consideration, to understand the nature of f near 0.

PROBLEMS

1. For Problems 1, 2, and 3 of Section 3, what can one say about the determinants of the forms (after reduction to sum-of-squares)? Compute the determinants directly, and check your result.

2. For Problems 1, 2, and 3 of Section 3, determine the type of the forms (positive definite, etc.). (Consider the forms after reduction to sum-of-squares.)

3. Decide whether the following are indefinite or not by trying out a number of vectors:

a. $$x^2+xy-2y^2$$

b. $$x^2+2y^2+z^2+2xy+2xz$$

4. Show: If A is positive definite (real symmetric), then it can be factored as $M^t \cdot M$, with an invertible M.

5. Prove the converse of Problem 4: If M is nonsingular (real), then M^tM is positive definite.

6. Extend Problem 5: If N is any real matrix (not even assumed square), then N^tN is positive semidefinite.

7. Prove the converse of Problem 6: If A is positive semidefinite, then it can be written as N^tN with square real N.

8. Show: If N is real and $N^tN=0$, then already $N=0$.

9. For $N=\begin{pmatrix} a & b \\ c & d \end{pmatrix}$ write out $\operatorname{tr} N^tN$. (tr means trace; see Chapter 9, Section 2.)

10. For $N=(n_{ij})$ write out $\operatorname{tr} N^tN$ in terms of the n_{ij}.

11. Let q be a quadratic form on V. Let T: $U \to V$ be a linear transformation. Define a quadratic form q' on U by $q'(u)=q(Tu)$. Show:
a. If q is positive definite, then q' is positive semidefinite.
b. If q is positive definite and T is injective, then q' is positive definite. Generalize to other types of forms.
c. Restate all this in matrix language.

5. QUADRICS AND CONES

We now come to the *geometry* of quadratic forms. Let X^tAX be a quadratic form on R^n (or abstractly, let q be a quadratic form on U, over R). For each scalar c we define a "surface" in R^n (or U), called the **quadric** Q_c of A (or q), by $Q_c=\{X\colon X^tAX=c\}$ (abstractly, $Q_c=\{u\colon q(u)=c\}$), the set of all vectors, at which the quadratic form takes the value c; for $c=0$ these are called **cones** instead of quadrics.

We shall discuss later in more detail what sense it makes to call the Q_c surfaces. For the moment we only note that they do in n dimensions what some familiar objects do in two and three dimensions: For $n=2$ the quadrics are the well-known conics, ellipse (e.g., $x^2+2xy+2y^2=1$) or hyperbola (e.g., $x^2-3xy+2y^2=1$) or "degenerate" conic (e.g., $x^2+2xy+y^2=1$, the two parallel lines $x+y=\pm 1$); the case "cone" appears here as a pair of lines through 0 (e.g., $x^2-y^2=0$, the two lines $y=\pm x$).

For $n=3$ we get ellipsoids, hyperboloids of one or two sheets, and degenerate cases and cones. We take the equations in sum-of-squares form. See Figure 18.

Note 1. For Figure 18(c) there is no point with $z=0$; the quadric does not meet the x-y-plane; it consists of two parts (sheets), one above and one below the x-y-plane. For Figure 18(d) we show only one type.

Note 2. The axes should be thought of as "general," that is, not at right angles, and the unit distances not equal to each other. Thus, the first figure is meant to suggest not a round sphere, but an ellipsoid; and the coordinate axes are not the "axes" of the ellipsoid known from analytic geometry. ■

We make a comment on cones: They have the special property of "consisting of lines through 0." Namely, if X satisfies the equation $X^tAX=0$, so does rX for every r. This is definitely not so in the case of the quadric Q_c with $c\neq 0$: If X satisfies the equation $X^tAX=c$, and rX, with $r\neq 1$, does too, then clearly r must be -1 (from $c=(rX)^tA(rX)=r^2X^tAX=r^2c$ we get $r^2=1$). In other words, a line through 0 meets the quadric, if at all, in exactly two points. Check all this for the quadrics in Figure 18. ■

For the general quadric we usually take $c=1$ (or -1). Changing c just expands or contracts the quadric, from 0 as center. For very small c the quadric is close to the corresponding cone; it is on one side of the cone or the other, depending on the sign of c (and the type usually changes; compare the hyperboloids of one or two sheets, Figures 18(b) and (c) above).

We might just as well take the equation of the quadric in sum-of-squares form; this will simplify the discussion. (But we should remember that this involves a coordinate change; see Note 2 above.)

We consider the positive definite case, $x_1^2+x_2^2+\cdots+x_n^2=1$. The quadric, the (hyper) ellipsoid, is contained in a finite portion of the space (does not stretch to "infinity"), since clearly every coordinate must be between $+1$ and -1. It "surrounds" the origin: If we take any vector X ($\neq 0$) and move in that direction by forming rX and letting r increase from 0, then pretty soon we get to a point on the quadric; namely, because of $X^tAX>0$ the equation $r^2X^tAX=1$ has a (real) solution. The quadric behaves in these respects like what we know in the special cases ellipse ($n=2$) and the ordinary ellipsoid ($n=3$).

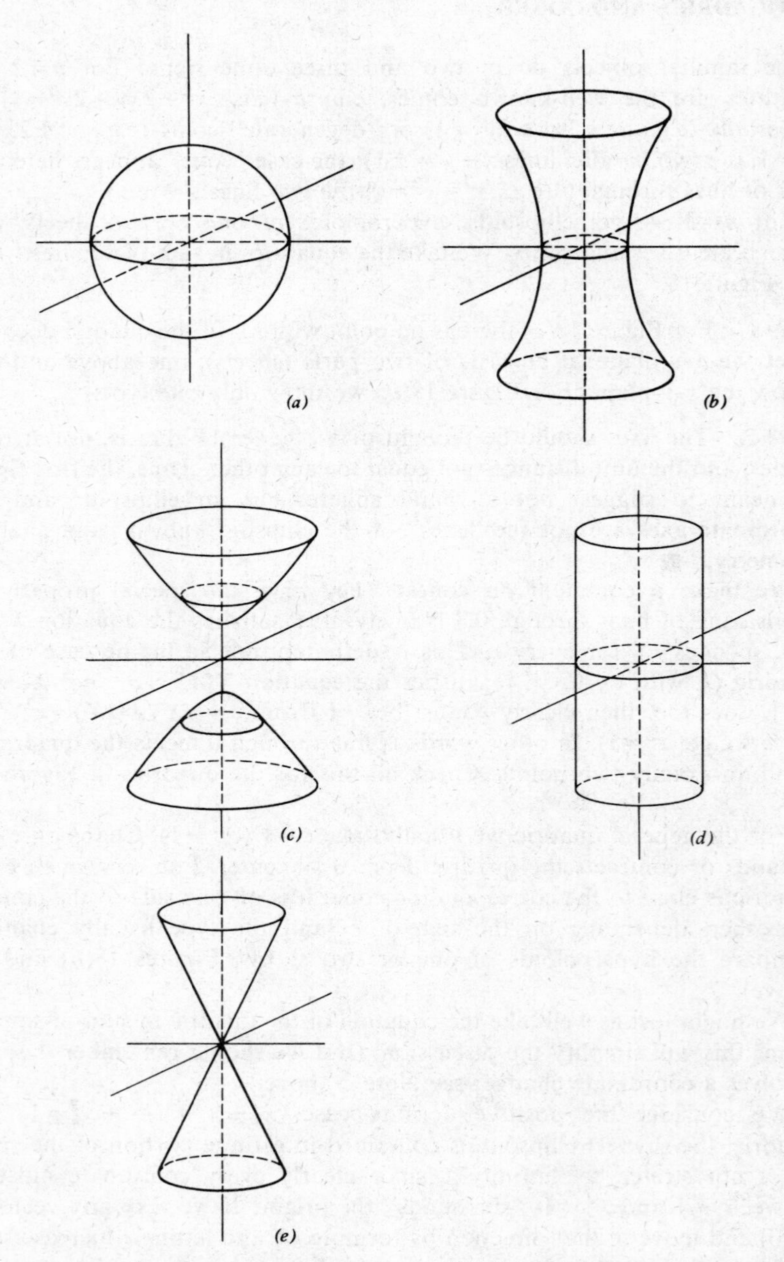

Figure 18. (*a*) $x^2+y^2+z^2=1$—ellipsoid; (*b*) $x^2+y^2-z^2=1$—hyperboloid of one sheet; (*c*) $-x^2-y^2+z^2=1$—hyperboloid of two sheets; (*d*) degenerate: $x^2 \pm y^2=1$ or $x^2=1$—cylinders; (*e*) $x^2+y^2-z^2=0$—cone.

To study this resemblance a bit more, we use the concept of **restriction** of a quadratic form q, defined on a vector space U, to a subspace V of U—this is the quadratic form on V, denoted by $q|V$ (or just q), whose value at any v in V is $q(v)$—thus $q|V(v)=q(v)$ for any v in V. In other words, it is just the given form q, except that we only use vectors in V and forget about the other vectors in U (compare the notion of restriction of a linear transformation to a subspace, (Chapter 8, Section 1). It is clear that $q|V$ is a quadratic form on V; indeed, restriction makes sense for (symmetric or not) bilinear functions on U and yields bilinear functions on V—conditions BL_1 and BL_2 of Section 1 remain valid.

In coordinates, we may assume that V is obtained by $x_{r+1}=x_{r+2}=\cdots=x_n=0$ (extend a basis of V to one of U). Then the form

$$q=\sum_{i=1}^{n} a_{ij}x_ix_j \qquad \text{(with } i \text{ and } j \text{ running from 1 to } n\text{)}$$

restricts to

$$q|V=\sum_{i=1}^{r} a_{ij}x_ix_j \qquad \text{(with } i \text{ and } j \text{ running from 1 to } r \text{ only).}$$

We look at a slightly different approach, with an example: The form $x^2+y^2-z^2$, restricted to the plane V with equation $z=2y$, restricts or reduces to $x^2+y^2-4y^2=x^2-3y^2$; here x and y serve as coordinates in V, corresponding to the basis $[1,0,0]$ and $[0,1,2]$ for V.

We apply this to quadrics. The main, and quite obvious, point is that the intersection of the quadric Q_c (equation $q(u)=c$) with V is the quadric Q_c' (equation $q|V(v)=c$) *in* V. For example, the intersection of a hyperboloid (of one sheet, say) with a plane (through 0) is a conic (ellipse or hyperbola, depending on the plane). Now if q is positive definite, so is, of course, $q|V$. This implies that a (hyper)ellipsoid in U is intersected by a subspace V in a (hyper)ellipsoid in V. For example an (ordinary) ellipsoid is cut by any plane (through 0) in an ellipse. In n dimensions, an ellipsoid is cut by any two-dimensional subspace in an ellipse, and by any three-dimensional subspace in an ordinary ellipsoid, etc. So much for the positive definite case. The quadric $q=1$ for $q=-x_1^2-x_2^2-\cdots-x_n^2$ is easier to describe: It is empty; there are *no* x_i to satisfy the equation.

The cone $q=0$ "collapses" in both cases; it contains only 0.

We turn to the in-between cases, the indefinite forms, with some $+1$'s and some -1's when reduced to sum-of-squares (for the time being we exclude degenerate forms—no 0's). We write such a form as $x_1^2+\cdots+x_r^2-x_{r+1}-\cdots-x_n^2$ and put $s=n-r$, which equals the number of minus signs; we assume $r>0$ and $s>0$.

The quadratic Q_c, a (hyper) hyperboloid, and the cone $q=0$ extend now "to infinity"; there are points on Q_c with arbitrarily large coordinates. This is familiar for the hyperbola in the plane and is true in the general case; the point is that the sums of the positive and the negative squares separately can be arbitrarily large, with their difference remaining equal to c. In more detail, the cone Q_0 plays the same role for the quadric Q_c that the asymptotes $y=\pm x$ play for the hyperbola $x^2-y^2=1$ in the plane: The quadric gets closer and closer to the cone, if one moves away from the origin. Let us take $c=1$, let X be a point on the quadric with a large x_1-entry (we are "moving to infinity"), and let Y be a point on the cone with $y_2=x_2,\ldots,y_n=x_n$, and with y_1 of the same sign as x_1 (this exists!). We have then $x_1^2-y_1^2=1$, and so $x_1-y_1=1/(x_1+y_1)$. Since x_1+y_1 is large, it follows that x_1-y_1 is small; thus, Y and X are very close to each other.

The fact that a form q is indefinite can also be expressed as follows: The cone $q=0$ divides R^n (or U, in the abstract case) into two nonempty parts, the *positive part* where $q(X)$ is greater than 0, and the *negative part* where $q(X)$ is less than 0. The two parts are on opposite sides of the cone. For the cone Figure 18(e) of Section 5, with $q(x,y,z)=x^2+y^2-z^2$, the negative part is "inside" the cone, surrounding the z-axis; the positive part is "outside" the cone; it contains the x-y-plane (except for 0) and any nearby plane. In the general case, it is difficult to say which part should be called "inside" and which "outside" the cone.

PROBLEMS

1. For Problem 4 of Section 4 determine the types of the restrictions of the form to the following subspaces: $x=0$, $y=0$, $x+y+z=0$, and $x+2y=y-z=0$.

2. For Problem 2 of Section 3, determine the types of the restrictions of the form to the following subspaces: $x=y-z$, $z=t=0$, $y+5t=z+2t=0$, and $x-y+z-t=y-3z-t=0$.

3. The form in Problem 2 of Section 3 is indefinite. Find some points in the positive and negative parts of R^4. Find a two-dimensional subspace on which the form is negative definite.

6. SYLVESTER'S THEOREM

To get a better picture of all this, we use the idea, introduced above, of restricting the form q to subspaces of U. We have to realize that now, in contrast to the positive definite case, there are different possibilities for the intersection of a hyperboloid with a subspace V of U: The intersection can, of course, be a hyperboloid in V (or degenerate), but it can also be an ellipsoid, or even empty. All these cases occur already for the ordinary hyperboloids Figures 18(b) and (c) of Section 5, for suitable planes (or lines). To put it a bit differently, the restriction of an indefinite form q to a subspace V can be indefinite, but it can also turn out positive or negative definite (and also degenerate). For instance, $x^2+y^2-z^2$, restricted to the x-y-plane $z=0$, becomes x^2+y^2, which is positive definite; restricted to the z-axis $x=y=0$ it becomes $-z^2=0$, which is negative definite. [Also, for $y=0$ we get x^2-z^2, an indefinite form, and with $x=0$ and $y=z$ we get $q|V$ identically 0, degenerate (very degenerate).]

Subspaces on which q is positive or negative definite are the same as subspaces that are completely (except for 0) contained in the positive part of U $(q>0)$, or in the negative $(q<0)$ part of U, determined by q. Of interest are subspaces of this kind that are as "big" as possible. This leads to a definition: Given a quadratic form q on a vector space U, we call **index** of **positivity** (or of **negativity**) of q the maximum of the dimensions of subspaces V or U such that $q|V$ is positive definite (or negative definite). It should be clear that for $x^2+y^2-z^2$ the index of positivity is 2, that of negativity 1. (The difference, index of positivity minus index of negativity, is often called the **signature** of q; for $x^2+y^2-z^2$ it is $+1$ $(=2-1)$.)

We come to the main fact about these concepts,

6.1. THEOREM. *Sylvester's Theorem—Law of Inertia. Let q be a quadratic form on U (over* R*) (nondegenerate); suppose q becomes, with respect to a suitable basis, $x_1^2+\cdots+x_r^2-x_{r+1}-\cdots-x_n^2$. Then r is the index of positivity of q, and $s=n-r$ is the index of negativity of q.*

The important fact here is that no matter how one reduces q to $\Sigma \pm x_i^2$, one always gets the same number of plus signs and of minus signs, since the indices of positivity and negativity are defined without any reference to coordinates. Instead of q we could talk here about a symmetric matrix A (nondegenerate) and attempt to make it $\mathrm{diag}(1,\ldots,1,-1,\ldots,-1)$ by going to P^tAP; the result is that one always gets the same number of -1's.

PROOF. Clearly q is positive definite on the r-dimensional subspace V, given (in the coordinates $x_1,\ldots,x_n$) by $x_{r+1}=\cdots=x_n=0$ (i.e., spanned by the first r basis vectors); this shows that the index of positivity is greater than or equal to r. Similarly, q is negative definite on the s-space, defined by $x_1=\cdots=x_r=0$; the index of negativity greater than or equal to s. Let now W be any subspace on which q is negative definite, of dimension t say. We claim that $V\cap W=0$. For the proof take u in $V\cap W$. Since u is in V, we have $q(u)\geqslant 0$ (in fact >0, unless $u=0$); since u is in W, we have $q(u)\leqslant 0$ (in fact <0, unless $u=0$). The only way out is $u=0$. With $V\cap W=0$ we can apply the dimension law (Chapter 3, Section 3): we have $\dim(V+W)=\dim V+\dim W=r+t$; we also have $\dim(V+W)\leqslant n$, since $V+W$ is a subspace of U. Thus, $r+t\leqslant n$ or $t\leqslant n-r=s$. This shows that the index of negativity is at most s; we say above that the opposite inequality also holds, and so the two are equal; similarly, the index of positivity is r. ■

For the missing case, where q is degenerate, we write the sum-of-squares form as $x_1^2 + x_r^2 - x_{r+1}^2 - \cdots - x_{r+s}^2$; degenerate means $r+s<n$. Sylvester's theorem goes through as before, with s the index of negativity (but no more $s=n-r$). In the proof we look at the space V given by $x_{r+1}=\cdots = x_{r+s}=0$, on which q is now only positive semidefinite; but the old argument still works. We add a new concept: The **degeneracy space** (sometimes called the null space) of q (or of A) is the space formed by all u with the property: $\varphi(u,v)=0$ for *all* v (φ is the associated bilinear form); in coordinates this means the vectors X with $X^tAY=0$ for all Y or simply, $X^tA=0$. Finally, by taking the transpose, $A\cdot X=0$. We see that this space is nonzero exactly if $\det A=0$, that is, if q is degenerate. In the sum-of-squares form the expression X^tAY is $x_1y_1+\cdots+x_ry_r-x_{r+1}y_{r+1}-\cdots - x_{r+s}y_{r+s}$. For this to be 0 for a given X and all choices of Y, we must have $x_1=\cdots=x_{r+s}=0$; that is, the degeneracy space is spanned by the vectors $E^{r+s+1},\ldots,E^n$. Put differently, the index of positivity plus the index of negativity plus dim degeneracy space equals n. By the way, the *rank* of a quadratic form is by definition the rank of any (symmetric) matrix representing it. This is always the same, from $A'=P^t\cdot A\cdot P$ with P of rank n. From sum-of-squares one sees that rank is the sum of index of positivity and index of negativity, and that rank less than n means degeneracy.

PROBLEMS

1. For Problems 1, 2, and 3 of Section 3 and Problems 3 and 4 of Section 4 find the indices of positivity and negativity, as well as those of nullity (i.e., the dimensions of the degeneracy spaces).

2. What is the index of negativity of a positive definite form?

3. Suppose q, defined on U, is positive definite on the subspace V and negative definite on the subspace W. What can one say about $V\cap W$?

4. Let q be a quadratic form in $x_1,\ldots,x_n$, and denote by q' the quadratic form in $x_1,\ldots,x_{n-1}$ obtained by putting $x_n=0$ (of course, any other x_i would do just as well). We know that if q is positive definite, then q' is also positive definite and $\det q$ is positive. Prove the converse: If q' is positive definite and $\det q$ is >0, then q is positive definite (*Hint.* What can one say about the index of positivity of q?)

5. (Continuation of Problem 4.) Write $q(X)=\Sigma a_{ij}x_ix_j$ with matrix $A=(a_{ij})$. Denote by D_r the det of the submatrix of A formed by the a_{ij} with i,j running from 1 to r. Prove that q is positive definite precisely if all the D_r (with $1\leqslant r\leqslant n$) are positive. This is the *det-criterion* for positive definiteness. (Note that $D_n=\det q$; one could, of course, take the variables in any other order.)

6. Can a negative definite or indefinite quadratic form have positive det?

7. As application of Problem 5, determine conditions on the coefficients a, b, and c under which the equation $ax^2+bxy+cy^2=1$ represents an ellipse (and not a hyperbola or a degenerate conic).

7. QUADRICS AS SURFACES; CONJUGACY

We discuss the "surface" nature of the quadrics and cones associated to $q(X)=X^tAX$ in some more detail. We recall the description of a *surface* in space by an equation $z=f(x,y)$, where $[x,y]$ can range through some portion of the x-y-plane, and for each such $[x,y]$ the equation yields the third coordinate of a point on the surface. (See Figure 19.) In our (higher dimensional) case, the equation $X^tAX=c$ can be considered as quadratic equation for any one of the x_i, and can be *solved* (quadratic formula) in the form $x_i=F\pm\sqrt{G}$, where F and G are certain expressions in the "other" x_j. We can now let the "other" x_j $(j\neq i)$ range through all those values for which G comes out positive; and get two values of x_i in each case. Thus, we get *two* "surfaces," extending over a certain region of the plane $x_i=0$, as parts of our quadric. The two surfaces get together where $G=0$, thus forming the whole quadric. (There is nothing wrong with the points $G=0$ on the quadric, they only behave peculiarly because we insisted on solving for x_i; if we solve for another x_j, these points will be quite regular, but some others will be peculiar.)

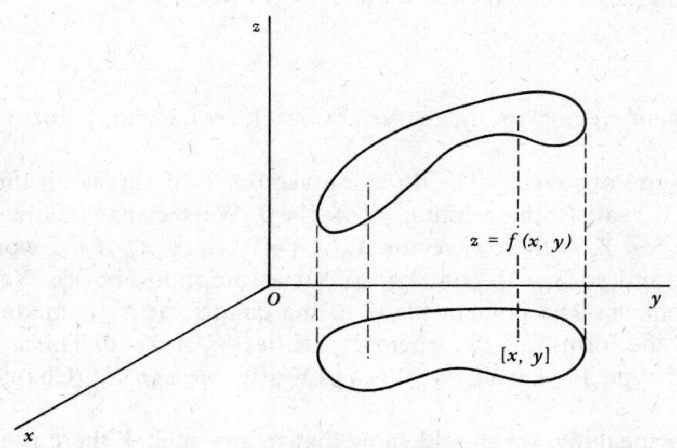

Figure 19.

Example

$x^2+y^2+z^2=1$. Solving for z, $z=\pm\sqrt{1-x^2-y^2}$. The allowable region in the x-y-plane is given by $x^2+y^2<1$; we get the upper and lower hemisphere. If we want to look at $[1,0,0]$, one of the points where the two hemispheres come together, we had better solve for x: $x=\pm\sqrt{1-y^2-z^2}$. Now we can let $[y,z]$ range through $y^2+z^2<1$, and taking the plus sign, we get $x=1$ for $y=z=0$ right in the middle of the hemisphere extending "over" the region $y^2+z^2<1$ in the y-z-plane in the positive x-direction.

In analogy to the tangent line to a curve in the plane, or the tangent plane to a surface in space, at any one of its points, there will be, at any point X_0 of our quadric, a "tangent plane," a linear variety of dimension $n-1$ which in some sense "touches" the quadric there. We consider two different approaches:

I. We use calculus. Let $X_0=[b_1,\dots,b_n]$ be a point on the quadric $X^tAX=c$. (For simplicity, take A nondegenerate and $X_0\neq 0$.) We consider *curves* on the quadric through X_0, that is, vector functions $X(s)$ of a variable s, such that (a) $X(0)=X_0$ and (b) $X(s)$ is on the quadric for all s. (b) means that $X^t(s)\cdot A\cdot X(s)=0$. We differentiate; by the product rule we get $dX^t(s)/ds\cdot A\cdot X(s)+X^t(s)\cdot A\cdot dX(s)/ds=0$; since A is symmetric, the two terms are equal, and we have $2X^t(s)\cdot A\cdot(dX(s)/ds)=0$. We put $s=0$ and write $(dX(0)/ds)=Y$, getting $X_0^t\cdot A\cdot Y=0$.

Now Y is the *direction* vector of the curve $X(s)$ at X_0: $dX(0)/ds=\lim_{s\to 0} 1/s\cdot(X(s)-X_0)$. Here $X(s)-X_0$ is the (small) vector from X_0 to the nearby point $X(s)$, and the factor $1/s$ restores it to good size. In the limit, for $s\to 0$, we get a tangent vector to the curve at X_0.

Example

The tangent vector to the curve $X(s)=[1,s+1,s^2,\sin s]$, for $s=0$, is $Y=[0,1,0,1]$.)

We restate our result. The direction vectors Y of curves on the quadric, through X_0, satisfy the relation $X_0^t\cdot A\cdot Y=0$. We see that this is *one* linear equation for Y, with row vector $X_0^t\cdot A$ ($\neq 0$, since $X_0^t\cdot A=0$ would imply $A\cdot X_0=0$, and so $X_0=0$, contrary to our assumption above). We interpret this as follows: The tangent plane to the quadric at X_0 is made up of all points of the form X_0+rY, where Y satisfies $X_0^t\cdot A\cdot Y=0$. This is the *linear variety* of type $V=\ker X_0^tA=\{Y\colon X_0^tAY=0\}$, *through* X_0 (Chapter 4, Section 4).

Strictly speaking, we should show that to any such Y there is a curve on the quadric that has Y as direction vector. To do this, we consider the 2-plane $((X_0,Y))=W$ and intersect it with the quadric; this amounts to restricting the form q to W. Then in W we simply get a conic with equation $Ax^2+Bxy+Cy^2=c$, which has the point X_0 on it; it is the required curve. (If $\dim W=1$, that is, if X_0 and Y are dependent, then the quadric must be a cone since now $X_0^tAX_0=0$; and then the line $((X_0))$ is the required curve.) The only point we have to stay away for all this from is the origin, $X_0=0$. This point is on the quadric only if the quadric is a cone; and the nature of a cone at 0 is "singular"; the cone is not a good surface there, and does not have a tangent plane. See the case $n=3$, Section 5, Figure 18(e).

The subspace $V=\ker X_0^tA$, abstractly given by $\{v: \varphi(u_0,v)=0\}$, where u_0 corresponds to X_0 and where φ is the symmetric bilinear form from which q derives, is called the **conjugate space** to u_0 (or X_0), or more precisely, the subspace of U conjugate to u_0 relative to q. More directly, two vectors u and v (or X and Y) are called **conjugate** with respect to the quadratic form, if $\varphi(u,v)=0$ (or $X^tAY=0$). This is symmetric: u is conjugate to v if and only if v is conjugate to u; in other words, if u lies on the conjugate space of v, then v lies on the conjugate space of u. We restate the result: The tangent space to the quadric $X^tAX=c$ at X_0 is the linear variety through X_0 whose type is the conjugate space $\ker X_0^t \cdot A$ of X_0. Here X_0 should not be 0, and we also assume q nondegenerate. ■

We consider a second approach: Again let u_0 (or X_0), $\neq 0$, be a point on the quadric Q_c: $q(u)=c$ (or $X^tAX=c$). Let v (or Y) be any vector ($\neq 0$), and consider the line through u_0 in direction v, consisting of all points $u_0+r\cdot v$, r in $\mathbf{R}$. The line meets the quadric Q_c in u_0, for $r=0$; we ask: Does it meet the quadric in some other point? This amounts to asking: For which r is $q(u_0+r\cdot v)=c$? We expand $q(u_0+r\cdot v)=q(u_0)+2\varphi(u_0,rv)+q(rv_0)=q(u_0)+2\varphi(u_0,v)\cdot r+q(v)\cdot r^2$. Putting this equal to c and recalling $q(u_0)=c$, we get $2\varphi(u_0,v)\cdot r+q(v)\cdot r^2=0$. One solution of this is $r_1=0$ (as it should be!), the other is $r_2=-2\varphi(u_0,v)/q(v)$ (assuming $q(v)\neq 0$).

We see that "in general" our line will meet Q_c in a second point, corresponding to r_2. However, the two points coincide, and so the line meets Q_c in only one point, if $\varphi(u_0,v)=0$, that is, if v is conjugate to u_0. This checks with what we know: The lines through u_0 corresponding to such v form the tangent plane to Q_c at u_0.

As to the exceptional case $q(v)=0$, that is, v lying on the cone Q_0 (q must be indefinite for this, since $v\neq 0$): The equation above reduces to $2\varphi(u_0,v)\cdot r=0$. Thus either it has only $r=0$ as solution. Or we have $\varphi(u_0,v)=0$ (u_0 and v conjugate), and then the equation holds for all r, so that the *whole line* lies on Q_c. This last result means that the intersection of the tangent plane at u_0 with Q_c can contain whole straight lines, namely those whose direction vectors v lie on the intersection of the cone Q_0 and the conjugate space of u_0. One can see this on the usual hyperboloid of one sheet in space (Section 5, Figure 18(b)), where the tangent plane at any u_0 has two straight lines in common with the surface. (*Warning*. There is more to this story; usually there are linear varieties of higher dimension involved.) ■

We note one last geometrical fact: Let v be any vector (not on the cone Q_0). We consider *all* lines of direction v in our vector space. Each one either meets our Q_c in two points, which might however "coincide," or it does not meet Q_c at all. In fact, if the line meets Q_c, we can take one point of intersection as u_0; we saw above that then there is either *one* or *no* other point of intersection. The second case (no other point) corresponds to $\varphi(u_0,v)=0$. In the first case, the other point of intersection is $u_0-2\varphi(u_0,v)/q(v)v$, using the value for r_2 we found. The midpoint of these two points of intersection is $1/2(u_0+u_0-2\varphi(u_0,v)q(v)v), = u_0-\varphi(u_0,v)/q(v)v = w$ say. We compute $\varphi(w,v)=\varphi(u_0,v)-\varphi(u_0,v)/q(v)\varphi(v,v)$; since $\varphi(v,v)\overset{\text{def}}{=}q(v)$, this is 0! Thus the midpoints of the pairs of points, in which lines parallel to a given direction v (with $q(v)\neq 0$) meet the quadric, all lie on the plane (subspace of codim 1) conjugate to v. (See Figure 20.) (For a line that happens to be tangent to the quadric this midpoint is the point of contact.) ■

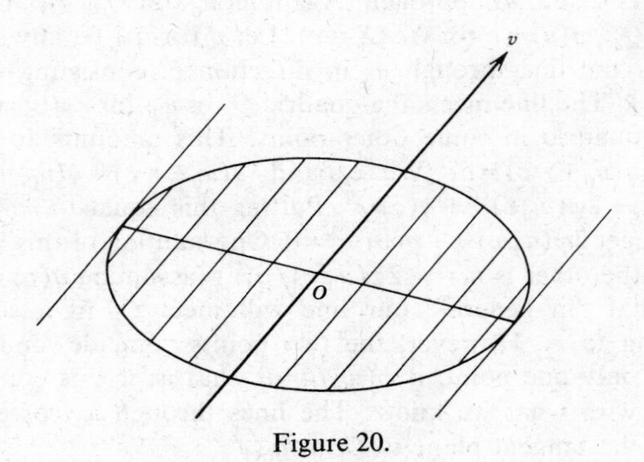

Figure 20.

Reduction to sum of squares can be interpreted in terms of the conjugate notion. Namely, that the matrix of the (symmetric) bilinear form φ with respect to a basis $\{u_1,\ldots,u_n\}$ is of sum-of-squares type means that the u_i are pairwise conjugate ($a_{ij}=\varphi(u_i,u_j)=0$, $i\neq j$) and further each $a_{ii}=q(u_i)=\varphi(u_i,u_i)$ is ± 1 or 0. This suggests another approach to reduction of theoretical and also of practical value: With φ (or A) given on U (or $\mathbb{R}^n$), first find a vector u_1 (or X_1) such that $q(u_1)\neq 0$ (or $X_1^tAX_1\neq 0$) (if there is no such vector, then φ is identically 0, by (P), Section 2); by adjusting a factor, we can make $q(u_1)=\pm 1$ ("normalization"). Next consider the conjugate space $\{v\colon \varphi(u_1,v)=0\}$ (or $\{Y\colon X_1^tAY=0\}$; all other basis vectors will be chosen in that space. So find *a* solution u_2 (or X_2) of the linear equation $\varphi(u_1,u_2)=0$ (or $X_1^tAX=0$), satisfying also $q(u_2)\neq 0$; again we can take $q(u_2)=\pm 1$. Continue: Having found $u_1,\ldots,u_k$ with $\varphi(u_i,u_j)=0$ for

$i \neq j$ and $q(u_i) = \pm 1$, find, if possible, a solution u_{k+1} of the linear equations $\varphi(u_1, v) = 0, \ldots, \varphi(u_k, v) = 0$, satisfying also $q(u_{k+1}) \neq 0$; make $q(u_{k+1}) = \pm 1$. (Note that the $u_1, \ldots, u_k$ here are automatically independent. Proof?) Either this continues all the way to n and the matrix of φ is now $\operatorname{diag}(\pm 1, \ldots, \pm 1)$, or at some k one cannot satisfy the requirement $q(u_{k+1}) \neq 0$. This means that q, restricted to the space of solutions of $\varphi(u_1, v) = \cdots = \varphi(u_k, v) = 0$, is identically 0. We then take any basis $u_{k+1}, \ldots, u_n$ of that space and find that the matrix of φ with respect to $\{u_1, \ldots, u_n\}$ is $\operatorname{diag}(\pm 1, \ldots, \pm 1, 0, \ldots, 0)$. The computations (starting from X^tAX) can be arranged quite compactly. Note that there are *choices* to be made at each stage; the process is far from unique. To sum up once more, sum-of-squares form means that the basis vectors at hand are pairwise conjugate (and normalized by $q(u_i) = \pm 1$ or 0).

PROBLEMS

1. Given the quadratic form $q = x^2 + z^2 + 2xy + 2xz + 4yz$ in $\mathbf{R}^3$.
a. Find the conjugate plane to $[1,0,0]$, to $[1,1,1]$, and to $[-3,2,1]$. (In each case find a basis for the plane in question.) Can a vector lie in its own conjugate plane?
b. Choose some Y (not on the cone $q = 0$); determine three chords of the quadric $q = 1$ parallel to Y; find their midpoints and check that they lie in the conjugate plane of Y.

2. Consider Problem 3.a, Section 4. Solve the equation $q = 1$ for x. Determine the region in the y-z-plane, where one gets (real) points on the quadric.

3. For the quadric of Problem 2, determine the tangent plane (a linear variety!) at the point $[-2,2,1]$.

4. For the hyperboloid $x^2 + y^2 - z^2 = 1$, determine the tangent plane at the point E^1. Show that the intersection of the hyperboloid with the tangent plane consists of two (straight) lines through E^1.

5. Reduce the form $x^2 - y^2 + 2xt - 2yt + 4zt - 2xz + 2yz$ to sum-of-squares form, using conjugacy as described at the end of this section. (*Advice*. Do not normalize the vectors, if awkward numbers appear.)

8. THE CASE F=C

This is our last section. So far we always had $\mathbf{F} = \mathbf{R}$ the real numbers. We now consider what happens on spaces over the *complex numbers* when $\mathbf{F} = \mathbf{C}$; x^2 can be negative now, and so all notions of positive definiteness, index of positivity, etc., go out. We bring them back in by modifying the idea of quadratic form: Basically, we replace x^2 by $|x|^2$, square of the absolute value, which also equals $\bar{x} \cdot x$. We define a **sesquilinear** (sesqui means one-and-a-half, "not quite *bi*linear") form on U, in coordinates as

$\overline{X}^tAY$ or $\Sigma a_{ij}\bar{x}_i y_j$ (complex conjugates of the x_i), or abstractly as a function φ of two variables u and v in U (values $\varphi(u,v)$), such that φ is linear in v, but *conjugate*-linear in u. Thus BL_1 is as before in Section 1 (additivity), but BL_2 is replaced by $\overline{BL}_2$: $\varphi(cu,v)=\bar{c}\varphi(u,v)$ and $\varphi(u,cv)=c\varphi(u,v)$ (note the $\bar{c}$ here). [One could even talk about sesquilinear forms on $U\times V$, for two vector spaces U and V (over C)]. The notion of symmetric form gets replaced by that of **Hermitean** (also Hermitian or hermitian) form; this is defined as a sesquilinear form φ, such that

$$\text{(H)} \qquad \varphi(u,v)=\overline{\varphi(v,u)} \qquad \text{for all } u \text{ and } v$$

(interchanging the variables yields the complex conjugate value); in matrix form we must have $\overline{X}^tAY=\overline{\overline{Y}^tAX}$ for all X,Y; this easily reduces (via $\overline{Y}^tAX=X^tA^t\overline{Y}$) to

$$\text{(H}'\text{)} \qquad A=\overline{A}^t \quad \text{or} \quad a_{ij}=\overline{a_{ji}}\,.$$

For $\overline{A}^t$ (complex conjugate transpose) one uses the symbol A^* and the term **adjoint** of A. Thus a Hermitean matrix A is one that equals its adjoint, $A=A^*$; such matrices or forms are also called **self-adjoint**. (If A is real, this is the same as symmetric.)

We can write X^*AY for $\overline{X}^tAY$. Under change of variables $X=P\cdot X'$ and $Y=P\cdot Y'$, the matrix of a sesquilinear form X^*AY changes to $P^*\cdot A\cdot P$; the det changes by $|\det P|^2$; the rank does not change.

One also calls Hermitean form the function q of *one* variable u, obtained from a Hermitean form φ by $q(u)=\varphi(u,u)$, or in coordinates (it is customary to write z_i instead of x_i) $q(Z)=Z^*AZ=\Sigma a_{ij}\bar{z}_i z_j$. The Babylonian method, completing squares, goes through (of course, one has to be careful that one is forming $\bar{z}\cdot z$ and not z^2; in the sum $\Sigma a_{ij}\bar{z}_i z_j$ terms like $a_{12}\bar{z}_1 z_2$ and $a_{21}\bar{z}_2 z_1$ cannot be combined): Any Hermitean form can be reduced to a sum-of-squares

$$|z_1|^2+\cdots+|z_r|^2-|z_{r+1}|^2-\cdots-|z_{r+s}|^2,\ r+s\leqslant n.$$

The notion of positive (or negative) definite or semidefinite and indefinite apply. (One should note that all values $q(u)$ are *real*, by

$$q(u)=\varphi(u,u)\overset{\text{(H)}}{=}\overline{\varphi(u,u)}=\overline{q(u)}\,.)$$

All the earlier notions, restriction to a subspace, index of positivity, etc., and theorems (Sylvester and complements) go through unchanged.

Example

$n=3$.

$$q(z_1,z_2,z_3)=|z_1|^2-i\bar{z}_1z_2+i\bar{z}_2z_1-\bar{z}_1z_3-\bar{z}_3z_1-2i\bar{z}_2z_3+2i\bar{z}_3z_2.$$

Collect terms in z_1 and complete the square:

$$|z_1|^2-i\bar{z}_1z_2+i\bar{z}_2z_1-\bar{z}_1z_3-\bar{z}_3z_1=|z_1-iz_2-z_3|^2-|z_2|^2-|z_3|^2+i\bar{z}_2z_3-i\bar{z}_3z_2.$$

Thus

$$q=|z_1-iz_2-z_3|^2-|z_2|^2-|z_3|^2-i\bar{z}_2z_3+i\bar{z}_3z_2.$$

Complete the square for z_2:

$$|z_2|^2+i\bar{z}_2z_3-i\bar{z}_3z_2=|z_2+iz_3|^2-|z_3|^2,$$

and so

$$q=|z_1-iz_2-z_3|^2-|z_2+iz_3|^2.$$

($|z_3|^2$ disappears.) Changing variables by $z_1'=z_1-iz_2-z_3$, $z_2'=z_2+iz_3$, and $z_3'=z_3$, we get $q=|z_1'|^2-|z_2'|^2$. Rank is 2, index of positivity is 1, index of negativity is 1, and index of degeneracy is 1.

PROBLEMS

1. Write out the matrix for the form in the example in the text.

2. Develop a polarization formula, reconstructing the Hermitean (sesquilinear) form $\varphi(u,v)$ from the "quadratic" function $q(u)=\varphi(u,u)$. Compare Formula (P) in Section 2; you will have to use not only $u+v$ but also $u+iv$.

3. Write out the Hermitean form on $\mathbf{C}^3$ with matrix

$$\begin{pmatrix} 1 & -i & 1 \\ i & 0 & -i \\ 1 & i & -2 \end{pmatrix}.$$

Reduce it to sum-of-squares form; determine rank and indices of positivity and negativity; find appropriate subspaces on which the form is positive definite or negative definite.

12

INNER PRODUCTS

In our whole development so far we have barely mentioned (Chapter 5, Section 6) or used the concepts length, distance, and angle. It is, in fact, somewhat surprising that one can do so much geometry without them. But, of course, they are important, and it is now time to consider them. The central definition is that of **inner product**, a bilinear form with certain special properties (symmetric and positive definite); length, distance, and angle are derived from the inner product. The **Schwarz inequality** is introduced; it reflects a simple geometric fact but finds application in unexpectedly many situations. The geometric idea of rotating a body around a fixed point leads to the notion of **isometry** (linear transformation that preserves the inner product) and of **orthogonal matrix (unitary** in the case of complex numbers). The main construction is the **Gram–Schmidt** process, which transforms an arbitrary basis into an orthogonal one.

1. INNER PRODUCT; SCHWARZ INEQUALITY

We start from the well-known formula (Pythagoras) giving the distance of a point P from the origin as $\sqrt{x^2+y^2}$ (in the plane) or $\sqrt{x^2+y^2+z^2}$ (in space), it being understood that the axes are "rectangular." We recognize the expression under the root sign as a positive definite quadratic form, reduced to sum-of-squares. This suggests our basic definition: An **inner product space** is a vector space U (over $\mathbb{R}$) *together with* ("*equipped with*") a positive definite quadratic form on U. To emphasize that one has committed oneself to a specific quadratic form, one calls the corresponding symmetric bilinear form the **inner product** and denotes it with a special symbol $\langle \cdot,\cdot \rangle$—we write $\langle u,v \rangle$ for the value of the form at u and v, where earlier we wrote $\varphi(u,v)$ for the values of a form φ—in the literature there are variations of this such as $\langle \cdot|\cdot \rangle$, $(\cdot,\cdot)$, and $(\cdot|\cdot)$. The value of the quadratic form at u is then $\langle u,u \rangle$; this is nonnegative, and equal to zero only for $u=0$. We write $|u|$ for the (positive) square root of this, $|u|=\sqrt{\langle u,u\rangle}$ —call it the length of the *norm* of the vector u and think of it as the analog of the ordinary length of a vector in the plane or space—given by the $\sqrt{x^2+y^2}$ or $\sqrt{x^2+y^2+z^2}$ above. Thus, it makes sense now to say that two vectors u and v of U are equally long: $|u|=|v|$ or that one is longer than the other: $|u|>|v|$. Note that this depends on our having adopted some particular p.d. form as inner product in U. How does all this relate to the usual notion of distance in physical 3-space? As earlier, we make space into a vector space by choosing an origin (see Chapter 1, Section 1). For every vector (arrow) OA we then have the *length*, which we write as $|OA|$. It is now a fact supported by long experience that the function whose value for any OA is $|OA|^2$ (the square of the length) is a quadratic function on our vector space, in the sense of Chapter 11; it is clearly positive definite (no negative values possible; 0 only for the 0-vector OO). Actually, that experience is not so easy to make explicit. The best way to do it might be to point out that if one refers space to "rectangular axes," the expression $x^2+y^2+z^2$ seems to represent $|OA|^2$ very well.

There is a fairly natural generalization to complex vector spaces: There an inner product means a positive definite Hermitean form. One uses the same notation as before: $\langle u,v\rangle$ for the inner product and $|u|=\langle u,u\rangle$ for the norm. One should realize that $\langle u,v\rangle$ is in general now a complex number; however, $\langle u,u\rangle$ is real and nonnegative, so $|u|$ makes good sense.

Incidentally, the term "inner product" is rather accidental and does not mean anything by itself.

Examples for inner products are simply examples of positive definite symmetric quadratic forms [Hermitean forms in the case $\mathsf{F}=\mathsf{C}$]. (From here on, we carry the case $\mathsf{F}=\mathsf{C}$ along in square brackets, to simplify statements that apply, with the obvious minor changes, to both cases.) Thus, using coordinates relative to a basis, we have $\langle u,v\rangle = X^tAY$ [or X^*AY] and $|u|^2=X^tAX$ [or X^*AX], with a positive definite symmetric [or Hermitean] matrix A.

Of importance is the **standard** inner product X^tY $(=\Sigma x_iy_i)$ on R^n with $|X|=\sqrt{x_1^2+\cdots+x_n^2}$. (This is the straightforward generalization to R^n of what we started from for $n=2$ and 3; we described it briefly in Chapter 4, Section 6.) It is frequently denoted by $X\cdot Y$ and referred to as the "dot product"; the similarity with the earlier $A\cdot Y$, with X in R^n and A in $(\mathsf{R}^n)'$, is intentional; compare Chapter 4, Section 6. Similarly, there is the **standard** inner product X^*Y $(=\Sigma\bar{x}_iy_i)$ in C^n; here the norm $|X|$ is given by $\sqrt{|x_1|^2+\cdots+|x_n|^2}$ (note $\bar{x}\cdot x=|x|^2$). We write again $X\cdot Y$.

The theorem on reduction of a quadratic form to sum-of-squares says that any inner product on any (real) vector space, when referred to a suitable basis, "looks like" the standard inner product on R^n; this is similar for the complex case.

To go on, we need a fundamental and omnipresent inequality for positive definite forms, called the **Schwarz** inequality.

1.1. PROPOSITION. *Let φ be a positive definite symmetric [or Hermitean, if $\mathsf{F}=\mathsf{C}$] form, with quadratic form q. Then, for any two vectors u and v, the inequality*

$$\text{(S)} \qquad |\varphi(u,v)| \leqslant \sqrt{q(u)} \cdot \sqrt{q(v)}$$

holds; equality holds if and only if u and v are dependent.

Formula (S) can also be written as $|\varphi(u,v)|^2 \leqslant q(u)\cdot q(v)$ or, in the real case, $\varphi(u,v)^2 \leqslant q(u)\cdot q(v)$.

Before the proof we restate this for an inner product space with inner product $\langle\cdot,\cdot\rangle$ and norm $|\cdot|$:

$$\text{(S)} \qquad |\langle u,v\rangle| \leqslant |u|\cdot|v|.$$

"The absolute value of the inner product is at most equal to the product of the two norms." Equality holds only if the two vectors are dependent. ($\mathsf{F}=\mathsf{R}$ *or* C.)

PROOF (in inner product notation). (1) $\mathsf{F}=\mathsf{R}$. With $w=|v|u-|u|v$, we have $0\leqslant\langle w,w\rangle$. Expanding by bilinearity and using $\langle u,v\rangle=\langle v,u\rangle$, we get $0\leqslant|v|^2\cdot\langle u,u\rangle-2|u||v|\langle u,v\rangle+|u|^2\langle v,v\rangle$. With $\langle u,u\rangle=|u|^2$ and $\langle v,v\rangle=|v|^2$ this turns into $0\leqslant 2\cdot(|u|^2\cdot|v|^2-|u|\cdot|v|\cdot\langle u,v\rangle)$. Dividing, we obtain $\langle u,v\rangle \leqslant |u|\cdot|v|$. This is almost (S), except that we need the absolute value $|\langle u,v\rangle|$ instead of $\langle u,v\rangle$ itself. If we start from $w'=|v|u+|u|v$, we get $-\langle u,v\rangle \leqslant |u|\cdot|v|$; that together with the first inequality is indeed (S).

(2) For $\mathsf{F}=\mathsf{C}$. With w as before we get this time $0\leqslant|v|^2\cdot\langle u,u\rangle - |u|\cdot|v|\cdot(\langle u,v\rangle+\langle v,u\rangle)+|u|^2\cdot\langle v,v\rangle$, since we can no longer use $\langle u,v\rangle =\langle v,u\rangle$. However, we *do* have $\langle v,u\rangle=\overline{\langle u,v\rangle}$, and therefore $\langle u,v\rangle+\langle v,u\rangle=2\,\mathrm{Re}(\langle u,v\rangle)$, where $\mathrm{Re}(z)$ means real part of the complex number z (with $z=a+ib$, $\mathrm{Re}(z)=a$, and $z+\bar{z}=a+ib+a-ib=2a=2\,\mathrm{Re}(z)$). Thus, dividing by $2|u|\cdot|v|$, we find $\mathrm{Re}(\langle u,v\rangle)\leqslant|u|\cdot|v|$. To get from here to (S) we use the following trick: If we replace u by $u'=e^{i\alpha}\cdot u$, then $\langle u,v\rangle$ becomes $\langle u',v\rangle=e^{-i\alpha}\langle u,v\rangle$, but $|u|$ does not change:

$$|u'|=\sqrt{\langle u',u'\rangle}=\sqrt{\langle e^{-i\alpha}u,e^{-i\alpha}u\rangle}=\sqrt{e^{i\alpha}\cdot e^{-i\alpha}\langle u,u\rangle}=\sqrt{\langle u,u\rangle}=|u|$$

(note the conjugate linearity in the first variable). Therefore, $\operatorname{Re}(e^{-i\alpha}\langle u,v\rangle) \leqslant |u|\cdot|v|$ for *any* $e^{i\alpha}$. Now by choosing $e^{i\alpha}$ right, we can make $e^{-i\alpha}\langle u,v\rangle = |\langle u,v\rangle|$ (if $z=|z|\cdot e^{i\alpha}$, then $e^{-i\alpha}z=|z|$): this yields $|\langle u,v\rangle| = \operatorname{Re}(|\langle u,v\rangle|) = \operatorname{Re}(e^{-i\alpha}\langle u,v\rangle) \leqslant |u|\cdot|v|$, establishing (S).

We consider equality in (S). In the real case $|\langle u,v\rangle| = |u|\cdot|v|$ means $\langle u,v\rangle = \pm|u|\cdot|v|$. Say $+$ here. Then in the computation above we get $\langle w,w\rangle = 0$; so $w=0$ because $\langle\cdot,\cdot\rangle$ positive definite, but this means $|v|u - |u|v = 0$, a dependence relation (the factors $|v|$ and $|u|$ are 0 only if v and u are 0).

For $\mathsf{F}=\mathsf{C}$ we have essentially the same argument; namely, $w=|v|\cdot e^{i\alpha}\cdot u - |u|\cdot v$, with $e^{i\alpha}$ chosen as above, has $\langle w,w\rangle = 0$, and so $w=0$.

PROBLEMS

1. For the vectors $X=[1,-2,4,2]$ and $Y[1,5,-3,1]$ of R^4 compute $X^2 = X\cdot X$, $Y^2 = Y\cdot Y$, and $X\cdot Y$. Verify that the Schwarz inequality holds.

2. Let an inner product $\langle,\rangle$ be defined by R^3 by the quadratic form with matrix $Q=\begin{pmatrix} 1 & 1 & -1 \\ 1 & 2 & -2 \\ -1 & -2 & 3 \end{pmatrix}$. (This is positive definite !) For the vectors $X=[1,-1,1]$ and $Y=[1,2,3]$ compute $|X|^2$, $|Y|^2$, and $\langle X,Y\rangle$. Check the Schwarz inequality. Find a (nonzero) vector Z such that $\langle X,Z\rangle = 0$.

3. In R^n, with standard inner product, let $X_1,\dots,X_n$ be a basis. Let $c_1,\dots,c_n$ be any n numbers. Show (using your knowledge of linear equations) that there exists a unique vector Y such that $Y\cdot X_1 = c_1$, $Y\cdot X_2 = c_2,\dots,$ and $Y\cdot X_n = c_n$.

4. Two bases $X_1,\dots,X_n$ and $Y_1,\dots,Y_n$ for R^n are called dual or reciprocal to each other if $X_i\cdot Y_j = \delta_{ij}$ (Kronecker delta). Show that it follows from Problem 3 that, given a basis, there is a unique dual basis.

5. In R^3, find the dual basis to $[1,1,1]$, $[1,2,2]$, and $[1,2,1]$.

6. Let U be a vector space (over R, say) with inner product $\langle,\rangle$. Let $v_1,\dots,v_k$ be any vectors in U. The matrix G, whose entries are the inner products $\langle v_i,v_j\rangle$, is called the Gram matrix or Gramian of the vectors. Show that $\det G$ is always nonnegative and is 0 exactly if the v_i are dependent. (*Hint*. With variables $x_1,\dots,x_k$ form $v=\Sigma x_i v_i$, and consider $\langle v,v\rangle$, (a quadratic form in the x_i.)

2. NORM, ANGLE, ORTHOGONALITY

We come back to our geometry, with consequences of the Schwarz inequality. First, there are some important properties of the norm:

1. $|u| \geqslant 0$ for all u ($=0$ only for $u=0$);

(N) 2. $|cu| = |c|\,|u|$ (here c is real or complex, as needed);

3. $|u+v| \leqslant |u| + |v|$ for any u, v; and $=$ only if u, v go in the same direction, that is, one is a nonnegative multiple of the other.

Property 3 is the **norm-** or **triangle-inequality**; it says that the diagonal of a parallelogram is at most as long as the sum of the two sides.

PROOF. (N) 1 and (N) 2 are trivial: Property 1 comes from the positive definiteness of $\langle \cdot \rangle$, in particular the fact that $\langle u,u \rangle = 0$ happens only for the vector 0; Property 2 comes from (part of) the bilinearity of $\langle , \rangle$: $\langle ru, ru \rangle = r^2 \langle u,u \rangle$ (or $\langle cu, cu \rangle = \bar{c} \cdot c \langle u,u \rangle$ in the complex case). (N) 3 comes from the Schwarz inequality: $|u+v|^2 = \langle u+v, u+v \rangle = |u|^2 + 2\langle u,v \rangle + |v|^2 \leqslant |u|^2 + 2|u| \cdot |v| + |v|^2$ (by Schwarz) $= (|u| + |v|)^2$. (There are some minor changes in the complex case: instead of $2\langle u,v \rangle$ we have $\langle u,v \rangle + \langle v,u \rangle = 2\,\mathrm{Re}(\langle u,v \rangle)$, which is $\leqslant 2|\langle u,v \rangle|$, and this in turn is less than or equal to $2|u| \cdot |v|$ by (S).) The "$=$"-case is relegated to Problem 1. ∎

To make this more geometric, we introduce the notion "distance": the distance $d(u,v)$ from (the tip of) u to (the tip of) v is the norm $|v-u|$. (See Figure 21.) And then, if u, v, and w are three vectors, we have the general

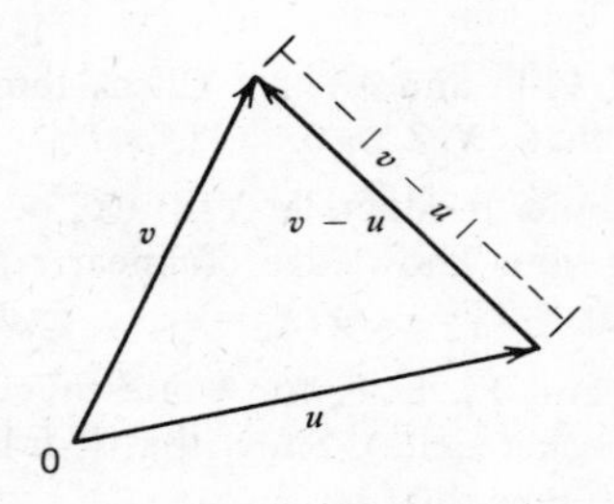

Figure 21.

triangle-inequality:

$$\text{(TI)} \qquad d(u,w) \leqslant d(u,v) + d(v,w).$$

A side in a triangle has length equal to, at most, the sum of the other two sides. (What happens in the case of "$=$"?)

PROOF. This is simply (N) 3 above, with u and v replaced by $v-u$ and $w-v$, and consequently $u+v$ replaced by $v-u+w-v=w-u$.

Next, we look at the concept "angle." For this we have to take $\mathsf{F} = \mathsf{R}$. We start from the cosine law in the plane: For a triangle we know, $c^2 = a^2 +$

$b^2-2ab\cdot\cos\gamma$. We compare this with Figure 22 in U: If γ in this abstract situation makes sense, we should have $|v-u|^2=|u|^2+|v|^2-2|u||v|\cdot\cos\gamma$. Expanding $\langle v-u,v-u\rangle=|u|^2-2\langle u,v\rangle+|v|^2$ (here is where we use $\mathsf{F}=\mathsf{R}$),

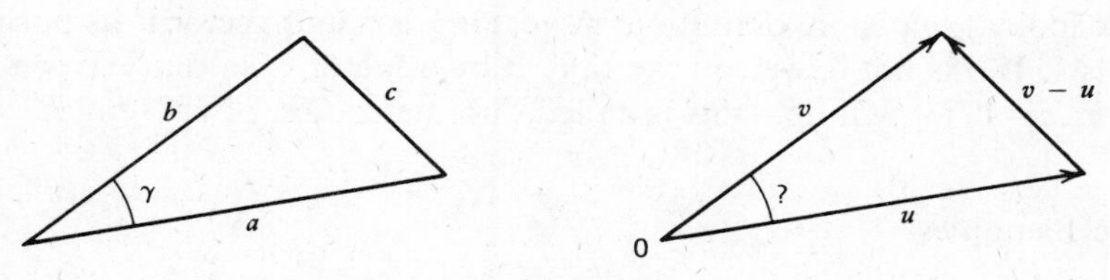

Figure 22.

we are lead to $\langle u,v\rangle=|u|\cdot|v|\cdot\cos\gamma$ (assuming, of course, $u\neq 0$ and $v\neq 0$):

$$\text{(A)} \qquad \cos\gamma=\frac{\langle u,v\rangle}{|u|\cdot|v|}$$

The only thing wrong here is that for an abstract U and $\langle,\rangle$ we do not really know what angle means; we should and will use (A) to *define* γ: The angle between u and v is measured by that number γ, $0\leqslant\gamma\leqslant\pi$, that satisfies (A). And for that to make sense (i.e., to get a γ from the value $\langle u,v\rangle/|u|\cdot|v|$ for $\cos\gamma$), we must be sure that values lies between -1 and $+1$ (endpoints included). But that is just what the Schwarz inequality says: $|\langle u,v\rangle|\leqslant|u|\cdot|v|$ is the same as $-1\leqslant(\langle u,v\rangle/|u|\cdot|v|)\leqslant+1$.

So now we can talk about the angle $\angle\,(u,v)$ formed by two nonzero vectors. Note some special cases before we go on: $\gamma=0$ means $\cos\gamma=1$ or $\langle u,v\rangle=|u|\cdot|v|$; by the "equality" part of the Schwarz inequality this can happen only if u and v have the same direction, that is, $u=r\cdot v$ with $r>0$. Similarly, $\gamma=\pi$ and $\cos\gamma=-1$ mean that u and v have opposite direction, $u=r\cdot v$ with $r<0$. Finally, $\gamma=\pi/2$, "u and v at right angles," means $\langle u,v\rangle=0$. This is an important relation; we say that u and v are **orthogonal** (also perpendicular) to each other; we write this $u\perp v$. Thus, $u\perp v$ amounts to $\langle u,v\rangle=0$. (By the way, then also $\langle v,u\rangle=0$, so that being perpendicular is a symmetric relation.) In particular we count the vector 0 as orthogonal to any vector—even though then "angle" strictly speaking does not make sense, since we cannot use formula (A). Note that in the case $\mathsf{F}=\mathsf{C}$ the concept of angle goes out: Since $\langle u,v\rangle$ is normally a complex number, we do not get a real angle γ from (A). However, "orthogonal" makes sense with the same definition: $u\perp v$ means $\langle u,v\rangle=0$.

Example

We consider a simple example: In $\mathbf{R}^3$, standard inner product, the vector $X=[1,2,3]$ is perpendicular to $Y=[4,1,-2]$, since $X\cdot Y=X^tY=1\cdot 4+2\cdot 1+3\cdot -2=0$.

We look at one more definition: A vector u is a **unit** vector if its norm $|u|$ equals 1. If v is not 0, we can multiply it by a factor c, so that $c\cdot v$ is a unit vector; $c=1/|v|$ will do—this is called "normalization of v."

Some Examples

$U=\mathbf{R}^3$, with $\langle u,u\rangle$ or $\langle X,X\rangle$ given by $x_1^2+2x_2^2+3x_3^2+2x_1x_2-2x_1x_3-4x_1x_3$ (this is *not* the standard inner product). (Check, by reducing to sum-of-squares, that this is positive definite.) We consider two vectors. $X=[1,1,1]$, $\langle X,X\rangle=\cdots=2$, $|X|=\sqrt{2}$; $Y=[1,2,2]$, $\langle Y,Y\rangle=\cdots=5$, $|Y|=\sqrt{5}$; $\langle X,Y\rangle=\cdots=3$. (Careful, we have to polarize the form, or better, get the matrix $A=\begin{pmatrix}1 & 1 & -1\\ 1 & 2 & -2\\ -1 & -2 & 3\end{pmatrix}$ for $\langle X,Y\rangle=X^tAY$.) So $\cos\gamma=\langle X,Y\rangle/|X|\cdot|Y|=3/\sqrt{10}$; from this (and trigonometric tables) $\gamma\approx 18°26'$. Find a vector orthogonal to X: $Z=[z_1,z_2,z_3]$ must satisfy $\langle X,Z\rangle=\cdots=z_1+z_2=0$; this is a space of dimension two, basis $Z_1=[-1,1,0]$, $Z_2=[0,0,1]$; *any* vector in $((Z_1,Z_2))$ is orthogonal to X.

Here is a more computational approach to the Schwarz inequality: By choosing a basis our $\langle\cdot,\cdot\rangle$ becomes X^tAY with a certain (symmetric) matrix; by changing to new coordinates we can reduce to sum-of-squares, which, because of positive definiteness, must be $x_1^2+\cdots+x_n^2$; the new A is I. The inner product $\langle u,v\rangle$ becomes then simply $X^t\cdot Y=x_1y_1+x_2y_2+\cdots+x_ny_n$. The Schwarz inequality becomes the **Cauchy** inequality

$$\text{(C)}\qquad (x_1y_1+\cdots+x_ny_n)^2\leqslant(x_1^2+\cdots+x_n^2)\cdot(y_1^2+\cdots+y_n^2)$$

for *any* real $x_1,\ldots,x_n$, $y_1,\ldots,y_n$.

For $n=2$ one verifies that $(x_1^2+x_2^2)\cdot(y_1^2+y_2^2)-(x_1y_1+x_2y_2)^2=(x_1y_2-x_2y_1)^2$; this shows $\geqslant 0$ ($=0$ only if $x_1/x_2=y_1/y_2$, i.e., if X and Y are dependent). For general n one can show $(x_1^2+\cdots+x_n^2)\cdot(y_1^2+\cdots+y_n^2)-(x_1y_1+\cdots+x_ny_n)^2=\Sigma_{i<j}(x_iy_j-x_jy_i)^2$, so again $\geqslant 0$.

For the complex case (C) becomes $|\bar{x}_1y_1+\cdots+\bar{x}_ny_n|^2\leqslant(|x_1|^2+\cdots+|x_n|^2)\cdot(|y_1|^2+\cdots+|y_n|^2)$.

Note that (C) is purely algebraic; there is no mention of geometry there; just an inequality for (sets of) numbers.

PROBLEMS

1. Carefully complete the argument for the "=" case in (N) 3.
2. Find $\cos\alpha$ for the angle α between X and Y in Problems 1 and 2 of Section 1.
3. Find Z in $\mathbf{R}^4$, orthogonal to X and Y of Problem 1 in Section 1.
4. Find Z in $\mathbf{R}^4$, orthogonal to X and Y of Problem 2 in Section 1.
5. In $\mathbf{R}^n$, with standard inner product, normalize the vector $[1,1,\ldots,1]$.
6. Verify the statement in the text about the Cauchy inequality for $n=2$.
7. Work out the Cauchy inequality for $n=3$: Show that the difference "right side minus left side" reduces to $(x_1y_2-x_2y_1)^2$ plus two similar squares.
8. Work out the Cauchy inequality in the complex case for $n=2$.

3. SOME EXAMPLES

1. A mechanical system consisting (in an idealized version) of a number of moving mass points can be described by a number of coordinates $x_1,\ldots,x_n$ (the x_i might be positions or angles or other parameters). The system moves, that is, the x_i are functions of t (time). The (generalized) velocity of the system, at any moment t is given by the n-vector $\dot{X}=[\dot{x}_1,\ldots,\dot{x}_n]$. The kinetic energy of the system, at any moment t, usually turns out to be given by a quadratic expression $\frac{1}{2}\Sigma a_{ij}\dot{x}_i\dot{x}_j$ (generalizing the usual $mv^2/2$ of a one-dimensional motion of one particle); the a_{ij} could depend on position (x_i) and time (t), but are often constant. The form $K(\dot{X}) = \frac{1}{2}\Sigma a_{ij}\dot{x}_i\dot{x}_j$ is positive definite, by its very meaning (energy 0 only for "no motion," all $\dot{x}_i=0$). It is natural to consider the vector space of all possible velocity vectors $\dot{X}=[\dot{x}_1,\ldots,\dot{x}_n]$ with $2K(\dot{X})$ (or rather the associated bilinear form) as inner product.

Example (a bit artificial)

A (one-dimensional) table of weight m moves on the x_1-axis; the center of the table is at x_1. On the table is an object, weight p, its center at distance x_2 from the center of the table. The kinetic energy is

$$\frac{m\dot{x}_1^2}{2}+\frac{p(\dot{x}_1+\dot{x}_2)^2}{2}=\frac{m+p}{2}\dot{x}_1^2+p\dot{x}_1\dot{x}_2+\frac{p}{2}\dot{x}_2^2.$$

2. We look at the moment of inertia. In ordinary 3-space consider a body, rotating at constant speed about a fixed axis through a fixed point of the body, located at O. We represent the angular velocity by a vector OA, usually denoted by ω, along the axis of rotation, with the length of OA

equal to the magnitude $|\omega|$ of the angular velocity (number of angle units per time unit). We shall show that the kinetic energy of the body, in its dependence on ω, is a quadratic form in ω; we write $K = \frac{1}{2}J(\omega)$, with the factor $\frac{1}{2}$ suggested by the analogy with $\frac{1}{2}mv^2$. This analogy becomes stronger by "factoring out" $|\omega|^2$, that is, by writing $K = \frac{1}{2}I(\omega)\cdot|\omega|^2$, where now $I(\omega)$, the *moment of inertia* of the body relative to ω, depends only on the direction of ω (and not on its length). In coordinates K becomes $\frac{1}{2}\omega^t J\omega$, with a certain symmetric matrix J.

We derive this now; for the details we need Proposition 4.2 below. For a particle of the body, of mass m, at distance r from the axis, the speed is $r\cdot|\omega|$, by definition of angular velocity, and so the kinetic energy is $\frac{1}{2}mr^2|\omega|^2$. We have to take the sum of all these over the whole body (for a "continuous" body the sum is an integral).

We regard space as $\mathbf{R}^3$, with the standard inner product and norm (strictly speaking the axes should be fixed to the body). ω is then given as $[\omega_1, \omega_2, \omega_3]$, and the position of the particle as $X = [x_1, x_2, x_3]$. The square of the distance from particle to axis is then $r^2 = (X - (\omega\cdot X/\omega\cdot\omega)\omega)^2$, by Proposition 4.2. And the kinetic energy becomes (after squaring out) $\frac{1}{2}\cdot m(\omega^2\cdot X^2 - \omega\cdot X^2)$. This is quadratic in the components of ω, and the same holds after we sum over all the particles. [Thus $J(\omega) = \Sigma m(\omega^2\cdot X^2 - (\omega\cdot X)^2]$.

Example

Suppose the "body" consists of four unit masses placed at the corners of the unit square in the horizontal plane. We have thus four X's, 0, E^1, E^2, and $E^1 + E^2$. With $(E^i)^2 = 1$, $(E^1 + E^2)^2 = 2$, and $\omega\cdot E^i = \omega^i$, the sum becomes $\frac{1}{2}(4\omega^2 - (\omega_1^2 + \omega_2^2 + (\omega_1 + \omega_2)^2) = \omega_1^2 + \omega_2^2 + 2\omega_3^2 - \omega_1\omega_2$. This is positive definite!

PROBLEMS

1. Work out the moment of inertia (kinetic energy of rotation) for a cube in space, with one vertex at 0, its faces parallel to the coordinate planes, length of side 1, and density 1.

4. ORTHOGONAL SPLITTING

We go back to mathematics. Let U be a vector space with inner product $\langle,\rangle$, over $\mathbf{R}$ or $\mathbf{C}$. Intuitively two vectors u and v are "as far apart as possible in direction," if they are orthogonal. This can be strengthened to a very important fact.

4.1. PROPOSITION. *Nonzero, pairwise orthogonal vectors are linearly independent. In detail, if* $u_1,\ldots,u_k$ *are nonzero and satisfy* $\langle u_i,u_j\rangle=0$ *for* $i\neq j$, *then they are independent.*

Even the method of proof is important: Let $w=\Sigma a_i u_i$ be any linear combination of the u_i. Then for each j we have $\Sigma_1^k\langle u_j,w\rangle=\Sigma_1^k\langle u_j,a_iu_i\rangle$ $=\Sigma_1^k a_i\langle u_j,u_i\rangle=a_j\cdot|u_j|^2$. Note that in the sum all $\langle u_j,u_i\rangle$, except the jth one, are 0; and the jth is $\langle u_j,u_j\rangle=|u_j|^2$. In particular, if $w=0$ (i.e., if we have a *relation* of the u_i), then we find $0=a_j\cdot|u_j|^2$. Furthermore, $|u_j|^2$ is not zero by hypothesis ($u_j\neq 0$), and so $a_j=0$. This is true for all $j=1,\ldots,k$, and shows that the u_i are independent. Be sure to understand this proof. ■

We consider a special case: An **orthonormal** set (ON-set) is, by definition, a sequence $u_1,\ldots,u_k$ of pairwise orthogonal unit vectors: $|u_i|=1$ and $\langle u_i,u_j\rangle=0$ for $i\neq j$. Similarly, an **orthogonal** set is a sequence $u_1,\ldots,u_k$ with $u_i\perp u_j$, that is $\langle u_i,u_j\rangle=0$ for $i\neq j$; we drop the normalization condition; in particular, we allow 0-vectors as u_i. We also talk about orthogonal and ON bases for U.

Proposition 1 implies the following.

4.1′. PROPOSITION. *ON-sets are automatically independent. In particular, the number k of vectors in such a set is at most* $n\,(=\dim U)$. ■

Note an important fact ("Pythagoras"): Let u and v be orthogonal to each other. Then $|u+v|^2=|u|^2+|v|^2$. This can be proved by $|u+v|^2=\langle u+v,u+v\rangle=\langle u,u\rangle+\langle u,v\rangle+\langle v,u\rangle+\langle v,v\rangle$. But $\langle u,v\rangle=0$, and also $\langle v,u\rangle=0$. Further $\langle u,u\rangle=|u|^2$ and $\langle v,v\rangle=|v|^2$. (See Figure 23.) ■

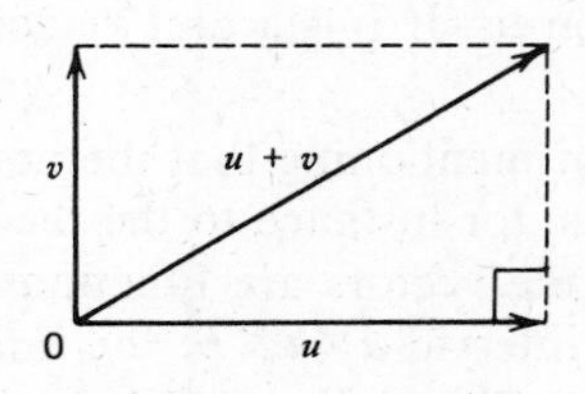

Figure 23.

Next we show that in a certain sense independent sets can always be replaced by ON-sets; this is the **Gram–Schmidt** orthogonalization process. We begin with the case of two vectors.

4.2. PROPOSITION. *Let* u *and* v *be two vectors in* U, *with* $v\neq 0$. *Then* u *can be "resolved" or "split" uniquely into two components* $u_{\|}$ *and* $u_{\perp}$, *where* $u_{\perp}$ *is orthogonal to* v, $u_{\|}$ *is a multiple of* v *("parallel to* v*"), and* $u=u_{\|}+u_{\perp}$.

(*See Figure* 24.) *And we have the formula*

$$u_{\|}=\frac{\langle v,u\rangle}{\langle v,v\rangle}v=\cos\phi\cdot|u|\frac{v}{|v|}$$

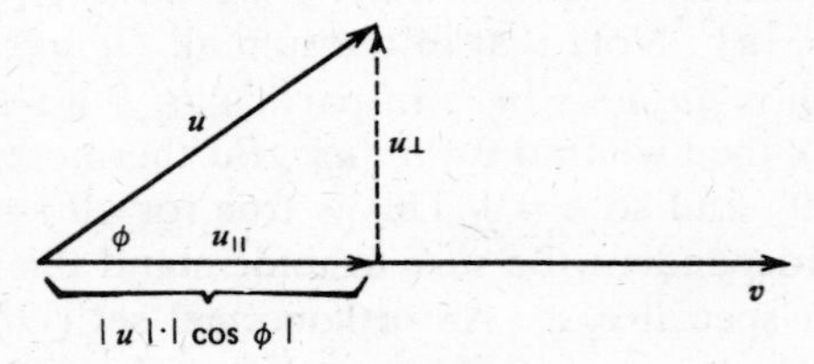

Figure 24.

The numerical factor $\langle v,u\rangle/\langle v,v\rangle$ occurring here is called the coefficient of u with respect to v. The vector $u_{\|}$ is called the component of u in direction v, or the (orthogonal) projection of u onto v.

PROOF. For existence of the splitting we just have to show that $u-u_{\|}$ (with $u_{\|}$ as described) is orthogonal to v. So we "take the inner product," we compute $\langle v,u-u_{\|}\rangle=\langle v,u\rangle-\langle v,u_{\|}\rangle$. The second term is $\langle v,(\langle v,u\rangle/\langle v,v\rangle)v\rangle$, which is by linearity L_2 equal to $(\langle v,u\rangle/\langle v,v\rangle)\langle v,v\rangle$, which is equal to $\langle v,u\rangle$, and so the difference of the two terms is 0. For uniqueness, say we have $u=u_{\|}+u_{\perp}$ with $u_{\|}=av$ and $u_{\perp}\perp v$. Again we take $\langle v,-\rangle$. We get $\langle v,u\rangle=\langle v,av\rangle+\langle v,u_{\perp}\rangle=a\cdot\langle v,v\rangle+0$. We find $a=\langle v,u\rangle/\langle v,v\rangle$: the coefficient is *forced* on us. If v is a unit vector, then $u_{\|}$ is just $\langle v,u\rangle v$. ■

This is a good point for mentioning that the notion "inner product" has significant generalizations, for instance to the theory of Fourier series and Fourier integrals. There the vectors are functions, generally complex valued, defined on, say, an interval $a\leqslant t\leqslant b$; the inner product $\langle f,g\rangle$ of two functions is by definition the value $\int_a^b\bar{f}(t)g(t)dt$. The constructions of Propositions 4.2, 5.1, 5.2, and 5.3, turn out to be very important.

PROBLEMS

1. Work out the details for the following alternate proof of Proposition 4.1: Compute $\langle w,w\rangle$. Conclude from the result: If w is 0, then all a_i must be 0. (Treat the cases $\mathsf{F}=\mathsf{R}$ or $\mathsf{F}=\mathsf{C}$ separately.)

2. Let $u_1,\ldots,u_k$ be an *ON*-set. For a vector $u=\sum a_iu_i$ (linear combination of the

u_i), compute the expressions for $|u|$ in terms of the a_i. (Distinguish the cases $\mathbf{R}$ and $\mathbf{C}$.)

3. Let $u_1,\ldots,u_k$ be vectors in U, over $\mathbf{R}$ say, with inner product $\langle,\rangle$. Suppose for any linear combination $u=\Sigma a_i u_i$ we have $|u|^2=\Sigma a_i^2$. Show that the u_i form an ON-set.

4. Prove: If $u_1,\ldots,u_k$ are pairwise orthogonal, then $|u_1+\cdots+u_k|^2=|u_1|^2+\cdots+|u_k|^2$.

5. Consider $\mathbf{R}^3$ with the quadratic form of Problem 2 of Section 1 as inner product. For $Y=[1,2,3]$ find the two components $Y_\parallel$ and $Y_\perp$ parallel and perpendicular to $X=[1,-1,1]$.

6. Given U with $\langle,\rangle$, let v be a nonzero vector, and let $u=u_\parallel+u_\perp$ be the splitting of a vector u with respect to v.

a. Show that $|u|\geqslant|u_\parallel|$ and $|u|\geqslant|u_\perp|$.

b. Under what circumstances can one have $|u|=|u_\parallel|$ or $|u|=|u_\perp|$?

7. Show that the two expressions for $u_\parallel$ in Proposition 4.2 are equal.

5. THE GRAM—SCHMIDT PROCESS

We extend Proposition 4.2.

First, we consider a definition: A vector u is orthogonal to a subspace V of U (we write $u\perp V$) if u is orthogonal to *every* vector in V.

Clearly, it is enough if u is orthogonal to a basis or a spanning set for V. Similarly, two subspaces V and V' of U can be orthogonal to each other—$V\perp V'$ means that every v in V is orthogonal to every v' in V'.

5.1. PROPOSITION. *Let $v_1,\ldots,v_k$ be pairwise orthogonal, nonzero vectors. Then any vector u can be "split," uniquely, into $u_\parallel+u_\perp$, where $u_\parallel$ lies in the span $V=((v_1,\ldots,v_k))$ and $u_\perp$ is orthogonal to V. In fact, $u_\parallel$ is given by*

$$\frac{\langle v_1,u\rangle}{\langle v_1,v_1\rangle}v_1+\cdots+\frac{\langle v_k,u\rangle}{\langle v_k,v_k\rangle}v_k.$$

Note that the coefficients $\langle v_i,u\rangle/\langle v_i,v_i\rangle$ appear; also note that the v_i must be pairwise orthogonal.

PROOF. This is similar to the proof of Proposition 4.2. The $u_\parallel$ described lies in V; we have to show that $u-u_\parallel$ is orthogonal to V. For this we take $\langle v_i,u-u_\parallel\rangle$, for $i=1,\ldots,k$; using the expression for $u_\parallel$ and the fact that $\langle v_j,v_i\rangle=0$ if $j\neq i$, we get $\langle v_i,u\rangle-(\langle v_i,u\rangle/\langle v_i,v_i\rangle)\langle v_i,v_i\rangle$ (of the k terms that make up $u_\parallel$ only the ith survives), and this is again zero. For uniqueness we also argue as before: If $u-(a_1v_1+\cdots+a_kv_k)$ wants to be orthogonal to V, it must have inner product zero with each v_i, but that inner product reduces to $\langle v_i,u\rangle-a_i\langle v_i,v_i\rangle$, and so $a_i=\langle v_i,u\rangle/\langle v_i,v_i\rangle$ is forced on us. The formula is again simpler if the v_i are unit vectors, that is, form an *ON* system: $u_{\|}=\langle v_1,u\rangle v_1+\cdots+\langle v_k,u\rangle v_k$. We call $u_{\|}$ the component of u parallel to V, and $u_{\perp}$ the component of u orthogonal to V. ■

In Proposition 5.1 the space V was *assumed* to have an orthogonal basis. We show next that orthogonal bases always exist, and give a procedure how to construct one starting from any basis.

5.2. THEOREM. (*Gram–Schmidt Orthogonalization Process*). *Let $u_1,\ldots,u_k$ be independent vectors. Then the vectors $v_1,\ldots,v_k$, defined inductively by taking for v_i the component of u_i orthogonal to $((v_1,\ldots,v_{i-1}))$ (beginning with $v_1=u_1$), form an orthogonal basis for $((u_1,\ldots,u_k))$.*

COROLLARY. *Any inner product space has ON bases.*

PROOF OF COROLLARY (from Theorem 5.2). Take any basis and orthogonalize it, then normalize the vectors (replace v_i by $(1/|v_i|)v_i$). ■

PROOF OF THEOREM 5.2. We start with $v_1=u_1$. Next, using Proposition 4.2, we split u_2 as $u_{2\|}+u_{2\perp}$, parallel and orthogonal to u_1, and put $v_2=u_{2\perp}$. We claim that $((u_1,u_2))=((v_1,v_2))$. We have $u_1=v_1$; we also have $u_2=a\cdot u_1+v_2$, since $u_{2\|}$ is a multiple of u_1 and $u_{2\perp}=v_2$; finally, $v_2=-a\cdot u_1+u_2$. This shows that the two spaces are equal. Now we look at the general step: Suppose we have $v_1,\ldots,v_{i-1}$, pairwise orthogonal, so that $((v_1,\ldots,v_{i-1}))=((u_1,\ldots,u_{i-1}))$. We split u_i into its components parallel and orthogonal to this space, using Proposition 4.2 again; and call the second component v_i; so that $u_i=u_{i\|}+v_i$ with $u_{i\|}$ in $(u_1,\ldots,u_{i-1}))$. It is clear then that $((u_1,\ldots,u_i))=((v_1,\ldots,v_i))$ since $u_i=u_{i\|}+v_i$ and $v_i=-u_{i\|}+u_i$! And so the induction continues. Note that the v_i must be nonzero, since they are k vectors spanning the k-dimensional space $((u_1,\ldots,u_k))$. We can write out formulae for the process: By Proposition 5.1 we have

$$v_1=u_1; \qquad v_2=u_2-\frac{\langle v_1,u_2\rangle}{\langle v_1,v_1\rangle}v_1;\ldots;$$

$$v_i=u_i-\frac{\langle v_1,u_i\rangle}{\langle v_1,v_1\rangle}v_1-\cdots-\frac{\langle v_{i-1},u_i\rangle}{\langle v_{i-1},v_{i-1}\rangle}v_{i-1};\ldots.$$

These formulae actually *compute* the v_i; note that to find v_i we use the previously found $v_1,\ldots,v_{i-1}$. ■

Just to make sure, we state a proposition formally.

5.3. PROPOSITION. *Let V be a subspace of U. Any vector u in U can be split, uniquely, into $u_{\|}+u_{\perp}$, with $u_{\|}$ in V and $u_{\perp}$ orthogonal to V.*

PROOF. This follows from Proposition 5.1 and Theorem 5.2. ■

Example and Exercise

$U=\mathsf{R}^3$; $\langle X,X\rangle = x^2+2y^2+3z^2+2xy-2xz-4yz$. $X_1=[1,0,0]$, $X_2=[1,1,0]$, and $X_3=[1,1,1]$. The new vectors Y_1, Y_2, and Y_3 are found as follows:

$$Y_1=X_1;\qquad Y_2=X_2-\frac{\langle Y_1,X_2\rangle}{\langle Y_1,Y_1\rangle}Y_1=[1,1,0]-2/1[1,0,0]=[-1,1,0]$$

$$Y_3=X_3-\frac{\langle Y_1,X_3\rangle}{\langle Y_1,Y_1\rangle}Y_1-\frac{\langle Y_2,X_3\rangle}{\langle Y_2,Y_2\rangle}Y_2=\cdots.$$

Complete this. (Answers do not always come in integers.) ■

An an application of Proposition 5.3, we consider a geometrical problem: Let V be a subspace of U, and let u_0 be a vector of U, preferably not in V. The *problem* is to find the nearest point to u_0 *in* V. (See Figure 25.) We claim that this is solved by splitting u_0 parallel and perpendicular to V; with $u_0=u_{0\|}+u_{0\perp}$, the "nearest point" is simply $u_{0\|}$.

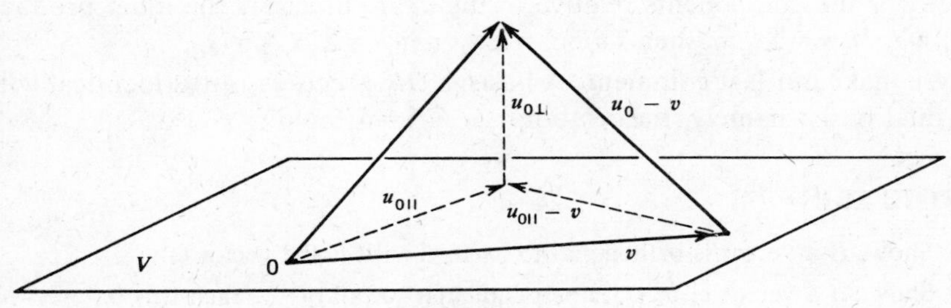

Figure 25.

PROOF. Let v be any vector in V. We have $u_0-v=u_0-u_{0\|}+u_{0\|}-v$ $=u_{0\perp}+(u_{0\|}-v)=u_{0\perp}+w$. Clearly, w is in V, and so it is orthogonal to $u_{0\perp}$. It follows (see Section 4) that $d(u_0,v)^2=|u_0-v|^2=|u_{0\perp}|^2+|w|^2$. Since $|w|^2\geqslant 0$, we find $d(u_0,v)^2\geqslant |u_{0\perp}|^2$ or $d(u_0,v)\geqslant |u_{0\perp}|$, and moreover, the equal sign holds exactly if $w=0$, that is, if $v=u_{0\|}$. This says that the smallest value of $d(u_0,v)$, as v ranges over V, occurs for $v=u_{0\|}$; and that minimum distance is $|u_{0\perp}|$. ■

We note, parenthetically, two reasons why *ON* bases are important: Let $\{u_1,\ldots,u_n\}$ be such a basis. First, for any vector u its components x_i with respect to it are simply the $\langle u_i,u\rangle$, by $\langle u_i,u\rangle=\langle u_i,\Sigma x_j u_j\rangle=\Sigma x_j\langle u_i,u_j\rangle$ $=\Sigma x_j\delta_{ij}=x_i$. (For a *unit* vector v the coefficients $\langle u_i,v\rangle$ are called the

direction cosines of v relative to the given ON basis. By Formula (A) of Section 2 they are indeed the cosines of the angles between v and the u_i. We assume $\mathsf{F}=\mathsf{R}$ here.) Second, we have $|u|^2=\langle u,u\rangle=\Sigma x_i x_j\langle u_i,u_j\rangle =\Sigma x_i x_j \delta_{ij}=\Sigma x_i^2$. [In the complex case we get $\Sigma|x_i|^2$.] In other words, "Pythagoras" holds: The norm of a vector is the square root of the sum of the squares of the coordinates. In the form $|u|^2=\Sigma|\langle u_i,u\rangle|^2$ this is often called the **Parseval identity**. We can write this also as $|u|^2=X^tX$ [or X^*X for $\mathsf{F}=\mathsf{C}$], whereas for a general basis we would have $|u|^2=X^tAX$ [or X^*AX]. To say that the basis is orthonormal ($\langle u_i,u_j\rangle=\delta_{ij}$) is to say that $A=I$, and conversely. The theorem on reduction to sum-of-squares gives therefore a second proof for the fact that ON bases exist. ∎

Two bases $\{u_1,\dots,u_n\}$ and $\{v_1,\dots,v_n\}$ of U are called dual (or reciprocal) to each other, if the relations $\langle u_i,v_j\rangle=\delta_{ij}$ (Kronecker delta) hold; this is related to, although not quite identical to, our earlier concept of dual bases for a space and its dual space. It is easy to see that the dual basis to a given basis exists and is unique: v_i is $1/\langle u_i,u_{i\perp}\rangle\cdot u_{i\perp}$, with $u_{i\perp}$ the component of u_i orthogonal to the span of the other u_j, i.e., with $j\neq i$. See Problem 4, Section 1.

The usefulness of dual bases comes from the following remark: For any vector w the components relative to the u_i are precisely the inner products $\langle v_i,w\rangle$: If $w=\Sigma x_j u_j$, then $\langle v_i,w\rangle=\Sigma x_j\langle v_i,u_j\rangle=\Sigma_j x_j\delta_{ij}=x_i$.

We make our last comment: A basis is ON precisely if it is identical with its dual basis; namely, the relations $\langle u_i,u_j\rangle=\delta_{ij}$ hold.

PROBLEMS

1. Show: If a vector is orthogonal to itself, then it is the vector 0.

2. Show: If a vector u, in U, is perpendicular to (all of) U, then it is 0.

3. Show: If two subspaces V and W of U are orthogonal to each other ($V\perp W$), then their intersection $V\cap W$ is 0.

4. Complete the example in the text.

5. In R^5, with standard inner product, compute the distance of the point $[0,2,0,2,1]$ from the plane spanned by $[1,1,1,1,1]$ and $[1,2,1,0,1]$.

6. Continuation of Problem 5 of Section 4: For the vector $Z=[-6,5,5]$ find the components parallel and perpendicular to the plane $((X,Y))$.

7. R^4 with standard inner product. With $V=(([1,-1,3,-3],[5,-5,1,-1]))$, and $Y=[3,4,-4,6]$, consider the linear variety $L=Y+V$, and find the distance of the point $Z=[3,6,5,1]$ from L. (*Hint*. For any X we have $d(Z,Y+X)=d(Z-Y,X)$, namely both $=|Z-Y-X|$.)

8. Orthogonalize the basis $\{[2,-1,2],[1,1,4],[6,3,9]\}$ of R^3.

9. Orthogonalize the set $\{[1,1,1,1],[0,2,0,2],[-1,1,3,-1]\}$ in R^4.

10. In $\mathbf{R}^3$, with the inner product of Section 1.2, find the dual to the basis $\{[1,2,-1],[0,1,1],[1,2,0]\}$.

11. Let $u_1,\ldots,u_k$ be an *ON* set. Show that for any vector v one has $|v|^2 \geqslant \Sigma_i^k|\langle u_i, v\rangle|^2$ (The **Bessel** Inequality). Can equality hold?

6. ORTHO-COMPLEMENTS

We defined in Section 5 the notions $u \perp V$, and $V \perp V'$. In the latter case we have $V \cap V' = 0$ (a vector in the intersection would be orthogonal to itself, but then $|u|^2 = 0$, and so $u = 0$). To a given subspace V we define $V^{\perp}$, its **ortho-complement** (in U), as the subspace of U of *all* vectors orthogonal to V. Clearly, this is a sub*space*, and we have $V \cap V^{\perp} = 0$. We make the following claim

6.1. PROPOSITION. (a) $U = V \oplus V^{\perp}$ and (b) $(V^{\perp})^{\perp} = V$.

The point of this is the following: We know from way back (Chapter 3, Section 2.5) that a subspace V has complements (expand a basis $u_1,\ldots,u_r$ of V to a basis $u_1,\ldots,u_r,\ldots,u_n$ of U; take $((u_{r+1},\ldots,u_n))$). Here we get a particular complement, the ortho-complement, which is adapted to the inner product.

Note that the Gram–Schmidt process together with 5.1 is a *constructive* approach to this; it produces to any u its components in V and $V^{\perp}$.

PROOF. Let $u_1,\ldots,u_r$ be an *ON* basis for V. Extend it to a basis $\{u_1,\ldots,u_r,\ldots,u_n\}$ of U. Applying Gram–Schmidt we can replace the vectors $u_{r+1},\ldots,u_n$ by vectors $v_1,\ldots,v_{n-r}$ so that the whole family $\{u_1,\ldots, u_r, v_1,\ldots,v_{n-r}\}$ is an *ON* basis for U. It should now be obvious that $\{v_1,\ldots,v_{n-r}\}$ is a basis for $V^{\perp}$, and that (a) and (b) of Proposition 6.1 hold. ■

The innocent looking Part (b) of Proposition 6.1 $(V^{\perp})^{\perp} = V$, means the following (compare Chapter 4, Section 6.4) for the case U equal to $\mathbf{R}^n$, with standard inner product: Let V be a subspace. If $X_1,\ldots,X_r$ span V, then $V^{\perp}$ is by definition the solution space of the r linear equations $X_1^t Y = 0,\ldots,X_r^t Y = 0$ (which we could also write as $X_1 \cdot Y = 0$, etc.). And now Proposition 6.1(b) says: If $Y_1,\ldots,Y_s$ span $V^{\perp}$, then the solution space of the linear equations $Y_1 \cdot X = 0,\ldots,Y_s \cdot X = 0$ is V. This is another version of the fact (known from Chapter 4, Section 6) that a subspace of $\mathbf{R}^n$ can be described by a system of linear equations. We now recognize the coefficient row vectors of these equations as the transposes of a set of vectors that span the ortho-complement. To extend this a little, suppose X_0 is a given vector in $\mathbf{R}^n$. Then the linear variety of type V, passing through X_0, is given by the nonhomogeneous system $Y_1 \cdot X = Y_1 \cdot X_0,\ldots,Y_s \cdot X = Y_s \cdot X_0$. The argument should be clear.

We note a special case which generalizes the well-known notion of normal form of a line in the plane: Let X_0 be any vector of $\mathbf{R}^n$, and let Y be a unit vector. Then the hyperplane (codim = 1) through X_0 orthogonal to Y is given by the one equation $Y\cdot(X-X_0)=0$. And one sees easily, using 5.3, that for any Z in $\mathbf{R}^n$ the value $Y\cdot(Z-X_0)$ is the distance (directed, i.e., with a sign) of Z from the hyperplane in question. In particular, $-Y\cdot X_0$ is the distance from 0 to the plane (this is positive if 0 lies on that side of the plane toward which Y "points").

We stop for a digression: There is a more sophisticated aspect to all this. The expression $\langle u,v\rangle$ for *fixed u* and *variable v* is a linear function on U (in components, it reads $X\cdot Y$, i.e., X^tY for $\mathbf{R}$ and X^*Y for $\mathbf{C}$); this linear function is determined by u; we call it (temporarily) f_u. By following the definitions one finds that this construction is linear, at least if $\mathbf{F}=\mathbf{R}$ [for $\mathbf{F}=\mathbf{C}$ see the note below]; to the vector $r\cdot u$ corresponds to the linear function $r\cdot f_u$ or $f_{ru}=r\cdot f_u$; and to a sum u_1+u_2 corresponds the sum $f_{u_1+u_2}=f_{u_1}+f_{u_2}$. To $u=0$ corresponds the linear function 0. But *conversely*, if f_u is 0, then u must have been 0—$f_u=0$ means $f_u(v)=0$ for all v, or $\langle u,v\rangle=0$ for all v; this implies $u=0$ by taking $v=u$; then $\langle u,u\rangle=0$, $|u|=0$, and so $u=0$.

Abstractly, we think of this as a linear map from U to the dual space U'; it sends u to f_u. And since the kernel of this map is 0 (as just seen) and $\dim U=\dim U'$, it is an *isomorphism.* In particular, to *any* linear function φ there is a unique vector u with $\varphi=f_u$, that is, with $\varphi(v)=\langle u,v\rangle$ for all v. The null space $\ker\varphi$ is just $((u))^\perp$. All this is easily understood with *ON* bases.

6.2. PROPOSITION. *Let $\{u_1,\ldots,u_n\}$ be an ON basis for U; let φ be a linear function on U. The vector u that represents φ as described above ($\varphi=f_u$, or $\varphi(v)=\langle u,v\rangle$ for all v) is given by $u=\Sigma a_iu_i$ with $a_i=\overline{\varphi(u_i)}$ [$=\varphi(u_i)$ in case $\mathbf{F}=\mathbf{R}$].*

PROOF. We only have to check $\langle u,u_j\rangle=\varphi(u_j)$, $j=1,\ldots,n$. We get $\langle u,u_j\rangle=\langle\Sigma_i a_iu_i,u_j\rangle=\Sigma_i\bar{a}_i\langle u_i,u_j\rangle=\Sigma_i\bar{a}_i\delta_{ij}=\bar{a}_j$. (Note: [In the complex case our map from U to U', $u\to f_u$, is still injective and surjective, also additive ($f_{u_1+u_2}=f_{u_1}+f_{u_2}$); but now $f_{cu}=\bar{c}\cdot f_u$ from $\langle cu,v\rangle=\bar{c}\langle u,v\rangle$—the map is only *conjugate-linear* and so not strictly an isomorphism. This slight trouble is unavoidable. By changing the definitions one can shift it to another place, but one cannot get rid of it. One just has to learn to live with it.] ■

The construction of the ortho-complement $V^\perp$ (proof of Proposition 6.1) can now be described as finding the nullspace of the linear functionals $f_{u_1},\ldots,f_{u_r}$. They (as elements of U') are just as independent as the u_i, and so $\dim V^\perp=n-r$ by rank-nullity. In Chapter 5, Section 1 we introduced $V^\perp$ as the space of all linear functionals that vanish on V; this is a subspace of the dual space U'. Our present definition puts $V^{\perp}$ into U. These two definitions seem to be in conflict. We resolve this by noting that our map $u \to f_u$ from U to U' maps the $V^{\perp}$ in U onto the $V^{\perp}$ in U'. This ends our digression.

PROBLEMS

1. Find a vector in R^4, whose inner products with E^1, E^2, and E^3 are 2, -1, and 3, respectively.

2. In R^5 let $V = (([1,2,1,0,1],[2,-1,0,1,0]))$. Find $V^{\perp}$.

3. Let $X_1,\ldots,X_t$ be vectors in R^n. Show that there exists a nonzero vector Y, orthogonal to all the X_i if and only if the rank of the matrix $(X_1,\ldots,X_t)$ is less than n. State the abstract version of this result.

4. Let $X_1,\ldots,X_{n-1}$ be $n-1$ independent vectors in R^n. Show that there exists a *unique* vector X_n, orthogonal to $X_1,\ldots,X_{n-1}$ and with $\det(X_1,\ldots,X_n)=1$. (For $n=3$ one writes $X_3 = X_1 \times X_2$, and calls it the cross- or vector-product of X_1 and X_2; if X_1 and X_2 are dependent, one puts $X_1 \times X_2 = 0$.)

5. A subspace of R^5 is defined by the equations $2x_1+3x_2-x_4+2x_5=0$, $x_1-2x_2+x_3+2x_4-6x_5=0$, $8x_1+5x_2+2x_3+x_4+6x_5=0$. Determine the ortho-complement (find a basis, preferably an orthogonal one).

6. Write out the system of equations for the linear variety L in R^4, passing through $[1,0,-1,0]$, whose type is the ortho-complement of the plane spanned by $[1,1,3,-1]$ and $[1,2,3,0]$; solve the equations.

7. ORTHOGONAL AND UNITARY MATRICES

ON bases are the appropriate bases for an inner product space; and usually, once one has an inner product, one uses only *ON* bases (using Gram–Schmidt to get them). The transition matrix between two *ON* bases has a special character: Let $\{u_1,\ldots,u_n\}$ and $\{u'_1,\ldots,u'_n\}$ be *ON* bases for U (over R). Then $u'_j = \sum_i p_{ij} u_i$ defines P (incidentally, we know $p_{ij} = \langle u_i, u'_j \rangle$, see Section 5). And now: $\delta_{rs} = \langle u'_r, u'_s \rangle = \langle \sum_i p_{ir} u_i, \sum_j p_{js} u_j \rangle = \sum_{i,j} p_{ir} p_{js} \langle u_i, u_j \rangle$; with $\langle u_i, u_j \rangle = \delta_{ij}$, this contracts to $\sum_i p_{ir} p_{is}$. The ensuing relation $\sum_i p_{ir} p_{is} = \delta_{rs}$ holds for any r and s and can be expressed matricially as

$$P^t \cdot P = I. \qquad \text{(O)}$$

We can also write this as $P^{-1} = P^t$ or $P \cdot P^t = I$ (Theorem 4.3.3). Such a matrix is called **orthogonal**. [In the Hermitean case the condition is $P^* \cdot P = I$ or $P \cdot P^* = I$ or $P^* = P^{-1}$. Such matrix is called **unitary**. Note that for a real matrix orthogonal is the same as unitary.]

Example

$$\begin{pmatrix} \frac{1}{\sqrt{2}} & -\frac{\sqrt{3}}{2} \\ \frac{\sqrt{3}}{2} & \frac{1}{\sqrt{2}} \end{pmatrix}; \quad \frac{1}{3}\cdot\begin{pmatrix} 1 & -2 & 2 \\ 2 & 2 & 1 \\ 2 & -1 & -2 \end{pmatrix}.$$

A 4×4 *Case.*

$$\frac{1}{2}\begin{pmatrix} 1 & -1 & -1 & -1 \\ 1 & 1 & 1 & -1 \\ 1 & -1 & 1 & 1 \\ 1 & 1 & -1 & 1 \end{pmatrix}.$$

Warning: The entries here are deceptively simple. Usually orthogonal matrices involve at least-square roots.

Example of a unitary matrix: $1/\sqrt{2}\begin{pmatrix} 1 & i \\ i & 1 \end{pmatrix}$ — Let $X_1,\dots,X_n$ be the columns of P. The entries of $P^t\cdot P$ [or of $P^*\cdot P$] are then the inner products $X_i^t\cdot X_j$ [or $X_i^*\cdot X_j$]. We see that orthogonality [or unitarity] can be expressed by saying that the columns of the matrix are an *ON* basis for R^n [or C^n] with standard inner product. Instead of columns we could say rows; each implies the other, since $P^t\cdot P=I$ has $P\cdot P^t=I$ as consequence (and similarly for the Hermitean case).

The usual rules about determinants and $P^t\cdot P=I$ (or $P^*\cdot P=I$) give $\det P=\pm 1$ for orthogonal P (or $|\det P|=1$ for unitary P, i.e., $\det P$ is a complex number of absolute value 1, $=a+ib$ with $a^2+b^2=1$); namely, $\det P^t\cdot\det P=1$ [or $\det P^*\cdot\det P=1$], which together with $\det P^t=\det P$, [$\det P^*=\overline{(\det P)}$] gives $(\det P)^2=1$ [or $|\det P|^2=1$]. Note some new terminology: An orthogonal P with $\det P=1$ is called **proper** or **special** orthogonal (also rotation), one with $\det P=-1$ is **improper** orthogonal (also reflection); a unitary P with $\det P=1$ is **special**-unitary.

We give more details on 2×2 examples: Any proper orthogonal matrix is of the form $\begin{pmatrix} a & -b \\ b & a \end{pmatrix}$ with $a^2+b^2=1$; any improper one is of the form $\begin{pmatrix} a & b \\ b & -a \end{pmatrix}$, still with $a^2+b^2=1$, but with $\det=-1$. (This is true because the first column must be a unit vector; thus $a^2+b^2=1$. The second column is orthogonal to the first, and therefore a solution of $ax+by=0$; there are two possible unit vector solutions $[-b,a]$ and $[b,-a]$.) We introduce an angle ϕ by $a=\cos\phi$ and $b=\sin\phi$. The matrices become $\begin{pmatrix} \cos\phi & -\sin\phi \\ \sin\phi & \cos\phi \end{pmatrix}$ for the rotations and $\begin{pmatrix} \cos\phi & \sin\phi \\ \sin\phi & -\cos\phi \end{pmatrix}$ for the reflections. (The latter is indeed a reflection, across the line with angle $\phi/2$.)

The 2×2 special-unitary matrices have the form $\begin{pmatrix} a & -\bar{b} \\ b & \bar{a} \end{pmatrix}$ with (complex), with a and b satisfying $|a|^2+|b|^2=1$ by similar reasoning. ∎

An obvious way to construct orthogonal [or unitary] matrices is to start from any nonsingular matrix and to convert the columns (or rows) into an ON set by the Gram–Schmidt process. This simple remark also leads to a remarkable fact.

7.1. PROPOSITION. *Any real nonsingular matrix M can be factored into a product $M=AB$ of an orthogonal matrix A and an (upper) triangular matrix B whose diagonal terms are positive; this splitting is unique. Similarly for complex matrices, with orthogonal replaced by unitary.*

The main point of the proof is the following: Let U be a vector space with inner product; let $\beta=\{u_1,\ldots,u_n\}$ be a basis. Change β to an ON basis, where $\gamma=\{v_1,\ldots,v_n\}$ by Gram–Schmidt and normalization. Then the transition matrix T_γ^β and its inverse T_β^γ (see Chapter 7, Section 4) are upper triangular: The equations in Section 5 show that each u_i is a linear combination of $v_1,\ldots,v_i$. Furthermore, both transition matrices clearly have positive diagonal terms.

For the proof of Proposition 7.1 we operate in $\mathbf{R}^n$ [or $\mathbf{C}^n$] with standard inner product. Let β be the basis, formed by the columns X_i of the given matrix M; let $A=(Y_1,\ldots,Y_n)$ be the orthogonal [or unitary] matrix obtained via Gram–Schmidt from β with the Y's forming an ON basis γ. As noted above, the transition matrix T_β^γ, which we shall also write as B, is upper triangular with positive diagonal terms. We write σ for the standard basis as usual; from Chapter 7, Section 4 we have the product relation $T_\beta^\sigma=T_\gamma^\sigma\cdot T_\beta^\gamma$ of transition matrices. But T_β^σ is, of course, just M, and T_γ^σ is A. Thus we have $M=AB$, and this is the splitting we wanted.

Now we look at the uniqueness statement: Two such factorizations $M=A_1B_1=A_2B_2$ lead to $(A_2)^{-1}A_1=B_2(B_1)^{-1}$. But if a matrix is simultaneously orthogonal (or unitary) and triangular with positive diagonal, then it must be the identity I (see Problem 11).

PROBLEMS

1. Show: If A, B are orthogonal (or unitary) matrices, then so are A^{-1}, A', A^*, and AB.

2. Which diagonal matrices are orthogonal (or unitary)?

3. Show that the product of two improper orthogonal matrices is proper orthogonal.

4. Write out the 2×2 special orthogonal matrices corresponding to the angles 30°, 45°, and 60°.

5. Starting from $A=\begin{pmatrix} -1 & 5 & 14 \\ 2 & 5 & 12 \\ 0 & 2 & -11 \end{pmatrix}$ and using Gram–Schmidt, construct an orthogonal matrix.

6. What value of b (if any) will make the matrix $\begin{pmatrix} 1/2+i/2 & b \\ 1/2-i/2 & 1/2-i/2 \end{pmatrix}$ unitary?

7. Show that the following matrix is orthogonal:

$$\begin{pmatrix} 1/3\sqrt{2} & -1/\sqrt{2} & 2/3 \\ 1/3\sqrt{2} & 1/\sqrt{2} & 2/3 \\ -4/3\sqrt{2} & 0 & 1/3 \end{pmatrix}.$$

8. Write $M=\begin{pmatrix} 3 & 1 \\ 2 & 1 \end{pmatrix}$ as product of an orthogonal and a triangular matrix.

9. Show that any real nonsingular matrix M can be written as CD or EF or GH, where C, F, and H are orthogonal, D and G are lower triangular, and E is upper triangular. (This is similar for complex M.)

10. a. In $\mathbf{R}^n$, with standard inner product, let $X_1,\ldots,X_n$ be n pairwise orthogonal vectors. Prove that $|\det(X_1,\ldots,X_n)|=|X_1|\cdot|X_2|\cdot\cdots\cdot|X_n|$. (*Hint.* Put $X_i=|X_i|X_i'$ so that the X_i' form an *ON* set.)
b. Let $Y_1,\ldots,Y_n$ be any n vectors in $\mathbf{R}^n$. Prove that $|\det(Y_1,\ldots,Y_n)| \leqslant |Y_1|\cdot|Y_2|\cdot\cdots\cdot|Y_n|$ (the "Hadamard inequality"). (*Hint.* Orthogonalize the Y_i to get $Z_1,\ldots,Z_n$. Compare the det of the Y_i with that of the Z_i, and compare $|Y_i|$ with $|Z_i|$. Use a.) Explain the case $n=2$ with a figure. Interpret the result as a statement about the volume. When does equality occur?

11. Suppose the matrix M is orthogonal [or unitary] and triangular. Show that M is a diagonal matrix.

8. ISOMETRIES

Orthogonal (or unitary) matrices come up in another context. We start with an important definition.

8.1. **DEFINITION.** *Let U and V be inner product spaces. A linear transformation T: $U \to V$ is an isometry if it is (a) an isomorphism (invertible) and (b) inner-product-preserving, that is, satisfies $\langle u_1, u_2 \rangle = \langle Tu_1, Tu_2 \rangle$ for any u_1, u_2 in U. (Note that the first $\langle , \rangle$ is that of U, the second that of V.)*

Instead of isometry one uses also the term "orthogonal" linear transformation for $\mathsf{F} = \mathsf{R}$ ["unitary" for $\mathsf{F} = \mathsf{C}$].

The definition is slightly redundant: (*b*) implies injectivity; if $Tu = 0$, we have $|u|^2 = \langle u, u \rangle = \langle Tu, Tu \rangle = 0$, and so $u = 0$. Also, instead of (*b*) it is enough to require "preservation of norm," that is, (*b'*) $|u| = |Tu|$ for any u in U (by polarization (Chapter 11, Section 2(P)) we can express $\langle , \rangle$ in terms of $| \, |$, and see that $\langle , \rangle$ is preserved by T). Spaces with an isometry between them "behave alike" as far as their inner products go. For instance, if $u_1 \perp u_2$, then $Tu_1 \perp Tu_2$. (Of course, $\dim U = \dim V$.)

Trivial Example

$U = V$ and $T = \mathbf{1}$.

Less Trivial Example

U is 3-space, and T is any rotation of space. In fact, the geometric idea of rotating a body around a fixed center, from an initial to a final position, is the motivation for our concept. Note that under rotation the distance of a point from the center does not change; similarly any angle subtended at the center does not change. This is equivalent to the inner product being preserved.

8.2. **THEOREM.** (*Construction Principle*). *If T: $U \to V$ is an isometry, then any ON basis of U is sent to an ON basis of V. Conversely, if $\dim U = \dim V$, and $\beta = \{u_i\}$, $\gamma = \{v_i\}$ are any ON bases of U and V, then the (unique) linear transformation T, defined by $Tu_i = v_i$, is an isometry of U with V.*

PROOF. The first part is quite obvious: Unit vectors go to unit vectors, since norm is preserved by T, and two orthogonal vectors go to two orthogonal vectors. For the second part, we work with coordinates: If $u \underset{\beta}{\leftrightarrow} X$, then $|u| = |X| = \sqrt{X^t X}$, and similarly $v \underset{\gamma}{\leftrightarrow} Y$, $|v| = \sqrt{Y^t Y}$ since β and γ are *ON*; if $v = Tu$, then $Y = X$ (the β–γ-matrix of T is, of course, I), and so $|v| = |u|$. ■

On the other hand, we can describe isometries with the help of arbitrary *ON* bases.

8.3. PROPOSITION. *Suppose* $\dim U = \dim V$. *Let* $\beta = \{u_i\}$, $\gamma = \{v_i\}$ *be arbitrary ON bases for U, V. A linear transformation T: $U \to V$ is an isometry exactly if its matrix with respect to β and γ is orthogonal* [*or unitary*].

PROOF. Let A be the matrix of T. With $u \underset{\beta}{\leftrightarrow} X, \ldots$ as before, we have $Tu \underset{\gamma}{\leftrightarrow} A \cdot X$ (i.e., $Y = A \cdot X$), and $|u| = |Tu|$ becomes $X^t X = Y^t Y = X^t A^t \cdot AX$. Now $A^t \cdot A$ is symmetric, and we know that a quadratic form determines its *symmetric* matrix uniquely. In the present case we get from the identity $X^t \cdot I \cdot X = X^t \cdot A^t \cdot A \cdot X$ the relation $I = A^t \cdot A$, that is, A is orthogonal. Conversely, if A is orthogonal, we find $|u| = |Tu|$, that is, T is an isometry (it is invertible, since A has A^t as inverse!). [The unitary case is practically the same.] In particular, by taking $U = V$, holding β fixed, and letting γ vary, we see that there are as many isometric operators on U as there are *ON* bases (γ) as there are orthogonal [or unitary] matrices. We emphasize once more that all this is based on *having* inner products in U and V.

PROBLEMS

1. Show: If T: $U \to V$ is an isometry, so is T^{-1}: $V \to U$.
2. Show: If T: $U \to V$ and S: $V \to W$ are isometries, so then is $S \cdot T$: $U \to W$.
3. Let $\sigma = \{i_1, i_2, \ldots, i_n\}$ be a permutation of $\{1, 2, \ldots, n\}$. Define T_σ: $\mathbf{R}^n \to \mathbf{R}^n$ by $[x_1, x_2, \ldots, x_n] \to [x_{i_1}, x_{i_2}, \ldots, x_{i_n}]$. Show that T_σ is an isometry and that its matrix relative to the standard basis is orthogonal (this kind of matrix is called a permutation matrix). Which T_σ are proper?
4. Let U be the space of solutions of $y'' + y = 0$. We define an inner product on U by $\langle f, g \rangle = \int_0^{2\pi} f(x) \cdot g(x)\, dx$. For each real α we define the "translation operator" T_α on U by $T_\alpha f$ is the function whose value at x is $f(x - \alpha)$. Show that T_α is an isometry of U.

9. ADJOINT

Remark. Let $\beta = \{u_1, \ldots, u_n\}$ be an *ON* basis for U; let T be an operator on Y. Then the matrix A of T relative to β is given by $a_{ij} = \langle u_i, Tu_j \rangle$.

PROOF. The $\langle u_i, Tu_j \rangle$ are the components of Tu_j relative to β (see note after Proposition 5.3). But that is how the representing matrix A is defined. Note that the above formula for a_{ij} does not hold for general bases. What takes its place? ■

We come to a concept which to some extent explains the importance of the notion of transpose or adjoint of a matrix.

9.1. DEFINITION. *Let U be a vector space with inner product* $\langle , \rangle$; *let T be* *an operator on U. An operator S is called adjoint (sometimes, in the real case, transpose) to T if it satisfies the relation $\langle Su,v\rangle=\langle u,Tv\rangle$ for all u and v. (By $\langle u,v\rangle=\langle v,u\rangle$ [or $\overline{\langle v,u\rangle}$] one could equally well require $\langle Tu,v\rangle=\langle u,Sv\rangle$.)*

We prove immediately the following proposition.

9.2. PROPOSITION. (*and notation*). *To any T the operator S of Definition 9.1 exists and is unique. It is denoted by T^* (sometimes, in the real case, by T^t).*

PROOF (via matrices). Let $\{u_1,\ldots,u_n\}=\beta$ be an *ON* basis. It is, of course, enough to require $\langle u_i,Su_j\rangle=\langle Tu_i,u_j\rangle$. We define S by prescribing its matrix A' with respect to β: $a'_{ij}=\langle Tu_i,u_j\rangle$; we have then, by the remark above, $\langle u_i,Su_j\rangle=a'_{ij}=\langle Tu_i,u_j\rangle$, as needed. (A more sophisticated proof can be constructed using the correspondence between vector and linear functions described above.)

By the symmetry of the definition of S mentioned in Definition 9.1 we get at once that $(T^*)^*=T$ (and $(T^t)^t=T$). We note briefly that the T^t, defined here, is a more "concrete" version of the transposed operator defined in Chapter 8, Section 8. Compare the "digression" in Section 6. ■

The relation to the matrix notions comes in the next proposition which again shows the importance of *ON* basis.

9.3. PROPOSITION. *Let $\beta=\{u_1,\ldots,u_n\}$ be an ON basis. Then S is the adjoint of T precisely if the matrix of S with respect to β is the adjoint (transpose in the real case) of the matrix of T with respect to β.*

PROOF. The condition on S is $\langle u_i,Tu_j\rangle=\langle Su_i,u_j\rangle$. The left side is the matrix entry a_{ij} for T (by the remark above); the right side equals $\overline{\langle u_j,Su_i\rangle}$ (or $\langle u_j,Su_i\rangle$), which is $\overline{a'_{ji}}$ (or a'_{ji}).

Our earlier description of isometries, that is, of orthogonal [unitary] operators (via their matrices with respect to *ON* bases), now simply becomes: Orthogonal [unitary] means $T^t\circ T=\mathbf{1}$, or also $T\circ T^t=\mathbf{1}$ [$T\circ T^*=\mathbf{1}$], or finally $T^{-1}=T^t$ [or $T^{-1}=T^*$].

PROBLEMS

1. With the data of Chapter 12, Section 1, Problem 2, find the adjoint T^* of the operator $T=T_M$ with $M=\begin{pmatrix}1&0&2\\2&1&1\\3&2&1\end{pmatrix}$; that is, find N so that $(NX)^tQY=X^tQMY$.

2. Let A be an $n \times n$ positive definite matrix; define an inner product $\langle , \rangle$ on $\mathbf{R}^n$ as $X^t A X$. Let $T = T_M$ be an operator on $\mathbf{R}^n$, defined by the matrix M. Find the operator T^* adjoint to T with respect to $\langle , \rangle$ (i.e., find its matrix).

3. Let T be an operator (on U with inner product $\langle , \rangle$; let T^* be its adjoint. Show that the kernel of T is the ortho-complement of the image of T^*, and the same with T and T^* interchanged.

4. Refer to the geometrical problem of Section 5. The assignment $u_0 \mapsto u_{0\|}$ defines an operator E: $U \to U$. Show that $\operatorname{im} E = V$ and $\ker E = V^{\perp}$; E is a projection (Chapter 8, Section 8), that is, $E^2 = E$; finally, the transpose of E equals E itself, that is, $\langle Eu, v \rangle = \langle u, Ev \rangle$ for all u and v in U. (E is called the orthogonal projection of U onto V.)

5. Work out Problem 4 for $U = \mathbf{R}^5$, standard inner product, $V = (([1,2,1,0,1],[1,-2,-2,1,-2]))$. (To find E means to find the matrix A with $E = T_A$. You will need an orthogonal basis for V.)

10. ISOMETRIES, NORMAL FORM

Isometries correspond to the intuitive idea of rotating plane or space (around the origin) to some new position without changing any distances or angles. (There are also *reflections*, with $\det A = -1$.)

Occasionally the following **normal form** for isometries is used.

10.1. THEOREM. *Let U be an inner product space; let T be an isometry of U with itself.*

a. *In the complex case there exists an ON basis β of eigenvectors of T; the matrix of T with respect to β is diagonal.*

b. *In the real case there exist two-dimensional subspaces $V_1, \ldots, V_r$ and one-dimensional subspaces $W_1, \ldots, W_s$, all pairwise orthogonal, such that the V_i and W_j are T-invariant and have U as direct sum. If in each V_i and W_j an ON basis is adopted, then the matrix of T is* $\operatorname{diag}(R_1, \ldots, R_r, 1, \ldots, 1, -1, \ldots, -1)$, *where each R_i is of the form* $\begin{pmatrix} \cos\phi_i & -\sin\phi_i \\ \sin\phi_i & \cos\phi_i \end{pmatrix}$ (i.e., *a rotation through ϕ_i*). (*The total number of* $+1$'s *and* -1's *is s; either r or s can be* 0; *two* $+1$'s *or two* -1's *can be combined to an R_i with $\phi_i = 0$ or π.*)

We restate this for matrices: *If M is a unitary matrix, there exists a unitary matrix such that $P^{-1} \cdot M \cdot P$ is diagonal. If M is real orthogonal, there exists a real orthogonal P such that $P^{-1} \cdot M \cdot P$ is of the form* $\operatorname{diag}(R_1, \ldots, R_r, 1, \ldots, 1, -1, \ldots, -1)$.

PROOF. We note first that eigenvalues of orthogonal and unitary operators (or matrices) have absolute value 1 (thus, *real* eigenvalues can only be ± 1): If $MX = \lambda X$, then $X^* M^* = \bar{\lambda} X^*$ and $X^* M^* M X = \lambda \bar{\lambda} X^* X$; by $M^* M = I$ this becomes $X^* X = |\lambda|^2 X^* X$, and so $|\lambda|^2 = 1$. Now let T be unitary, and let λ_1 be an eigenvalue and u_1 an eigen(unit)vector. Then by the unitary property the $(n-1)$-space $((u_1))^\perp$ is also T-invariant. By induction we may assume that in $((u_i))^\perp$ we have an ON basis $u_2,\ldots,u_n$ of eigenvectors. Together with u_1 we get the required basis for (a). For (b) the difficulty is that although the operator (or matrix) is real, the eigenvalues can be complex. We saw earlier (Chapter 9, Section 4) that such an eigenvalue gives a T-invariant 2-space. We push this a bit further omitting details: With each complex eigenvalue λ the conjugate $\bar{\lambda}$ also is eigenvalue since M is real. For each pair $(\lambda,\bar{\lambda})$ of complex eigenvalues find an orthogonal basis (in $\mathbf{C}^n$) of the λ-eigenspace of M. Each basis vector gives a T-invariant 2-space V_i in U. Together with an ON basis for the $+1$ and -1 eigenspace of T as the W_j this is what is promised in Theorem 10.1. ■

Example

Any rotation in 3-space (3×3 proper orthogonal matrix) has $+1$ as eigenvalue (see below), the corresponding eigenvector gives the *axis* of the rotation, and the 2-plane orthogonal to the axis is rotated in itself, that is, is T-invariant.

We consider Theorem 10.1.b once more. As noted, two -1's can be combined into an R_i with $\phi_i=\pi$; therefore, we may assume that there is either no -1 or exactly one -1 in the normal form. In the first case, we have $\det T=+1$, in the second, $\det T=-1$ (since each R_i has $\det=1$). The first case is what we called proper or special orthogonal or rotation; it acts on U by rotating through certain angles in a number of pairwise orthogonal 2-planes (together with the identity on the remaining axes). In the second case, the normal form matrix A can be factored as $\operatorname{diag}(1,\ldots,1,-1)\cdot A'$, where A' means A with the -1 replaced by 1 (we assume the -1 in A to be at the (n,n)-position). The operator acts by a rotation "around the x_n-axis" (points on this axis stay fixed) followed by a reflection across the subspace $x_n=0$ ($[x_1,\ldots,x_n]$ goes to $[x_1,\ldots,x_{n-1},-x_n]$; the last coordinates change sign). Hence, the name reflection for improper orthogonal operators or matrices.

PROBLEMS

1. Find the eigenvalues of the (orthogonal) matrix $\begin{pmatrix} \cos\phi & -\sin\phi \\ \sin\phi & \cos\phi \end{pmatrix}$.
2. Find the normal form for the 3×3 orthogonal matrix of Section 7.
3. Find the normal form for the 4×4 orthogonal matrix of Section 7.
4. Explain why every rotation in 3-space has an axis of rotation.

5. Let A be a proper orthogonal matrix, with dimension n *odd*. The normal form shows that A has 1 as eigenvalue. Prove directly that $\det(A-I)$ is 0 by substituting $A^t \cdot A$ for I and factoring A out.

6. Show: If A is at the same time triangular and unitary (or orthogonal), then it is a diagonal matrix.

7. Let A be an orthogonal $n \times n$ matrix (say a proper one); it defines a rotation of $\mathbf{R}^n$. Suppose it satisfies the relation $A^3 = I$ (the rotation, three times repeated, gives the identity transformation). What can one say about the angles ϕ_i that appear in the normal form for A? What if we had $A^6 = I$ instead?

11. THE CROSS-PRODUCT OR VECTOR-PRODUCT

We come to a phenomenon that is peculiar to dimension three. For simplicity we work with $\mathbf{R}^3$, standard inner product. To any two vectors X and Y we associate a third vector, called the **cross-** or **vector-** or **outer product** of X and Y, and denoted by $X \times Y$, as follows: It is orthogonal to both X and Y; it is so directed that X, Y, and $X \times Y$, in that order, form a "right-handed screw" (i.e., the det is positive); finally, its length equals the area of the parallelogram spanned by X and Y, that is, $|X| \cdot |Y| \cdot \sin\alpha$, where α is the angle between X and Y (given by $\cos\alpha = X \cdot Y / |X| \cdot |Y|$). This fairly complicated description leads to a quite simple formula; we simply write it down and then verify that it does what it should do: If $X = [x_1, x_2, x_3]$ and $Y = [y1, y_2, y_3]$, then

$$\text{(V)} \qquad X \times Y = [x_2 y_3 - x_3 y_2, x_3 y_1 - x_1 y_3, x_2 y_3 - x_3 y_2].$$

Example

$X = [1, -1, 2]$, $Y = [2, 3, 1]$; then $X \times Y = \cdots = [-7, 3, 5]$. Verify $X \cdot X \times Y = 0$ and $Y \cdot X \times Y = 0$.

One should note that the three entries of $X \times Y$ can be thought of as determinants; for instance, the first one is $\det\begin{pmatrix} x_2 & y_2 \\ x_3 & y_3 \end{pmatrix}$. In fact, the three entries are simply the cofactors of the (1,3)-, (2,3)-, and (3,3)-entries of any 3×3 matrix that has X and Y as first and second columns (see Chapter 8, Section 3.e). Laplace expansion, relative to the last column, tells us then that $X \times Y$ is so defined that for any vector Z we have the ("outer-inner") identity

$$\text{(OI)} \qquad \det(X, Y, Z) = X \times Y \cdot Z.$$

(One could well *define* $X \times Y$ by this equation.)

It follows at once that $X \times Y$ is orthogonal to X and Y: If we take either X or Y for Z, we get a matrix with two equal columns and, therefore, with $\det 0$. If we take Z as $X \times Y$, we find $\det(X, Y, X \times Y) = X \times Y \cdot X \times Y > 0$ (unless $X \times Y = 0$), so that the right-hand screw condition holds. Finally, to study the norm $|X \times Y|$, we note that the *square* of the area in question is $|X|^2 \cdot |Y|^2 \cdot \sin^2 \alpha$, which with the help of $\sin^2 = 1 - \cos^2$ and of $\cos \alpha = X \cdot Y / |X| \cdot |Y|$ turns into $|X|^2 \cdot |Y|^2 - (X \cdot Y)^2$. Thus we have to prove the formula $(X \times Y)^2 = X^2 \cdot Y^2 - (X \cdot Y)^2$. But this is quite simple; all one has to do is to write out what the various expressions mean in terms of the x_i and y_i (using Definition (V)) and to check that everything cancels. In fact, we did this, more or less, in Section 2 when we talked about the Cauchy inequality.

Can $X \times Y$ equal 0? The answer is yes, if and only if X and Y are linearly dependent. One can see this from (V): $X \times Y = 0$ means that the ratios x_1/y_1, x_2/y_2, and x_3/y_3 are equal, say equal k, whereupon $X = kY$ (one should really be a little more careful with this argument; as written, it assumes that the y_i are not 0). Or from $|X \times Y| = |X| \cdot |Y| \cdot \sin \alpha$; $X \times Y = 0$ amounts to $X = 0$ or $Y = 0$, or $\alpha = 0$ or π.

The operation $\times$ is *bilinear*: We have $(aX) \times Y = a(X \times Y)$ and $(X' + X'') \times Y = X' \times Y + X'' \times Y$, and the same thing is true for the second factor; this is clear from (V).

Finally, one verifies from (V) that $\times$ is "skew-symmetric":

(SK) $\qquad X \times Y = -Y \times X \qquad$ for any X and Y.

So far we have operated in $\mathbf{R}^3$. The definition for $\times$ makes sense in any three-dimensional vector space (over $\mathbf{R}$) with an inner product, except for the part about the "right-handed" screw—there is no det in an abstract vector space. What takes its place, and what really is behind this condition, is the notion of "orientation" of a vector space (see Chapter 8, Section 10): We choose, in advance, some basis for the space, and call it, and all the bases that one gets from it by a transition matrix with *positive* det, positively oriented (and the remaining bases negatively oriented). (In $\mathbf{R}^3$ we start, of course, with $\{E^1, E^2, E^3\}$ as positively oriented.) And now we take the direction of $u \times v$ orthogonal to u and v, so that u, v, and $u \times v$ form a positively oriented base. Everything goes through as before. One finds that a basis $\{u_1, u_2, u_3\}$ is *ON and* positively oriented exactly if the three relations $u_1 \times u_2 = u_3$, $u_2 \times u_3 = u_1$, and $u_3 \times u_1 = u_2$ hold. And, finally, if one uses coordinates with respect to such a basis, then Formula (V) gives the coordinates of $u \times v$.

We note once more that the cross-product is derived from inner product plus orientation in U. This has the following consequence: Let $T: U \to U$

be an orthogonal operator on U (so that it preserves the inner product) and suppose that it also preserves the orientation (sends any positively oriented basis to a positively oriented one; this means that $\det T$, relative to any basis, is positive). Then T must preserve the cross-product: $T(u\times v) = Tu\times Tv$; the T-image of the cross-product is the cross-product of the T-images. What happens if T reverses the orientation (has negative det)? The answer is: $T(u\times v)=-Tu\times Tv$; the T-image of the cross-product is the *negative* of the cross-product of the T-images. The reason is simply that in this case the operator $-T$ preserves the orientation, and therefore also the cross-product: We have $-T(u\times v)=(-Tu)\times(-Tv)=Tu\times Tv$. (This behavior of $u\times v$ under improper orthogonal transformations or reflections is usually expressed in physics by saying that $u\times v$ is only a "pseudo-vector"; what this means is that the definition of $u\times v$ involves an orientation of the space.)

PROBLEMS

1. Compute the cross-product of $[1,2,-1]$ and $[2,-1,3]$, and verify $\det(X,Y,X\times X)=(X\times Y)^2=X^2\cdot Y^2-(X\cdot Y)^2$.

2. Find the equation of the plane (in $\mathbf{R}^3$) spanned by the two vectors of Problem 1, using the fact that it is the ortho-complement of the line spanned by the cross-product.

3. Prove $X\times Y\cdot Z=Y\times Z\cdot X=Z\times X\cdot Y$ for any three X, Y, and Z in $\mathbf{R}^3$

12. SYMPLECTIC MATRICES

For the record we mention another kind of matrices. Let J_1 be the matrix $\begin{pmatrix} 0 & 1 \\ -1 & 0 \end{pmatrix}$. With a given n let J_n, or J in short, be the $2n\times 2n$ matrix $\operatorname{diag}(J_1,\ldots,J_1)$. A (possibly complex) $2n\times 2n$ matrix M is called **symplectic**, if the relation $M^tJM=J$ holds; the term **canonical** is also used. These matrices are characterized by the fact that they leave the skew bilinear form X^tJY invariant; that is, that they satisfy $(MX)^tJ(MY)=X^tJY$ for all X and Y in $\mathbf{C}^{2n}$. They appear in connection with canonical or Hamiltonian systems of differential equations. The determinant of a symplectic matrix is one, and with each eigenvalue λ the reciprocal $1/\lambda$ also appears as an eigenvalue, with the same algebraic multiplicity; this follows from $M^t=JM^{-1}J^{-1}$.—Abstractly, a symplectic operator is one that leaves a preassigned non-degenerate *skew* form φ invariant (meaning $\varphi(Tu,Tv)=\varphi(u,v)$ for all u and v); the dimension of U is necessarily even.

13 THE SPECTRAL THEOREM

Symmetric matrices (and operators) play a special role and deserve a separate chapter. They appear frequently "in nature" (an example is the stress tensor of elasticity discussed below); the symmetry usually comes from the existence of some conservation law (like that for angular momentum in physics). The central fact is that such a matrix can be made diagonal by a *rotation* of the coordinate system (similarity via *orthogonal* matrix). This is the **spectral theorem**; we discuss several variants of it. The name comes from the fact that the theorem often amounts to the description of a complex oscillation ("light") in terms of a number of simple oscillations whose frequencies ("colors") constitute the *spectrum* of the oscillation; these frequencies are equal to or at any rate closely related to the diagonal terms of the matrix (after the rotation). Geometrically this corresponds to the notion of **principal axes** of a quadric. Over $\mathbf{C}$ "symmetric" is replaced by **"self-adjoint."**

1. DEFINITIONS; THE SPECTRAL THEOREM

Let U be a vector space with inner product $\langle \cdot, \cdot \rangle$ (and norm $|\cdot|$) over $\mathbf{R}$ or $\mathbf{C}$. We recall the notions transpose and adjoint of Chapter 12, Section 9. With their help we define the following: An operator T: $U \to U$ is **symmetric** (real case) if $T = T^t$ or **self-adjoint** (SA; complex case) if $T = T^*$. Both cases mean that $\langle u, Tv \rangle = \langle Tu, v \rangle$ for any two vectors u and v in U. Note that this involves the inner product; one should really say that T is symmetric [or SA] with respect to (a given) $\langle\ ,\ \rangle$.

It is clear from Proposition 12.9.3 that T is symmetric [SA] if its matrix with respect to any *ON* basis (and then with respect to all such) is symmetric [SA]. In brief, symmetric [SA] operators are abstract versions of symmetric [SA] matrices. The main fact about these operators and matrices, the **spectral theorem**, says that they are diagonalizable (semisimple), so that the Jordan trouble (Chapter 10) does not arise for them, and even more: The transition matrix needed for diagonalization can be chosen orthogonal [unitary]. There are in fact three versions, one for operators, one for matrices, and one for quadratic [Hermitean] forms.

1.1. THEOREM. *(The Spectral Theorem). Let T: $U \to U$ by a symmetric* [SA] *operator on the inner product space U. Then the eigenvalues λ_i of T are real, and there exists an ON basis for U made up of eigenvectors of T.*

1.1′. THEOREM. *(Diagonalization of symmetric [or SA] matrices). Let A be a real symmetric [or complex self-adjoint] matrix. Then the eigenvalues λ_i of A are real, and there exists an orthogonal [unitary] matrix M such that $M^{-1} \cdot A \cdot M$ is diagonal,* $= \mathrm{diag}(\lambda_1,\ldots,\lambda_n)$.

1.1″. THEOREM. *(Principal Axes Theorem). Let $q(x_1,\ldots,x_n) = \Sigma a_{ij} x_i x_j$ [or $= \Sigma a_{ij} \bar{x}_i x_j$] be a real-quadratic [Hermitean] form with symmetric [Hermitean] matrix A. There exists a change of coordinates $X = PX'$ with P orthogonal [unitary] that transforms q to diagonal form $\Sigma \lambda_i (x_i')^2 [\Sigma \lambda_i |x_i'|^2]$; the λ_i are real and are the eigenvalues of A.* (This can, of course, be stated abstractly, for a form on a vector space with inner product.)

The eigenvalues $\lambda_1,\ldots,\lambda_n$ constitute the **spectrum** of T or A or q.

Before the proof, we comment on the relation between the three theorems; in fact we show that they are equivalent to each other.

Theorems 1.1 and 1.1′ are related by Proposition 12.9.3: If T is given, we take any ON basis $\beta = \{u_1,\ldots,u_n\}$ and get a symmetric [or SA] matrix A for it. Using the matrix M of Theorem 1.1′ as transition matrix, we get a new basis such that the matrix for $T(M^{-1} \cdot A \cdot M)$ is diagonal. The new basis vectors are then eigenvectors (9.3.1) and the fact that M is orthogonal [unitary] guarantees that the new basis is ON. Thus, we get Theorem 1.1. Conversely, if A is given, we consider $\mathbf{R}^n$ (or $\mathbf{C}^n$) with the standard inner product, and the usual operator T_A, given by $T_A(X) = A \cdot X$. The matrix of T_A with respect to $E^1,\ldots,E^n$ is just A. From Theorem 1.1 we get another ON basis $X_1,\ldots,X_n$, eigenvectors of T_A. The transition matrix is precisely $M = (X_1,\ldots,X_n)$; it is orthogonal [unitary]. The matrix for T_A with respect to the new basis, $M^{-1} \cdot A \cdot M$, is $\mathrm{diag}(\lambda_1,\ldots,\lambda_n)$, since the new basis vectors are eigenvectors of T_A.

Consider the relation between 1.1′ and 1.1″: The matrix of q changes under $X = P \cdot X'$ to $A' = P^t A P\,[P^* A P]$, see Formula 1.(C) of Chapter 11. We choose as P the M of Theorem 1.1′. Then, since $M^t = M^{-1} [M^* = M^{-1}]$ (this is the crucial point where orthogonality [unitarity] is used) we have $P^t \cdot A \cdot P = M^{-1} \cdot A \cdot M = \mathrm{diag}(\lambda_1,\ldots,\lambda_n)$ [or $P^* A P = M^{-1} \cdot A \cdot M$

$=\operatorname{diag}(\lambda_1,\ldots,\lambda_n)$]; and so $q=\Sigma\lambda_i(x_i')^2$ [or $\Sigma\lambda_i|x_i'|^2$]. Similarly from 1.1″ to 1.1′.

To restate the main point, for orthogonal [unitary] M the relation "similarity" $A'=M^{-1}\cdot A\cdot M$ is the same as the relation "congruence" $A'=M^t\cdot A\cdot M$ [or $=M^*\cdot A\cdot M$ in the complex case]. Note incidentally that under $X=PX'$ the determinant changes by $\det P^t\cdot\det P=\det P^t\cdot P=1$ [or $\det P^*\cdot\det P=1$]; this means that, as long as we allow only orthogonal [unitary] change of coordinates, the det of q is well defined (in the final diagonal form we have $\det q=\lambda_1\cdot\lambda_2\cdot\cdots\cdot\lambda_n$). ■

The case $n=2$ of the principal axes theorem should be familiar (and explains the name of the theorem): A quadratic expression $Ax^2+Bxy+Cy^2$ can be changed to $A'x'^2+C'y'^2$ (cross term absent!) by a rotation of the axes, $x=\cos\theta x'-\sin\theta y'$ and $y=\sin\theta x'+\cos\theta y'$ with suitable θ; the new axes give the principal axes (major and minor) of the conic in question. We will see below how the usual "formula" for θ, $\tan 2\theta=B/(A-C)$, fits into the present context.

We make some comments on the role of the spectral theorem. It is important as it stands because of many geometric and physical applications (see Section 3 for some examples). But beyond that, it has far reaching generalizations (related to those of the inner product, mentioned in Chapter 10, Section 4). Many ordinary and partial differential equations, particularly those that have to do with oscillating systems of any kind (this goes up all the way to quantum theory) are governed by extensions of the spectral theorem.

PROBLEMS

1. Show that every (real or complex) square matrix A can be written, uniquely, in the form A_1+iA_2, with A_1 and A_2 self-adjoint. (*Hint*. If $A=A_1+iA_2$ as described, what would A^* be?)

2. In Problem 1, what is the relation of A_2 to the "skew part" of A (see Chapter 6, Section 2, Problem 11)? If A is real, are A_1 and A_2 real?

3. Which diagonal matrices (over $\mathbf{C}$) are Hermitean?

4. Construct an example in dimension two or three of a symmetric matrix A and a nonorthogonal matrix M such that $M^{-1}AM$ is not symmetric.

5. Show from the spectral theorem that a real symmetric (complex self-adjoint) matrix (or operator) is positive definite exactly if all its eigenvalues are positive (and positive semidefinite exactly if all the eigenvalues are nonnegative).

6. Let A be real symmetric positive definite. Show, using the spectral theorem, that there exists B, also real symmetric positive definite, with $A=B^2$ (a "square root" of A). Can a symmetric B with $A=B^2$ exist, if A is not positive definite? (*Hint*. Suppose A were diagonal.)

7. A matrix (or operator) is called *normal*, if it commutes with its adjoint: $A \cdot A^* = A^* \cdot A$. Show that A is normal if and only if the two SA matrices A_1 and A_2, associated to A as in Problem 1, commute with each other. (*Hint*. How is the adjoint of iM related to that of M?)

2. SPECTRAL THEOREM, PROOF

We come to the proof of Theorems 1.1, 1.1′, and 1.1″. First, we must consider the reality of the eigenvalues.

2.1. PROPOSITION. *Let A be a self-adjoint matrix (this includes real symmetric). Then the eigenvalues of A are real.*

PROOF. Let λ be an eigenvalue, with (possibly complex) eigenvector X, so that $AX = \lambda X$. "Multiplying" by $X^* = \overline{X}^t$, we get $X^*AX = \lambda X^*X$. Here $X^*X = \Sigma |x_i|^2$ is real nonzero, in fact positive. And X^*AX is also real: For a 1×1 matrix $*$ equals $\bar{}$; so $\overline{(X^*AX)} = (X^*AX)^* = X^*A^*X^{**} = X^*AX$ (by $A = A^*$ and $X^{**} = X$), but a complex number that equals its conjugate is real. Finally, $\lambda = X^*AX / X^*X$ is real. ■

Next, we look at a proposition that will make an inductive proof possible.

2.2. PROPOSITION. *Let T: $U \to U$ be symmetric* [SA]. *If V is a T-invariant subspace of U, then the ortho-complement $V^\perp$ is also T-invariant.*

PROOF. (We know $U = V \oplus V^\perp$.) Take w in $V^\perp$. For any v in V we have $\langle w, Tv \rangle = 0$, since Tv is again in V by hypothesis, and w is orthogonal V. But $\langle Tw, v \rangle = \langle w, Tv \rangle$, and "$\langle Tw, v \rangle = 0$ for all v in V" implies $Tw \perp V$. ■

The proof of the spectral theorem is surprisingly short. We start with a (real) eigenvalue λ_1 and the corresponding eigenvector u_1; we normalize u_1 to a unit vector. The 1-space $((u_1))$ is T-invariant; so is then its ortho-complement $W = ((u_1))^\perp$ of dimension $n-1$ ($n = \dim U$). T, restricted to W, is still symmetric [SA], since the relation $\langle u, Tv \rangle = \langle Tu, v \rangle$ is unchanged. By induction (the theorem being trivial for $n = 1$) Theorem 1.1 holds for $T|W$, that is, there is an ON basis $u_2, \ldots, u_n$ made up of eigenvectors. Clearly, $u_1, u_2, \ldots, u_n$ do for U what Theorem 1.1 requires. ■

2.3. PROPOSITION. *With T as before, any two eigenvectors to two different eigenvalues are orthogonal to each other.*

Briefly, the eigenspaces are pairwise orthogonal.

PROOF. (This could be read off from the spectral theorem.) Let $\lambda \neq \mu$ be eigenvalues, with eigenvectors u and v; $Tu = \lambda u$ and $Tv = \mu v$. We start from the basic relation $\langle u, Tv\rangle = \langle Tu, v\rangle$. We get $\langle u, \mu v\rangle = \langle \lambda u, v\rangle$ or $\mu\langle u, v\rangle = \lambda\langle u, v\rangle$ [in the complex case we used that μ is real; otherwise we would have to write $\bar{\mu}$]. This means $(\lambda - \mu)\cdot\langle u, v\rangle = 0$. Since $\lambda \neq \mu$, we find $\langle u, v\rangle = 0$. ■

We comment on how this works out in practice: 1.1′ says that a real symmetric [SA] can be diagonalized, and in fact via an orthogonal [unitary] matrix M. To find M one proceeds as follows: First, one finds the eigenvalues, as usual (Chapter 10, Section 2); they are real; if one cannot find them, one is stuck, more or less. Next, to each eigenvalue λ one finds the eigenspace U_λ, solutions of $AX = \lambda X$. One finds a basis for U_λ, and orthonormalizes it via Gram–Schmidt (standard inner product). These *ON* bases for the various U, taken together, will form the required *ON* basis for $\mathbf{R}^n$ [or $\mathbf{C}^n$]; if they are $X_1, \ldots, X_n$, then the orthogonal [unitary] transition matrix M that transforms A to diagonal form is $M = (X_1, \ldots, X_n)$.

Remark. If A has no multiple eigenvalues, then the eigen(unit-)vectors are determined up to a factor ± 1 [or $e^{i\varphi}$ in the complex case].

Example

Reduce $q(x_1, x_2, x_3) = 2x_1^2 + 2x_2^2 + 2x_3^2 - 2x_1x_2 - 2x_1x_3 - 2x_2x_3$ to principal axes. Matrix $A = \begin{pmatrix} 2 & -1 & -1 \\ -1 & 2 & -1 \\ -1 & -1 & 2 \end{pmatrix}$. $\chi_A(x) = \det(A - xI) = \cdots - x^3 + 6x^2 - 9x$. Roots are $\lambda_1 = 0$ and $\lambda_2 = 3$ twice. We find eigenspaces. U_0 is given by $AX = 0$; the solution (row-echelon form, etc.) is $[1, 1, 1]$, or, normalized, $X_1 = (1/\sqrt{3})[1, 1, 1]$. U_3 is found from $(A - 3I)\cdot X = 0$; there are two solutions, for example, $[1, -1, 0]$ and $[1, 0, -1]$, which can be orthonormalized to $X_2 = 1/\sqrt{2}\,[1, -1, 0]$ and $X_3 = 1/\sqrt{6}\,[1, 1, -2]$, respectively.

$$M = \begin{pmatrix} \frac{1}{\sqrt{3}} & \frac{1}{\sqrt{2}} & \frac{1}{\sqrt{6}} \\ \frac{1}{\sqrt{3}} & -\frac{1}{\sqrt{2}} & \frac{1}{\sqrt{6}} \\ \frac{1}{\sqrt{3}} & 0 & -\frac{2}{\sqrt{6}} \end{pmatrix};$$

this *is* orthogonal. (Check!)

Finally, $M^{-1}\cdot A\cdot M = \operatorname{diag}(0, 3, 3)$ (Check!), and the substitution $X = M\cdot X'$ reduces the quadratic form to $3(x_2')^2 + 3(x_3')^2$ (Check!). (The quadric $q = c$, with $c = 3$ say, to make things simple, has equation $(x_2')^2 +$

$(x_3')^2=1$ in the new coordinate system; it is the cylinder in the x_1'-direction over the "unit circle" in the $x_2'-x_3'$-plane—"cylinder" means that through each point on that circle we draw the whole line parallel to the x_1'-direction; since x_1' does not enter the equation, such a line is completely on the surface if it has one point on it.)

A Hermitean Example

$$A=\begin{pmatrix} 1 & i \\ -i & 1 \end{pmatrix}.$$

$$\chi_A(x)=(x-1)^2-i\cdot-i=x^2-2x.$$

Eigenvalues $\lambda_1=0$, $\lambda_1=2$. U_0 is given by $A\cdot X=0$; $x_1+ix_2=0$ and $-ix_1+x_2=0$; solution $X_1=(1/\sqrt{2})[-i,1]$; U_2 is given by $(A-2I)\cdot X=0$; $-x_1+ix_2=0$ and $-ix_1-x_2=0$; and solution $X_2=(1/\sqrt{2})[i,1]$. Check $X_1\perp X_2$. $M=(X_1,X_2)$, and $M^{-1}\cdot A\cdot M=\text{diag}(0,2)$. (Check! Note $M^{-1}=M^*$.) ■

We add a standard "application" of the principal axes theorem: Let $p(x_1,\ldots,x_n)=\Sigma a_{ij}x_ix_j$ and $q(x_1,\ldots,x_n)=\Sigma b_{ij}x_ix_j$ be *two* quadratic forms with *p positive definite*. Then there is a change of coordinates $X=PX'$ (not orthogonal) such that *both p* and *q* become diagonal; in fact, we can arrange p to become $\Sigma(x_i')^2$; and then the constants λ_i in $q=\Sigma\lambda_i(x_i')^2$ are the "eigenvalues of B with respect to A," that is, the roots of $\det(B-xA)=0$. One can do this in two steps: A first change $X=P_1X'$ changes p to sum-of-squares $\Sigma(x_i')^2$ (p is positive definite) (Chapter 11, Section 6); that is, $P_1^tAP_1=I$; B changes to $B_1=P_1^tBP_1$. A second change, by Theorem 1.1″, changes B_1 to diagonal form via an *orthogonal* matrix $P_2, P_2^tB_1P_2=\text{diag}(\lambda_1,\ldots,\lambda_n)$. Since P_2 is orthogonal, *p remains* sum-of-squares. Thus $P=P_1\cdot P_2$ does the trick. As for the eigenvalue assertion, by Theorem 1.1″ the λ_i are the roots of $\det(B_1-xI)$. By the above relations we can rewrite the det as $\det(P_1^tBP_1-xP_1^tAP_1)=\det(P_1^t\cdot(B-xA)P_1)=\det P_1)^2\cdot\det(B-xA)$. Furthermore, by an eigenvector of B with respect to A with eigenvalue λ, one means a nonzero vector X with $(B-\lambda A)X=0$. The new basis that diagonalizes p and q can be shown to consist of n eigenvectors of B with respect to A, normalized relative to p, and chosen orthogonal relative to p in case of a multiple eigenvector. We note that one can interpret the whole thing as the principal axes theorem for q with p as inner product in $\mathbf{R}^n$ instead of the standard $\langle\, ,\rangle$.

PROBLEMS

1. Transform $A = \begin{pmatrix} 5 & 2 & 2 \\ 2 & 2 & -4 \\ 2 & -4 & 2 \end{pmatrix}$ to diagonal form with the help of an orthogonal matrix.

2. Transform $A = \begin{pmatrix} 3 & 2 & 4 \\ 2 & -4 & -4 \\ 4 & -4 & -3 \end{pmatrix}$ to diagonal form with the help of an orthogonal matrix.

3. The quadratic form $2x^2 + 5y^2 + 5z^2 + 4xy - 4xz - 8yz$ can be diagonalized, that is, transformed to $\Sigma\lambda_i x_i^2$ by an orthogonal change of variables. Find the coefficients λ_i.

4. Work out the spectral theorem (i.e., find the relevant unitary matrix M) for $A = \begin{pmatrix} 1/2 & \bar{c} \\ c & -1/2 \end{pmatrix}$ with $c = 1/2\sqrt{2} \cdot (1 + \sqrt{2} + i(1 - \sqrt{2}))$; this is in $\mathbf{C}^2$ with standard inner product.

5. Find eigenvalues and vectors of $B = \begin{pmatrix} 8 & 4 & -7 \\ 4 & 2 & -3 \\ -7 & -3 & 12 \end{pmatrix}$ with respect to $A = \begin{pmatrix} 6 & 4 & 1 \\ 4 & 3 & 3 \\ 1 & 3 & 17 \end{pmatrix}$. (Note that A is positive definite!)

3. SOME COMMENTS

Two General Examples

i. Many of the "symmetric tensors" that appear in physics or engineering are really symmetric operators in $\mathbf{R}^3$. We describe the stress tensor of elasticity: We are in an (idealized) elastic medium, at a point P that we take as origin. Let u be any vector; we make the following construction: We form the plane through P, perpendicular to u, and remove a part of the substance, behind the plane, such that the piece of the plane in the boundary of the removed part has area equal to $|u|$ (length of u). To keep equilibrium, we have to supply a force at P, which depends, of course, on u; we write Su for the vector representing this force. (See Figure 26.) Physical equilibrium considerations show that (a) S is a linear operator (this is, of course, "idealized:'; in reality the linearity relation would hold only for small vectors u—so that removal of a part does not upset things too much—and even then only approximately) and (b) the operator is *symmetric*—S is the so-called stress tensor. The spectral theorem says that there are three directions u_1, u_2, and u_3, such that Su_i goes in direction u_i (eigenvector) and that these three directions are mutually orthogonal. The Su_i are the "principal stresses."

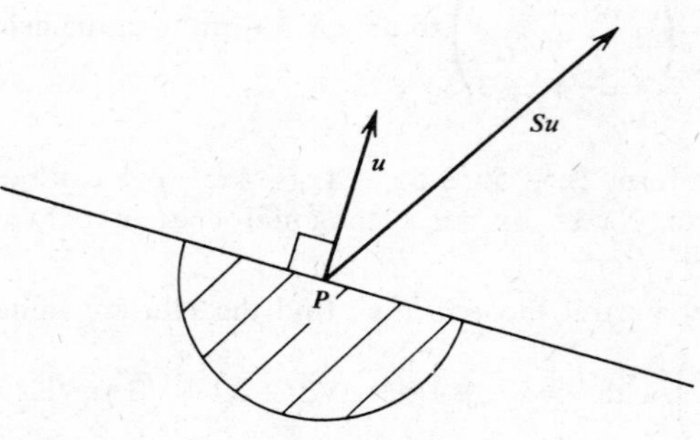

Figure 26.

ii. The application "simultaneous reduction of a pair of quadratic forms, if one is positive definite" comes up in mechanics: A mechanical system, with coordinates $x_1,\ldots,x_n$, near an equilibrium position is often characterized by having kinetic energy $E_k=\frac{1}{2}\Sigma a_{ij}\dot{x}_i\dot{x}_j$ and potential energy $E_p=\frac{1}{2}\Sigma b_{ij}x_ix_j$, with certain a_{ij} and b_{ij}. Here the kinetic energy (which generalizes $1/2mv^2$) is positive definite. By our result we can change to new coordinates $X=PX'$, such that $E_k=\frac{1}{2}\Sigma(\dot{x}_i')^2$ and $E_p=\frac{1}{2}\Sigma\lambda_i(x_i')^2$. The equations of mechanics say that the motion is given by the solutions of the differential equations $\ddot{x}_i'+\lambda_i x_i'=0$; the solutions are given by sin and cos, if λ_i is positive, and by exponentials, if $\lambda_i<0$; oscillation in the first case but not in the second. The point is that before our change of variables the equations of motion are $A\ddot{X}+BX=0$ or $\Sigma_j a_{ij}\ddot{x}_j+b_{ij}x_j=0$; the x_j are all mixed up with each other, and one cannot see what the solutions look like, whereas after the change each x_i has its own equation $\ddot{x}_i+\lambda_i x_i=0$ ("normal modes").

Correspondence Between Operators and Quadratic Forms

The spectral theorem (for symmetric operators) and the principal axes theorem (for quadratic forms) are in a strange relation: two theorems, for different objects, with the *same* proof. Behind this is the fact that in an inner product space there is a close relation between operators and bilinear forms; they are "almost the same thing." Strictly speaking, there are *two*, slightly different, ways of setting up this relation: We say that the (arbitrary, not necessarily symmetric) operator T *corresponds* to the bilinear form φ if $\varphi(u,v)=\langle u,Tv\rangle$ for all u and v in U (first or right correspondence), or again if $\varphi(u,v)=\langle Tu,v\rangle$ for all u and v (second or left correspondence). Instead of all u and v, it is enough to take basis vectors u_i and u_j.

The first correspondence simply means that φ and T have the same representing matrix ($\varphi(u_i,u_j)=\langle u_i,Tu_j\rangle$, relative to an ON basis, and for the second correspondence their matrices are transposes of each other ($\varphi(u_i,u_j)=\langle Tu_i,u_j\rangle=\langle u_j,Tu_i\rangle$, not $=\langle u_i,Tu_j\rangle$). (In the complex case, where φ is sesquilinear, we would get A and A^* for the first and second correspondence.) The matrix interpretation shows also that to each given φ we can *find* a (right) T_1 and (left) T_2, and similarly two φ's for a given T. (Abstractly, what one really should do first is verify that for a given T the expressions $\langle u,Tv\rangle$ and $\langle Tu,v\rangle$ are bilinear [or sesquilinear] forms.) The two operators, corresponding to a given form φ (by right and left correspondence) are in general not equal; they are in fact transposes [adjoints] of each other, since $\varphi(u,v)=\langle T_2u,v\rangle=\langle u,T_1v\rangle$. Equality, where $T_1=T_2$ (i.e., T_1 symmetric [or SA]), thus amounts to φ being symmetric [Hermitean]: $\varphi(v,u)=\varphi(u,v)$ [or $=\overline{\varphi(u,v)}$] via $\langle v,u\rangle=\langle u,v\rangle$ [or $=\overline{\langle u,v\rangle}$]. This correspondence means that any time one proves something about operators in an inner product space, one automatically gets something about bilinear forms: Theorems 1.1 and 1.1″ are an example of this.

Remarks to the Principal Axes Theorem

a. We saw in Chapter 11, Section 3, that any quadratic form can be reduced to a sum-of-squares by suitable change of variables. For the spectral theorem we allow only *orthogonal* change of variables—the new axes in $\mathbf{R}^n$ must be orthogonal. The theorem says that with this restriction we can still diagonalize the form, but we cannot reduce completely to sum-of-squares $\Sigma\pm(x_i')^2$; we can only get $\Sigma\lambda_i(x_i')^2$ with certain unavoidable factors λ_i; the λ_i are the eigenvalues of the original matrix. A step that we could do earlier, namely substitute $x_i''=\sqrt{|\lambda_i|}\;x_i'$, cannot be done now, since it is not an orthogonal change of variables (unless $|\lambda_i|=1$ already). In fact the λ_i are important; geometrically, they tell us something about the shape of the quadric $q=1$ (or $=-1$): We write $\lambda_i=\pm 1/a_i^2$, so that q takes the form $\Sigma\pm x_i'^2/a_i^2$. Then the x_i'-axis meets the quadric at distance $\pm a_i$. This should be familiar for $n=2$: We have changed, by rotation of axes, from $Ax^2+Bxy+Cy^2=1$ to $\pm x'^2/a^2\pm y'^2/b^2=1$; a and b are major and minor axis of the conic.

b. We describe a second aspect: "Principal axes" of the quadratic form X^tAX are the directions of the eigenvectors of A, or, in case several eigenvalues are equal, those of an *ON* basis of eigenvectors. That a quadratic form q has a diagonal matrix in a coordinate system means, by Chapter 11, Section 7, that the basis vectors are pairwise *conjugate* with respect to q. The special property that distinguishes the principal axes is that they are also pairwise orthogonal. Geometrically, this amounts to the following: Let Y_i be a point where the line (1-subspace) $((X_i))$ spanned by a principal axis vector X_i meets the quadric Q_c; then the tangent plane to Q_c at Y_i is *orthogonal* to Y_i, that is, equals $((Y_i))^\perp$. Namely, by Chapter 11, Section 7 the tangent plane is a translate of the conjugate space of Y_i (the nullspace of the equation $Y_i^tAX=0$). But Y_i, a multiple of X_i, is eigenvector of A, with eigenvalue λ_i. Thus, $Y_i^tA=(AY_i)^t=\lambda_i Y_i^t$, and so $Y_i^tAX=0$ becomes $\lambda_i Y_i^tX=0$. We shall assume $\lambda_i \neq 0$ (we might assume q nondegenerate, the most important case anyway; the case $\lambda_i=0$, where $A\cdot Y_i=0$, requires special handling); then the equation becomes $Y_i^tX=0$; but that describes precisely the space orthogonal to Y_i. ■

If the eigenvalues are pairwise different (the "good" case), there are exactly n such lines (along the eigenvectors), and so $2n$ points on Q_c with this orthogonality property. For any other line the tangent space to the quadric at the point of intersection with the line *is not* orthogonal to the line. (See Figure 27.) If an eigenspace has a dimension greater than one, then *any* vector in this subspace has the orthogonality property (e.g., ellipsoid of revolution). For Theorem 1″ we still have to pick out an *ON* basis in that space.

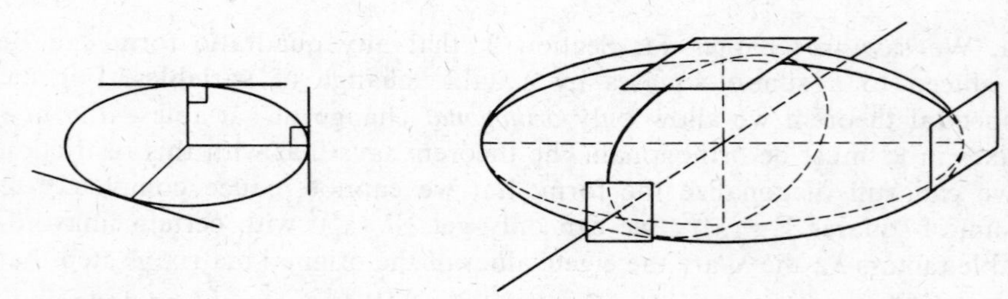

Figure 27.

For $n=2, ax^2+bxy+cy^2$ has matrix $A=\begin{pmatrix} a & b/2 \\ b/2 & c \end{pmatrix}$. Suppose λ is an eigenvalue and $[x,y]=[\cos\theta, \sin\theta]$ a normalized eigenvector. Thus, $ax+(b/2)y=\lambda x$ and $(b/2)x+cy=\lambda y$. We rewrite this as $a+(b/2)(y/x)$

$=\lambda,(b/2)(x/y)+c=\lambda$. Eliminating λ we find

$$a-c=\frac{b}{2}\left(\frac{x}{y}-\frac{y}{x}\right)=b\cdot\frac{x^2-y^2}{2xy}=b\frac{\cos^2\theta-\sin^2\theta}{2\sin\theta\cos\theta}=\frac{b}{\tan 2\theta};$$

we have found, by a slightly mysterious process, the usual formula $\tan 2\theta = b/(a-c)$ for the angle by which one should rotate the coordinate system in order to "eliminate the xy-term," that is, to diagonalize the matrix.

PROBLEMS

1. For $\mathbf{C}^3$ with standard inner product write out the sesquilinear form q that has the matrix $A=\begin{pmatrix} 2 & 1-i & i \\ 3i & 1+i & 2-i \\ -i & 1 & 2i \end{pmatrix}$ (or better, the operator T_A) as right correspondent; then write out the matrix for the operator that is left correspondent of q.

2. Find the principal axes of the quadric $-x^2-2y^2+3z^2+12xy-8xz+4yz=18$.

3. Find the principal axes of the quadric $5x^2+2y^2+2z^2+4xy+4xy-8yz=6$.

4. Rotate the coordinate system to remove the xy-term from the equation of the conic $x^2-2xy-3y^2$.

5. (Rayleigh's principle) Let U be a vector space (over $\mathbf{R}$, say) with inner product $\langle\,,\rangle$; let q be a quadratic form on U (or, equivalently, a symmetric operator T with $q(v)=\langle v, Tv\rangle$). Let $\lambda_1,\ldots,\lambda_n$ be the associated eigenvalues, arranged in order, so that $\lambda_1 \geqslant \lambda_2 \geqslant \cdots \geqslant \lambda_n$, and let $u_1,\ldots,u_n$ be the corresponding (unit) eigenvectors. Prove that λ_1 is the maximum value of $q(v)$ with v running over the *unit-sphere*—set of all unit vectors—or also the maximum value of $q(v)/\langle v,v\rangle$ with v running over all nonzero vectors. (*Hint*. Express q in coordinates with respect to the eigenvector basis.)

6. (Continuation of Rayleigh's principle.) Let U_2 be a subspace of U of codimension 1, and let q_2 be the restriction of q to U_2. Prove that the maximum of $q_2(v)$ with v running over the *unit sphere* in U_2 is always (i.e., for any U_2) greater than or equal to λ_2 (the second eigenvalue), and for a suitable U_2 it is equal to λ_2. (One says that λ_2 is characterized as a *minimax*—it is the minimum value of the maxima of q on the unit spheres in the various subspaces of codimension 1.) (*Hint*. Show that there is a unit vector in U_2 that also belongs to $((u_1,u_2))$.)

7. (Continuation of Rayleigh's principle.) State and prove a similar characterization of λ_i, for any k, in terms of subspaces U_k of codimension $k-1$.

8. Verify the inequality of Rayleigh's principle for the λ_2 of the form in Problem 2, Section 2, for the subspace $y+z=0$ (i.e., find a vector v in this subspace with $q(v)\geqslant\lambda_2$).

SOLUTIONS

(TO MOST OF THE NUMERICAL PROBLEMS)

INTRODUCTION

1. $x=-5, \quad y=17, \quad z=30.$
2. $(-1,-4)$ yes; $(1,1)$ and $(2,4)$ no.

CHAPTER 1,

Section 2

1. (a) and (d) no; (b and (c) yes.
2. VS_5–VS_9 fail.
3. $[3,1,-1,0], \quad [-1,3,-5,-2], \quad [2,4,6,-2], \quad [-1,8,-13,-5].$

Section 3

1. $-4x, \quad 25x^4+3x^2+6, \quad 252x^{17}-192x^{15}.$
2. $SX=[3,0], \quad SY=[0,3], \quad X+Y=[3,3,-3], \quad S(X+Y)=[3,3].$
3. x^3.

CHAPTER 2,

Section 1

1. $[3,22,5,-17], \quad [0,0,0,0].$
4. $X_1=E^1+E^2$, $X_2=E^1-E^2$; $E^1=1/2X_1+1/2X_2$, $E^2=1/2X_1-1/2X_2$.
5. (a) No (cannot "remove" x^6); (b) no (after subtracting p_1 cannot remove x^4); (c) yes, $=p_1+2p_2+p_3$.
7. For instance, $[1,0,0,0,0], \quad [0,0,0,1,0], \quad [0,0,0,0,1], \quad [0,1,1,0,0].$
8. $1, \quad x^2, \quad x^4, \quad x^6.$

Section 2

4. $x_2=2x_1$ and $x_2=3x_1$ together imply $x_1=x_2=0$. On the other hand, any x_1 and x_2 can be written as y_1+y_2 and $2y_1+3y_2$, with suitable y_1 and y_2. $X=[1,1,2,3]$ splits as $[2,4,2,3]+[-1,-3,0,0]$ or $[2,4,2,0]+[-1,-3,0,3]$, etc.

Section 3

5. $aX+bY+cZ=0$ means $a+b=0, \quad a-b=0, \quad c=0, \quad b+c=0$, which implies $a=b=c=0$.

Section 4

1. (i) $[1,0,0,0,0,0]$,

(ii) $\begin{pmatrix} 1 & 0 & 0 \\ 1 & 1 & 0 \\ 1 & 2 & 1 \end{pmatrix}$,

(iii) $\begin{pmatrix} 1 & 0 & 0 \\ 2 & 1 & 0 \\ -3 & -1 & 1 \\ 1 & 1 & 2/3 \end{pmatrix}$

(iv) $\begin{pmatrix} 1 & 0 & 0 & 0 \\ 2 & 0 & 0 & 0 \\ 1 & 1 & 0 & 0 \\ -4 & -3 & 0 & 0 \end{pmatrix}$

2. (i) $\begin{pmatrix} 1 & 0 & 1 & 3 \\ 0 & 1 & 1 & -1 \\ 0 & 0 & 1 & 2 \\ 0 & 0 & 0 & 1 \end{pmatrix}$

(ii) $\begin{pmatrix} 1 & 3 & 2 & 5 \\ 0 & 1 & 1 & 2 \\ 0 & 0 & 1 & 1 \end{pmatrix}$

Section 5

1. No.
2. No.
3. First three vectors.
4. $X_3 = 3X_1 - X_2$.
5. X_1 and X_2 are independent.
6. Independent.
7. $Y = 3X_1 - 2X_2$.

CHAPTER 3,

Section 1

1. 3, 3, 2.

2. $E^1 = 1/2(-X_1 + X_2 + X_3)$, etc.
3. X, E^2, E^3, E^4, for instance.
4. $x = a + b$ and $y = 2a + 3b$ can be solved to $a = 3x - y$ and $b = y - 2x$.
5. $6 + 4(x-1) + (x-1)^2$.

Section 2

2. X, Y, E^3.
3. X, Y, E^1, E^2.
4. No.
5. $X_2 = iX_1$.
6. Yes.
8. 3, 3, 2.
9. $((E^3, E^4, E^5))$, for instance.

Section 3

3. $\dim V + W = 4$; $\dim V \cap W = 0$.

Section 4

1. $Y_1 \leftrightarrow [10, -7]$, $Y_2 \leftrightarrow [-4, 3]$, $Y_3 \leftrightarrow [2, 0]$.
2. $x^2 = 2p_1 - p_3$, etc. $p \leftrightarrow [5, 0, -3]$.

CHAPTER 4,

Section 5

2. -17, 10, 15.

Section 6

1. [2,1,0,0,0,0], [−1,0,1,0,0,0], [3,0,0,1,0,0], [0,0,0,0,1,0], [−2,0,0,0,0,1].
4. (1, −1, 1) or any multiple.

Section 3

1. [8, −7, 1, 0], [−4, 3, 0, 1].
2. [−3, −1, 1, 0].
3. [−2, 1, 0, 0, 0], [1, 0, 3, 1, 0].
4. (a) [1, −1, 1, 0, 2], [2, −1, 0, 1, −1]; (b) [3, 1, −2]; (c) [0, 0, 0, 0]; (d) [3, 0, 4, 1, 0], [1, −1, 1, 0, 1]; (e) [1, 2, 0, −1], [2, 1, 1, 3]; (f) [2, 2, 1, 0, −1], [1, 2, 2, 1, 0], [1, −1, 1, 0, 1]; (g) [1, 1, 1, 1, 1, 1].

Section 4

1. [5/6, −1/6, 1/6, 0].
2. (a) [2, 1, 3, 0, 0] + (([−3, 1, 0, 0, 0], [−2, 0, −2, 1, 0], [−2, 0, 0, 0, 1])); (b) inconsistent; (c) [−1, 2, 0, 0] + a[−1, 1, −1, 1]; (d) [1, 1, −2, 1] + a[1, 2, 0, 1].
3. (a) $3a + b - 1 = 0$; (b) for $a = 0$, $b = 1$, [0, 1, 0, 0] + a[1, 0, −1, 1]; (c)

$[2a, a+b, -a, 0]+(([1,0,-1,1]))$, with $3a+b=1$.
4. (a) $[1,0,2]+a[1,1,-2]$; (b) inconsistent; (c) $[1,2,0,3]+(([1,1,-1,-1],$ $[2,1,3,-1]))$; (d) inconsistent.
5. (a) $a-b=-1$; (c) $[1,1,a+b,a-b]+(([1,1,-1,2],\quad [0,1,4,3]$.

Section 5

1. $Y=3X_1-X_2+2X_3$.
2. $V_1 \cap V_2=(([1,1,-1,1]))$.
3. No, inconsistent.
4. $(([-1,2,1,2]))$.
5. $(0,-1,0,-1,1),\quad (11/3,-2,-7/3,0,0)$, for instance.
6. $[-7,16,42]$.
7. $[1,2,1,0]$.
10. $[1,1,1,1]$.
12. $[1,1,1,1]+(([3,3,-2,3]))$.

Section 6

2. $[2,-1,2,3,0]$, for instance.
5. $(([1,-1,-1,1],\quad [2,0,0,-1]))$.
6. $[2,1,-3,1],\quad [-1,-1,4,2],\quad [1,0,1,3]$.
7. 5.
8. $3\pi/4$.

Section 7

1. 2.
6. $V^{\perp}$ spanned by $(1,-2,3,0)$ and $(2,-1,0,-3)$.
7. $x_1+x_2+x_3+x_4=3,\quad x_1-x_3+x_4=1$.

CHAPTER 5,

Section 2

1. $[2,-1,-1],\quad [-6,3,4],\quad [-3,2,2]$.
2. $(2,2,-10,-8),\quad (-1,-3,5,3),\quad (-1,-2,3,4),\quad (0,-1,1,1)$.

Section 3

1. $X_0=[3,6,7,4,1]$, $A=(1,-1/2,1/3,-1/4,1/5)$; $\eta(p_0)=23/15$.

CHAPTER 6

Section 1

2. $\begin{pmatrix} 1+i & i \\ 1+3i & 3 \end{pmatrix}$.

Section 3

1. $\begin{pmatrix} 1 & 0 & -1 \\ -2 & 1 & 3 \\ -3 & 1 & 5 \end{pmatrix}$.

2. $\begin{pmatrix} -10 & -3 & -2 & 5 \\ -26 & -8 & -5 & 13 \\ -5 & -2 & -1 & 3 \\ 18 & 6 & 4 & -9 \end{pmatrix}$.

8. $B = \begin{pmatrix} 1 & -2 & -3 \\ 1 & -3 & -5 \\ 1 & -4 & -6 \end{pmatrix}$.

14. $\begin{pmatrix} 0 & 0 & 1 \\ 0 & 1 & 0 \\ 1 & 0 & 0 \end{pmatrix} \cdot \begin{pmatrix} 1 & 0 & 1 \\ 0 & 1 & 0 \\ 0 & 0 & 1 \end{pmatrix}$.

$$\times \begin{pmatrix} -1 & 0 & 0 \\ 0 & 1 & 0 \\ 0 & 0 & 1 \end{pmatrix} \cdot \begin{pmatrix} 1 & 0 & 0 \\ 0 & 1 & -1 \\ 0 & 0 & 1 \end{pmatrix} \cdot \begin{pmatrix} 1 & 0 & 0 \\ 3 & 1 & 0 \\ 0 & 0 & 1 \end{pmatrix} \cdot \begin{pmatrix} 1 & 0 & 0 \\ 0 & 1 & 0 \\ 2 & 0 & 1 \end{pmatrix} \cdot \begin{pmatrix} 1 & 0 & 0 \\ 0 & 1 & 0 \\ 0 & -1 & 1 \end{pmatrix}.$$

15. $\begin{pmatrix} -1 & -2 \\ i & i \end{pmatrix}$.

Section 4

2. $P^{-1} = \begin{pmatrix} -2 & 2 & 1 \\ -2 & 1 & 2 \\ 3 & -2 & -2 \end{pmatrix}$, $\quad X' = [1, -1, 1]$.

3. $P = \begin{pmatrix} -1 & -1 \\ 3 & 2 \end{pmatrix}$, $\quad 3Y_1 - 4Y_2 = X_1 + X_2$.

CHAPTER 7

Section 2

1. $+1$.
2. $w = 6$.
4. (a) -36; (b) 6; (c) 1; (d) 0.

Section 3

1. 5!
2. det = 0.
3. −6.
4. $\{4,2,1,6,3,5\}$; $w=6$.
5. (a) $\det M = 1$; $\tilde{M} = \begin{pmatrix} 1 & -1 & 0 \\ 0 & 1 & -1 \\ -1 & 0 & 2 \end{pmatrix} = M^{-1}$;

(b) $\det = 2$; $\tilde{M} = \begin{pmatrix} 4 & -2 \\ -1 & 1 \end{pmatrix}$, $M^{-1} = \begin{pmatrix} 2 & -1 \\ -1/2 & 1/2 \end{pmatrix}$;

(c) $M^{-1} = 1/(ad-bc)\cdot\begin{pmatrix} d & -b \\ -c & a \end{pmatrix}$.

8. $\det = x^4 - 4x^3$; roots 0, 0, 0, 4.
9. $[2,1]$.
11. $\det\begin{pmatrix} 1 & 2 & x_1 \\ 2 & 3 & x_2 \\ 3 & 4 & x_3 \end{pmatrix} = -x_1 + 2x_2 - x_3$.
12. $\det = -11x_1 - 6x_2 - 4x_3 - x_4$.

Section 4

1. $\{t[3,1]+(1-t)E^1, t[2,1]+(1-t)E^2\}$; $\det = 1+2t-2t^2 \neq 0$ for $0 \leqslant t \leqslant 1$.
2. First step: $X_1(t) = X_1 - tX_2$, $0 \leqslant t \leqslant 1$, reduces to $[1,1,0]$, $[1,2,1]$, $[1,1,1]$.

CHAPTER 8

Section 1

1. $-2x+1$, $2x^2+x$, for instance.
2. $[4,-1]$, $[-6,0]$, $[-1,1]$, $[9,0]$.

Section 2.

1. $A' = \begin{pmatrix} 1 & 0 \\ 0 & 1 \\ 0 & 0 \end{pmatrix}$.

Section 3

3. ker = the constants; im $= P^2$, all polynomials of degree $\leqslant 2$.
4. $\ker = ((X_1 - X_2 + X_3))$; $\mathrm{im} = ((Y_1, Y_2))$.

Section 5

5. $$A=\begin{pmatrix} 4 & 4 & 11 & 23 \\ 0 & 4 & 10 & 33 \\ 0 & 0 & 4 & 15 \\ 0 & 0 & 0 & 4 \end{pmatrix}.$$

Section 6

8. No.

Section 7

1. $\ker T=((E^1-E^2))$, $\ker T^2=((E^1,E^2))$, $\operatorname{im} T=((E^1-E^2,E^2+E^3,E^4))$, $\operatorname{im} T^2=((E^2+E^3,[2,-1,1,1]))$.

CHAPTER 9

Section 2

1. $-(x+1)(x+1)(x-8)$, $[-1,2,0]$, $[0,2,-1]$, $[2,1,2]$; $-(x-1)(x-2)(x-3)$, E^1, $[2,1,0]$, $[9,6,2]$; $(x-1)(x+1)$, $[1,1]$, $[1,-1]$; $x^2+1=(x-i)(x+i)$, $[i,1]$, $[i,-1]$; $-(x-1)^2(x+1)$, $[2,1,1]$, $[1,1,1]$; $x(x-1)(x-2)^2$, $[3,0,-1,0]$, $[2,0,-1,0]$, E^2, $[1,0,-1,1]$; $\lambda_{1,2}=\sqrt{3}/2\pm i/2$, $[i,1]$, $[i,-1]$; $x^2-2\cos\alpha\cdot X+1$, $\lambda_{1,2}=\cos\alpha\pm i\cdot\sin\alpha$, $[i,1]$, $[i,-1]$.
3. i, $-i$, 1, $[i,1,0]$ and $[i,-1,0]$ not real, E^3.

Section 3

1. No for (*e*), only two eigenvectors.
3. Not possible.
4. Eigenvalues 0, 1, 2, 2; not possible.

Section 4

1. $-(x-1)(x-2)(x-3)$, $[1,0,-1]$, $[2,-1,0]$, $[0,1,-1]$; $-(x-1)(x-2)\cdot(x-3)$, $[i,1,0]$, $[0,0,1]$, $[i,-1,0]$.
2. -1, i, $-i$; $[0,-1,1]$, $(([2,-1,0],[-1,0,1]))$.
9. 1, 1, 0; 1, -1, 0; $[1,2,1]$, $[1,1,1]$, $[3,-2,2]$.

CHAPTER 10

Section 1

2. $(x-1)^2$.
5. $-(x-1)^2(x+1)$; x^2-1.

Section 2

1. No.
2. Yes.

Section 4

1. $U^0 = (([-1,1,0]))$, $U^1 = ((E^1,[0,-1,1]))$.

Section 6

1. (a) $(x-1)(x-2)$, $(([0,1,0]))$, $(([0,2,1],E^1))$; $-(x-1)^2(x-2)$, $(([5,0,-2],$ $[5,2,-5]))$, $(([1,1,-2]))$; $-(x+1)^3$, $((E^1,[1,1,0],[1,2,1]))$.
(b) $(x-2)(x-1)^3$, $(([-2,1,1,0]))$, $(([-1,1,1,0],[-1,0,1,0],[0,-1,0,1]))$.
4. One basic block; $m(x) = (x-1)^4$.

Section 8

1. With E^3 and E^4 for W, the columns of A are $[1,0,2,2]$, $[0,1,1,1]$, 0, 0.
2. $\ker = (([2,-1,0]))$, $\operatorname{im} = (([-1,1,0],[-2,1,1]))$.

Section 9

1. $[bte^{2t} + ce^{2t}, ae^t + 2be^{2t}, be^{2t}]$.

CHAPTER 11

Section 1

1. $\begin{pmatrix} 14 & 19 \\ 21 & 29 \end{pmatrix}$.
3. $a_{ij} = 1/(i+j)$; $a'_{ij} = (-1)^{i+j}/(i+j+1)(i+j+2)$.
4. For example, $q = 8 - 15t$.

Section 2

4. $A = A' = \begin{pmatrix} 0 & 1 \\ -1 & 0 \end{pmatrix}$.

Section 3

1. (a) $x'^2 + y'^2 - z'^2$; (b) $x'^2 + y'^2 - z'^2 - t'^2$; (c) $x'^2 + y'^2 + z'^2$; (d) $x'^2 + y'^2$.

Section 4

1. $\det = -1$, 1, 1 ($\det P = 1$ in all three cases).
2. Indefinite, indefinite, positive semidefinite.
3. (a) Indefinite; (b) indefinite.

Section 5

1. Positive definite, positive semidefinite degenerate, indefinite nondegenerate, negative definite.
2. Indefinite nondegenerate, positive definite, indefinite nondegenerate, identically 0.
3. Negative definite on $x - y + z = y - 2z + t = 0$.

Section 6

1. 2, 1, 0; 2, 2, 0; 3, 0, 1; 1, 1, 0; 2, 1, 0.

Section 7

1. $x+y+z=0$, $3x+3y+4z=0$, $y=2z$. Yes.
2. $y^2-2yz<1$ ("interior" of a hyperbola).
3. $x+2y-z=1$.
5. $x'^2-y'^2+z'^2-t'^2$.

Section 8

3. $|z_1'|^2-|z_2'|^2+|z_3'|^2$.

CHAPTER 12

Section 1

2. For example, 6, 10, 5; $[1,1,1]$.
5. $[2,-1,0]$, $[-1,0,1]$, $[0,1,-1]$.

Section 2

2. $-19/30$; $5/2\sqrt{15}$.
3. For example, $[-2,1,1,0]$.
4. $[8,-4,-1]$.

Section 3

1. With $\omega=[\alpha,\beta,\gamma]$, $J(\omega)=2/3(\alpha^2+\beta^2+\gamma^2)-1/2(\beta\gamma+\gamma\alpha+\alpha\beta)$.

Section 4

5. $[5/6,-5/6,5/6]$, $[1/6,17/6,13/6]$.

Section 5

5. 2.
6. $[2,1,4]$, $[-8,4,1]$.
7. $\sqrt{10}$.
8. $[2,-1,2]$, $[-1,2,2]$, $[2,2,-1]$.
9. $[1,1,1,1]$, $[-1,1,-1,1]$, $[-2,1,2,-1]$.
10. $[-5,3,0]$, $[-5,4,1]$, $[7,-4,0]$.

Section 6

1. For example, $[2,-1,3,17]$.
2. $(([1,2,-5,0,0], [-2,1,0,5,0], [1,2,0,0,-5]))$.
5. $(([2,3,0,-1,2], [3,1,1,1,-4]))$.
6. $[1,0,-1,0]+(([3,0,-1,0], [2,-1,0,1]))$.

Section 7

5. $$\begin{pmatrix} -1/\sqrt{5} & 6/7 & 4/7\sqrt{5} \\ 2/\sqrt{5} & 3/7 & 2/7\sqrt{5} \\ 0 & 2/7 & -15/7\sqrt{5} \end{pmatrix}$$

8. $$\begin{pmatrix} 3/\sqrt{13} & -2/\sqrt{13} \\ 2/\sqrt{13} & 3/\sqrt{13} \end{pmatrix} \cdot \begin{pmatrix} \sqrt{13} & 5/\sqrt{13} \\ 0 & 1/\sqrt{13} \end{pmatrix}.$$

Section 8

4. Note that all solutions of the DE have period 2π.

Section 9

1. $$\begin{pmatrix} 1 & 0 & 4 \\ 0 & -1 & 3 \\ 1 & 0 & 3 \end{pmatrix}$$

5. $EX = X - (X_1 \cdot X/7)X_1 - (X_2 \cdot X/7)X_2$ with $X_1 = [1,2,1,0,1]$, $X_2 = [2,0,-1,1,-1]$.

Section 10

1. $e^{\pm i\phi}$ $(= \cos\phi \pm i \cdot \sin\phi)$.

2. $$\begin{pmatrix} 2/3 & -\sqrt{5}/3 & 0 \\ \sqrt{5}/3 & 2/3 & 0 \\ 0 & 0 & -1 \end{pmatrix}$$

3. $\mathrm{diag}(R,R)$ with $R = \begin{pmatrix} 1/2 & -\sqrt{3}/2 \\ \sqrt{3}/2 & 1/2 \end{pmatrix}$.

Section 11

1. $[5,-5,-5]$.
2. $x - y - z = 0$.

CHAPTER 13

Section 2

1. $\lambda_{1,2} = 6$, $\lambda_3 = -3$; $1/\sqrt{2}\,[0,1,-1]$, $1/3\sqrt{2}\,[4,1,1]$, $1/3[1,-2,-2]$.

2. $\lambda_{1,2,3} = 5, 0, -9$; $1/\sqrt{5}\,[2,0,1]$, $1/3\sqrt{5}\,[2,5,-4]$, $1/3[1,-2,-2]$.
3. $1, 1, 10$.
4. $\lambda_{1,2} = 1, -1$; $[(\sqrt{2}-i)/2, (1-i)/2\sqrt{2}\,]$, $[-(1+i)/2\sqrt{2}\,, (\sqrt{2}+i)/2]$.
5. $1, 2, -1$; $[4,-6,1]$, $[1,-1,0]$, $[-5,8,1]$.

Section 3

2. $\sqrt{6}\,, \sqrt{3}\,, \sqrt{2}\,.$
3. $1, 1, \sqrt{2}\,.$
4. $\cos\theta = ((5-2\sqrt{5}\,)/10)^{1/2}$, $\sin\theta = ((5+2\sqrt{5}\,)/10)^{1/2}$; $-(\sqrt{5}+1)x'^2 + (\sqrt{5}-1)y'^2$.
8. $[0, 1, -1]$.

APPENDIX: SPECIAL SYMBOLS

INDEX